未解之谜

[美] 克雷格 · P. 鲍尔　著
（CRAIG P. BAUER）
鲁冬旭　译

UNSOLVED!

The History and Mystery of the World's Greatest Ciphers from Ancient Egypt to Online Secret Societies

下

中信出版集团 | 北京

图书在版编目（CIP）数据

未解之谜. 下 /（美）克雷格 · P. 鲍尔著；鲁冬旭译. -- 北京：中信出版社，2019.1（2019.4重印）
书名原文：Unsolved!: The History and Mystery of the World’s Greatest Ciphers from Ancient Egypt to Online Secret Societies
ISBN 978-7-5086-9286-9

I. ①未… II. ①克… ②鲁… III. ①密码学 – 普及读物 IV. ①TN918.1-49

中国版本图书馆CIP数据核字（2018）第253463号

未解之谜（下）

著　　者：[美] 克雷格 · P. 鲍尔
译　　者：鲁冬旭
出版发行：中信出版集团股份有限公司
（北京市朝阳区惠新东街甲4号富盛大厦2座　邮编　100029）
承 印 者：中国电影出版社印刷厂

开　　本：880mm × 1230mm　1/32　　印　　张：12.5　　字　　数：307千字
版　　次：2019年1月第1版　　印　　次：2019年4月第2次印刷
京权图字：01–2018–4344　　广告经营许可证：京朝工商广字第8087号
书　　号：ISBN 978-7-5086-9286-9
定　　价：56.00元

目录

下

第7章
来自坟墓的密码

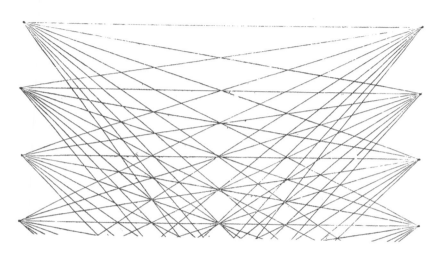

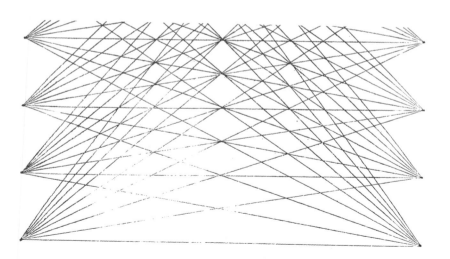

虽然我并不相信《未解之谜（上）》第6章中提到的那些已经死亡的受害者（或者任何已经死亡的人）能够通过某种方式与警方沟通，然而却有一些人试图用各种方式证明这种沟通是可能的。有趣的是，在证明死者确实能与这个世界沟通的过程中，有时会用到密码这一工具。一些新的未解之谜就这样产生了。

伟大的逃脱

第一个例子的主角是哈里·霍迪尼（Harry Houdini）。在他的职业生涯早期，他就建立了一套与妻子贝丝（Bess）交流的秘密代号。通过这套代号，霍迪尼夫妇二人可以互相传递信息，而周围人却没有办法知道他们在交流什么。如果哈里·霍迪尼死后还能与他的妻子交流，他也会使用这种密码的。这套密码的密钥是这样的：

Pray（祈祷）=1=A

Answer（回答）=2=B

Say（说）=3=C

Now（现在）=4=D

Tell（告诉）=5=E

Please（请）=6=F

Speak（说出）=7=G

Quickly（快速地）=8=H

Look（看）=9=I

Be quick（快点儿）=10 或者 0=J

在这套密钥中，英文字母表中的前10个字母，以及前10个数字分别用一个单词来替代。比如，“BAD”（坏）这个单词的密文是“Answer, Pray, Now（回答，祈祷，现在）”。从表面上来看，这套密钥似乎没有办法表示字母J之后的字母，但是因为这些单词不仅对应字母，还能对应数字，所以这套密钥其实是可以表示字母J之后的字母的。比如，如果霍迪尼要表示英语字母表中的第19个字母S的话，他只需要用上面这套密钥表示出1和9这两个数字就行了，也就是说密文的“Pray-Look（祈祷—看）”代表明文的S。哈里·霍迪尼于1926年10月31日去世。他死后，他的妻子贝丝仍然等待着丈夫用这套密钥发来加密的信息。

不少来源称，阿瑟·福特（Arthur Ford）能够与死去的哈里·霍迪尼建立联系。但是，福特的说法究竟有多少真实的成分呢？事实上，福特第一次“成功通灵”时并没有使用这套密钥，而是直接给出了一个明文的单词“FORGIVE”（原谅），据说这条信息来自霍迪尼的母亲。在霍迪尼去世之前，他的母亲就已经去世了。贝丝知道，对于霍迪尼来说“FORGIVE”是一个非常重要的词，因为霍迪尼一直希望母亲能原谅自己。有些人认为，既然福特能给出一个只有当事人知道的重要单词，就说明他确实具有通灵的本事。然而，在这件事情上，我们需要认真考虑“日期”这个重要因素。

福特给出“FORGIVE”一词的时间是1928年2月8日。早在大约一年前，也就是1927年的3月13日，《布鲁克林鹰报》（*Brooklyn Eagle*）就引用了贝丝的一段话。这段话称，任何来自她丈夫的真实的信息都肯定会包含“forgive”这个词。

后来，福特又给出了另一条号称是来自霍迪尼本人的消息。这条消息的内容是：

Rosabelle, answer, tell, pray-answer, look, tell, answer-answer, tell.

《罗萨贝尔》（*Rosabelle*）是霍迪尼的妻子经常唱的一首歌曲。而剩下的信息则使用了霍迪尼与妻子用过的那套密钥，翻译成明文就是“BELIEVE”（相信）一词。

贝丝相信，这确实是一条来自已故丈夫的信息。但是，我们仍然需要考虑日期这个重要因素。在霍迪尼死后，贝丝并没有将这套密钥向公众保密，她将这套密钥告诉了哈罗德·凯洛克（Harold Kellock），而后者则将这套密钥写进了一本经过授权的霍迪尼传记中。这本传记于1928年出版，而福特给出上面这条“通灵”信息的日期是1929年1月8日。

虽然贝丝后来改变了主意，不再相信福特真的能与她死去的丈夫交流，但是关于福特有通灵本事的故事已经在世间流传开了。

索利斯的三个密码

到了20世纪40年代，关于死者通过密码与世人交流的故事仍在继续。

这一次，故事的主角是剑桥大学特别研究员罗伯特·H. 索利斯（Robert H. Thouless）。索利斯创造了一条加密的信息，他认为，在没有密钥的情况下，任何人都无法破解这条信息。然后，索利斯计划在自己死后尝试将这套密钥传送给还活着的人。索利斯认为，如果自己死后有人能够获得这套密钥，就足以证明死者确实能与这个世界交流。

图7-1 罗伯特·H. 索利斯（1894—1984）

我将索利斯发表在《心灵学研究学会会刊》（*Proceedings of the Society for Psychical Research*）上的第一段加密信息称为"密码A"。下面我将密码A的全文复制如下[1]：

CBFTM HGRIO TSTAU FSBDN WGNIS BRVEF BQTAB QRPEF BKSDG MNRPS RFBSU TTDMF EMA BIM

关于这段密码，索利斯只透露了以下信息：

> 这段密码使用了一种著名的加密方式。这种加密方式包含一个关键词，我希望，在我死后，我还能记得这个关键词。我从未把这个关键词告诉任何人，也不准备在我的余生中告诉任何人。并且，在完成这段密码以后，我很快将所有相关文件都销毁了。[2]

索利斯并非不知道世界上有"密码分析"这种技术，在表达了对有人在他生前通过读心术来获得关键词的担忧以后，他写道：

> 进一步的怀疑是：是否无法通过理性的推断来找到这个关键词，因为人们普遍相信，技术高超的密码专家能够在不知道关键词的情况下破译任何加密信息，只要花上足够长的时间。然而，除非满足一些特定的条件，否则上述想法是没有用的。如果密码专家能够破译一段关键词未知的密码，那么要么这种密码是一种简单密码（比如单套字母替代密码），要么密文长度要足够长。事实上，非密码专家也能够破译用简单替代法加密的短信息，人们常常把这种信息作为一种谜题来逗孩子玩。而如果待破解的是一段用更复杂的加密方法加密的长段信息（或者若干段短信息），在不知道关键词的情况下破译这段信息就要复杂得多了。但是，据说只要给密码专家足够长的时间，他们就能够在不知道关键词的情况下破译出用大部分加密方法（注意并不是全部加密方法）加密的信息。我给出的这段密码既不是用简单的替代或者换位重排方法加密的，长度也不够长，所以没有办法通过上述针对简单密码的破译方式来破译。我认为，在不知道关键词的情况下，即使是密码专家也没有办法解开我的密码。而只要知道这个加密系统的关键词，任何有基础密码知识的人都能够很轻松地解开这段密码。[3]

虽然索利斯不相信简单的密码分析技术能够破解自己的密码，但他也考虑到了另一种可能性：也许某人可以通过某种与他的原意不同的方式将他的密码翻译为一段有意义的信息。[读者可以回忆一下，在《未解之谜（上）》中，伏尼契手稿、多拉贝拉密码，以及黄道十二宫杀手的340密码都出现过一些错误的解法。] 为了防止这种情况的发生，索利斯透露，他的这段密码“引自莎士比亚的一出戏剧”。[4]索利斯相信，即使有人能找到某种有意义的错误解法，这种解法也不太可能正好出自莎士比亚的戏剧。

在处理完上述所有不确定因素以后，索利斯还对另外一种可能性做好了准备，也就是：尽管索利斯尽一切努力要记住这个关键词，但他仍然有可能在死后记不起来。为了防止这种情况的发生，索利斯将密码的关键词放在一个密封的信封中，并让心灵学研究学会负责保存。按照索利斯的要求，只有在他死后没有人能通过通灵术解开这段密码的情况下，才可以打开信封查看里面的内容。

索利斯没有想到的是，这个信封根本就没有打开的必要。因为在索利斯还活着的时候，这段密码就已经被解开了。一位身份不明的“密码专家”将索利斯给出的密码A当作一项挑战，并利用业余时间解开了这段密码，只花了两个星期的时间。

事实上，这已经不是索利斯第一次在密码的问题上栽跟头了。在密码A之前，索利斯还创造过另外一段密码，这段密码被德尼斯·帕森斯（Denys Parsons）先生解开了。但是由于这段密码并没有公开发表，所以索利斯也就没有因为密码被破译而在公众面前出丑。[5]在索利斯的一篇论文中，他致谢了“D. 帕森斯”和另外一位不具名的人士，感谢他们告诉他滚动密钥密码是可以破译的。从上述内容我们可以看出，索利斯的第一段失败的密码是一种滚动密钥密码。在本书的第8章中，我们将详细讨论滚动密钥密码的解法。总之，密码专家只用了48小时就破译了他创造的滚动密钥密码。

虽然索利斯并没有明确提示密码A的加密方法，但是我们知道密码A属于“普莱费尔密码”（Playfair Cipher）。加密一段普莱费尔密码会用到一个5×5的网格，加密者要在这个网格中放入被打乱的英文字母表。为了将英文字母表打乱，加密者需要挑选一个关键词或者关键短语。比如，如果关键词是“MACHETE”（大砍刀）的话，字母表就是以下的样子：

M	A	C	H	E
T	B	D	F	G
I/J	K	L	N	O
P	Q	R	S	U
V	W	X	Y	Z

为了把26个英文字母塞进5×5的网格中，通常会将字母I和字母J合并在一起，或者干脆省略字母J。还有一种处理这个问题的方法是省略出现概率最低的字母Z。

为了更好地向读者解释如何用上述网格来加密信息，让我们来考虑以下的这条信息：

IT'S TOO HOT FOR CLOTHES.（天太热了，穿不住衣服。）

加密的第一步是，如果任何两个相同的字母连在一起，就必须在这两个字母之间加上一个X。在上面这段信息中，“TOO”中出现了“OO”两个相同字母连在一起的情况，因此我们将“TOO”变为“TOXO”。至于为什么要进行这种操作，继续读下去你就会明白了。

接下来，我们要把这段信息中的字母两两组合成对：

IT ST OX OH OT FO RC LO TH ES

接下来，我们用一种简单的规则来加密每一对字母。这种规则取决于字母在5×5的网格中的相对位置。在5×5的网格中，两个字母的位置关系一共有以下3种可能性：

1. 两个字母在同一行。
2. 两个字母在同一列。
3. 两个字母既不在同一行，也不在同一列。

我们的第一对字母IT属于第2种情况。在这种情况下，我们用每个字母下方的字母来替代原字母，也就是IT被加密为PI。

我们的第2对字母ST属于第3种情况。在这种情况下，我们首先在上述5×5的网格中找到包含字母S和字母T的最小矩形，然后再用矩形另外两个顶点上的字母来替代原始字母。我在下面的这个网格中用黑体字标出了这个矩形的4个顶点，并且在替代原始字母的两个字母下加了下划线，这样读者就能清楚地看到这个小小的矩形了。

M	A	C	H	E
T	B	D	**<u>F</u>**	G
I/J	K	L	N	O
<u>P</u>	Q	R	**S**	U
V	W	X	Y	Z

那么，替代原始字母ST的密文字母究竟应该是FP还是PF呢？普莱费尔密码的加密规则规定，与明文中的第1个字母在同一行的密文字母应该先出现。也就是说，ST应该被加密为PF。如果我们需要加密的这对明文字母是TS，那么对应的密文字母就应该是FP。

接下来，我们可以继续根据上述规则加密接下来的几对字母，得到密文LZ、NE、IG、GN，以及XD。

直到加密到LO这对字母时，才出现了以上的第1种情况。字母L和字母O出现在网格中的同一行，在这种情况下，我们用原始字母右侧的字母来代替原始字母，于是字母L就变成了字母N。但是在字母O的右侧已经没有字母了，因此我们只能再回到这一行的第一个字母（就像在“吃豆人”游戏里一样），也就是用字母I来代替字母O。这样，原始字母LO就被加密成了NI。

经过上述这些加密步骤以后，完整的密文信息如下：

PI PF LZ NE IG GN XD NI FM HU

我们之所以要在两个相同的字母之间插入字母X，是为了防止第4种情况的产生。两个相同的字母连在一起时，既符合第1种情况，又符合第2种情况，因此根据规则无法明确知道究竟是应该用哪种规则加密。此外，在加密的时候，消除两个相同字母连在一起的情况一般来说都是一种比较好的处理方式！

由于这种普莱费尔密码在加密的时候会以两个字母为单位进行替换，所以这种密码又叫作“双字母替代密码”（digraphic substitution cipher）。虽然双字母替代密码已经比单套字母替代密码前进了一大步，但这仍然是一种比较简单的加密方式。这种加密方式的弱势在于：虽然两个相同的字母对有时会被加密成不同的密文字母——是否会出现这种情况主要取决于相同字母对出现在信息的何种位置（参见以下的例子），但是，在大部分情况下，相同的字母对都会被加密成同样的密文字母。因此，就像我们可以通过单字母概率分析法来破译MASC密码一样，我们也可以通过双字母概率分析法来破译这种双字母替代密码。

例子：丹尼·特乔（Danny Trejo）有句名言——“EVERYTHING

GOOD THAT HAS HAPPENED TO ME HAS HAPPENED AS A DIRECT RESULT OF HELPING SOMEONE ELSE."（发生在我身上的所有好事都是我帮助别人所产生的直接结果。）

下面让我们来用双字母替代密码的方式给这句话加密。首先，我们在连续出现的两个相同字母之间插入字母 X，这样我们就得到：

EVERYTHING GOXOD THAT HAS HAPXPENED TO ME HAS HAPXPENED AS A DIRECT RESULT OF HELPING SOMEONE ELSE

接着我们把字母分成两两一对，得到：

EV ER YT HI NG GO XO DT HA TH AS HA PX PE NE DT OM EH AS HA PX PE NE DA SA DI RE CT RE SU LT OF HE LP IN GS OM EO NE EL SE

由于把字母两两组成一对的方式不同，在这段话中出现过两次的单词"HAS"被加密成了两种不同的形式。但是，单词"HAPPENED"也出现了两次，而且两次都被加密成了同样的密文。不需要完成对整段话的加密，我们就可以看出这是双字替代密码系统的一个弱点。如果破译密码的人能够准确地猜出某个多次出现的单词，他就非常有可能把这个词和重复出现的密文字符匹配起来。以上述发现作为基础，密文余下的部分也将被破解。

不管索利斯究竟对破译密码的技巧有多少了解，他都非常清楚地表示：因为他给出的密文信息太短，所以所有这些密码破译的技巧都不适用于他的密码。然而，索利斯的这种想法是错误的。虽然索利斯既没有透露破译密码的专家是谁，也没有透露该专家究竟是用何种技巧破译这段密码的，但是，通过观察，我们可以发现索利斯的这段密码中有一些弱点，这些弱点可能会帮助密码破译专家破解这段密码。

我们观察到的第一个问题是，虽然加密者可以用任意关键词打乱字母表，但是许多关键词中并不包含V、W、X、Y、Z这几个字母。如果加密者选择的关键词中不包含上述几个字母，那么网格的最后一行就会保留这几个字母的原始顺序。也就是说，当加密者使用不包含V、W、X、Y、Z这几个字母的单词作为关键词时，网格的形式如下：

?	?	?	?	?
?	?	?	?	?
?	?	?	?	?
?	?	?	?	?
V	W	X	Y	Z

打“?”的位置上的字母目前尚不明确，然而，我们已经知道了网格中20%的字母的正确位置。密码分析师有时会假设双字替代密码的网格以上述形式出现，这实际上是在赌加密者选择的关键词中不包含V、W、X、Y、Z这几个字母。在索利斯的密码中，这个赌注下得很对，因为索利斯选择的关键词中确实没有这几个字母。

事实上，仔细观察密码A以后，我们还能看出其他一些可能性。密码A的密文如下：

CBFTM HGRIO TSTAU FSBDN WGNIS BRVEF BQTAB QRPEF BKSDG MNRPS RFBSU TTDMF EMA BIM

我们把以上密文中的字母两两组成一对，就得到：

CB FT MH GR IO TS TA UF SB DN WG NI SB RV EF BQ TA BQ RP EF BK SD GM NR PS RF BS UT TD MF EM AB IM

我们可以看出，有些双字母组合（BQ、EF、SB以及TA）出现了两次。这些双字母组合很可能与正常英语中高频出现的双字母组合相对应。虽然只有经过多次尝试才能把这些双字母组合与正常英语中高频出现的双字母组合匹配起来，但密码分析师们是非常有耐心的。此外，在这段密码中至少还有另外一处弱点能够帮助密码分析师攻破这段密码。

在这段密码的密文中，出现了一些字母顺序相反的双字母组合。比如，既有BS，又有SB。在正常英语中，某些字母组合常常以一种顺序出现，却极少（或者从来不）以相反的顺序出现，比如QU/UQ这两种字母组合。而ER/RE这两种组合则都会在正常英语中高频出现，因此这是两种在英语中十分特别的字母组合。根据这一点，密码的破译者可以合理地假设这段密码中的BS代表ER，而SB代表RE。当然情况也可能正好相反（也就是BS代表RE，而SB代表ER）。事实证明，在这段密码中，第一种假设是正确的。研究字母顺序相反的字母组合是破解许多加密系统的一种很有用的方法。在《未解之谜（上）》的第5章中，我曾指出这种密码分析技巧可能会帮助我们破译亨利·德博斯尼斯留下的未解之谜。

虽然我们已经发现了索利斯密码中的3处弱点，但距离我们完全破译这段密码还有很远的距离。从这3处弱点到最终的答案之间有多条不同的路径，破译这段密码的专家很可能会因为一些错误的猜测而不得不走回头路（毕竟破译这段密码用了两个星期的时间）。以上这些分析只是为了向读者展示在破译这段密码时可能采取的初始步骤。从上面这些分析中，读者也可以看出，这段密码并没有索利斯想象的那么安全。

最终，这名密码专家发现，索利斯选择的关键词是“SURPRISE”（惊喜）——这就是索利斯打算在自己死后通过通灵的方式告诉世人的关键词。索利斯认为，有了这个关键词以后，世人就能够很容易地破译出自己创造的这段密码了。有了这个关键词，我们就可以构建出以下这个用于加密或解密的网格：

S　U　R　P　I（索利斯省略了字母J。）
E　A　B　C　D
F　G　H　K　L
M　N　O　Q　T
V　W　X　Y　Z

有了这个网格，我们就可以轻松地破译密码A。密码A的破译结果如下：

密文：CB FT MH GR IO TS TA UF SB DN WG NI SB RV EF BQ TA
明文：BA LM OF HU RT MI ND SG RE AT NA TU RE SX SE CO ND
密文：BQ RP EF BK SD GM NR PS RF BS UT TD MF EM AB IM
明文：CO UR SE CH IE FN OU RI SH ER IN LI FE SF EA ST

加入正确的空格和标点符号以后，我们就得到以下结果：

BALM OF HURT MINDS GREAT NATURE'S X SECOND COURSE CHIEF NOURISHER IN LIFE'S FEAST.

（抚慰伤痛思绪的香膏，大自然的第二道菜，生命筵席的主要营养。）

上面这段信息中的字母X只是为了分隔两个连续出现的字母S而加入的。果然，这段信息引自莎士比亚的戏剧——《麦克白》，我帮读者找到了这句话的上下文，并且摘录如下：

将劳心纠结的衣袖编结整齐的睡眠，
那每天生命的死亡，酸楚劳碌后的沐浴，
抚慰伤痛思绪的香膏，大自然的第二道菜，
生命筵席的主要营养。

索利斯知道，如果密文足够长，这种普莱费尔密码是比较容易破解的。但是他认为，自己给出的这段密码只有65个字母，这么短的信息应该是安全的。不管是谁破译了索利斯的密码，它都不是被成功破译的最短的普莱费尔密码。如果有"被成功破译的最短的普莱费尔密码"这项纪录的话，保持纪录的人应该是阿尔夫·蒙日（Alf Mongé）。

蒙日出生在挪威，他在挪威度过了生命的前19年岁月。然后，他来到美国并参了军。在第二次世界大战期间，蒙日在伦敦为英国政府工作了很长时间。为了表彰他的贡献，英国政府为他颁发了大英帝国勋章。在战争爆发前的1936年，蒙日曾经破译过一段只有30个字母的普莱费尔密码，这段密码是在1933年被当作一项密码挑战而公布出来的。此外，蒙日还能在30分钟内破译很短的普莱费尔密码，这段密码仅比索利斯的密码短一点儿（长度是50~60个字母）。

在第5章中，我们曾经解释过，对于一种给定的加密系统，密文必须足够长，才能保证解密的过程不会给出多种有意义的明文信息。读者可以回忆一下，在《未解之谜（上）》第5章中我们给出了“唯一点”的概念，唯一点的公式写作：$U = \log_2(K)/D$，其中U代表唯一点的数值，K代表可能密钥的数量，而D代表这段信息中单位字母的冗余度。要把这个公式运用到普莱费尔密码系统中，我们首先需要决定究竟可以以多少种方法将字母表打乱并放入5×5的网格中。如果我们规定关键词必须是一个真实的英文单词，那么K的值取决于加密者在选择关键词时使用的词典中究竟有多少个单词。在极端情况下，我们假设加密者选择的词典是1989年出版的《牛津英语词典》第二版，这一版的《牛津英语词典》共有20卷，全书总重62.6千克。这本词典中一共收录了615 100个单词。[6]我们把615 100这个数字代入上述公式，就得到$U = \log_2(615\,100)/0.7 \approx 27$个字母。事实上，普莱费尔密码的实际唯一点数值比27还要小，因为一些不同的关键词会构建出完全一样的加密网格。比如，Care和Career这两个关键词对应的网格就是完全一样的。假设网格中的字母可以按任意顺序排列，考虑字母顺序的所有情况，我们则会得到：$U=\log_2(26!)/0.7 \approx 126$个字母。由此我们可以看出，密钥空间的大小是非常重要的！

在设计现代军事密码时，最重要的一条设计标准就是：密码的唯一点数值必须超过用这种加密系统加密的信息的长度。

备用计划

虽然索利斯的密码被破译了，但他并不打算就此放弃对通灵术的测试。事实上，索利斯已经对这种情况做好了准备。在索利斯的论文中，还有一个名为“附属测试”的附录部分，索利斯在该部分写下了以下内容：

> 虽然我不相信会出现这样的情况，但是从理论上讲，仍然不能排除以下可能：如果一位破译技术足够高的密码专家愿意花极大的精力来攻克这段密码，他有可能会在不知道关键词的前提下就破解我的这段密码。因为，虽然我的密码很短，从而给破译带来了难以逾越的障碍，但是这种加密系统在理论上并不是“不可破译的”。因此，我认为有必要增添第二段密码，这段密码是用一种绝对不可能被破译的加密系统加密的。也就是说，在不知道关键词的情况下，即使完全了解这种加密系统，也不可能破译出密码的内容。因此，我需要一个满足上述条件的加密系统，同时还要保证这个加密系统的关键词传递起来非常容易，满足这两个条件的加密系统不可能非常简单。接下来，我会给出加密的信息，并解释这种加密系统的原理。我还会描述我死后将会通过通灵的方式向大家提示关键词的性质。[7]

索利斯在密码B中使用的加密系统与被破译的密码A中使用的加密系统完全不同。我将密码B的密文复制如下[8]：

INXPH CJKGM JIRPR FBCVY WYWES NOECN SCVHE GYRJQ
TEBJM TGXAT TWPNH CNYBC FNXPF LFXRV QWQL

公布密码A的时候，索利斯并没有透露他加密的方法，但是在公布密码B的时候，他却解释了密码B的加密方式。密码B的加密方式基于一种16世纪的加密系统，被称为“维吉尼亚密码”（Vigenère cipher），但布莱兹·德维热内尔（Blaise de Vigenère，1523—1596，也被译为布莱兹·德·维吉尼亚）却并不是这种密码的真正发明者。我们将在本书的第9章中详细讨论维吉尼亚密码的解法，但在本章中，读者只要理解经索利斯修改后的维吉尼亚密码就足够了。

在经索利斯修改后的系统中，他将某种已经发表的作品作为密钥。索利斯从这部作品中挑选出一串单词，如果这串单词的后部有前部已出现过的单词，就将这些重复的单词删去。索利斯挑选的单词串必须足够长，使删除重复的字母以后至少剩余74个不同的单词——这是因为索利斯想要加密的信息一共有74个字母，而每加密一个字母需要用到单词串中的一个单词。

那么，如何用一个单词来加密一个字母呢？第一步是把每个单词转化成一个数字。这一步需要我们把这个单词中的每一个字母都转化成一个数字，转化的规则是A=1，B=2，C=3，…，Z=26。然后，我们把单词中每个字母对应的数值加在一起，得到一个总和，比如“BACON”（熏肉）这个单词对应的数字就是 2+1+3+15+14=35。如果某个单词对应的数值大于或等于26，索利斯就把这个数值减去26。因此，“BACON”这个单词对应的最终数值是35-26=9。当一个单词比较长时，可能需要多次从这个总和中减去26。用数学术语来描述，这个过程叫作用单词对应的总数去“模除26”，简称“模26”（mod 26）。用一个数字去模除26相当于把这个数字除以26，然后取余数（除法的商则不是我们关心的问题）。在对每一个单词进行上述操作以后，索利斯就会得到一串数字，在这串数字中，每一个数字的值都在0~25之间（包括0和25）。假设索利斯想要加密的原始信息是：“GREETINGS FROM THE DEAD”（来自死者的问候），而他从某部作品中取得的单词串所产生的密钥是：14，8，21，5，2，11……那么在加密的时候就要把这段信息和这串密钥一一对齐如下：

G　R　E　E　T　I　N …

14,　8,　21,　5,　2,　11,　17 …

接下来，我们把这段信息中的每个字母都往字母表的后方移动几步，具体步数等于这个字母下方的数字。完成这个步骤以后，我们就得到：

U　Z　Z　J　V　T　E …

在上面列出的这串字母中，最后一个字母的加密方式是N+17=E，对于这种操作，我有必要做一些解释。如果我们从字母N开始数，数到12的时候就已经到达字母表的最后一个字母Z了。在Z之后再也没有其他字母了，于是，我们就重新回到字母表的开头，从第一个字母A开始继续数下去，一直数到17，我们正好到达字母E处，所以N+17=E。我们还可以通过另一种方式来理解上述操作，那就是把整个字母表想象成一个巨大的圆环，就像一面钟上的数字一样。当我们在某个字母上加上一个数字时，钟面上的指针就从这个字母开始向前移动一定的步数，指针停止的地方就是这个字母对应的密文。

为了将加密的规则定义得更加明确，索利斯规定，在产生密钥的过程中，“用连字符（-）连在一起的两个单词一律被视作两个分开的单词”。[9]这条规则无意中为密码的破译者提供了一条小小的线索。如果索利斯选作密钥的这串单词中根本没有连字符的话，索利斯就完全没有必要引入这条规定了。因此，这条规定是否意味着他所选择的密钥单词串中包含连字符呢？

此外我们还可以假设，密码B中的信息可能也引自莎士比亚的作品。但是，在打开《莎士比亚全集》寻找含有连字符的句子之前，我们首先看看索利斯论文中的以下段落：

> 我没必要针对原始信息给出任何提示。一段密码的破译方法是无穷的，但是以下事实可以帮助我们区分正确和错误的解法：在正确的解法中，作为密钥的单词串必须是能够从一本印刷作品中找到的一段话（当然，接下来需要对这段话进行上述一系列操作）。我认为，在印刷作品中无法找到多条能将密码破译成有意义信息的密钥串——这种情况发生的概率几乎为零。我的目标是：我在死后会通过通灵术告诉大家密钥串究竟来自哪一部作品中的哪一段话。[10]

从上面这段话中，我们可以看出索利斯不相信密码B会产生多种解法，因为“这种情况发生的概率几乎为零”。这就意味着密码B的明文不应该引自莎士比亚的作品。但是索利斯也不太可能挑选一本特别不出名的作品作为密钥串的来源。因为如果索利斯打算在死后将这部作品的题目告诉某人的话，他肯定希望此人能够比较轻松地找到它，解开索利斯留下的密码，从而证实死者确实可以与这个世界交流。

索利斯希望其他人也用和他一样的方法创造密码，以便共同验证通灵术是否真的存在。然而，德尼斯·帕森斯认为，密码B所用的加密方法计算量太大，许多人根本没有能力完成这种加密工作。因此，帕森斯建议普通人使用一种更加简单的加密方法。在这种加密方法下，只需要选择某部作品中的一段话作为密钥，但不需要用到密钥中的所有字母（在第8章中，我将向读者介绍这种“滚动密钥加密系统”）。加密者只需要每隔两个字母挑出一个字母来，然后把这些字母转化成数字，就可以用与密码B最后一步的相同方式来加密信息了。

索利斯认识到，如果从一段有意义的文字中每隔两个字母挑出一个字母来，所得到的结果不会是随机的字母串，而会是一串与正常英语的字母概率一致的字母串。然而，索利斯错误地认为，双字母组合或者三字母组合的概率会“接近随机，因此不能帮助破译者破译密码”。[11]而此前破译过索利斯密码的那位不愿具名的密码专家则认为，这种简化的加密系统“安全性相当低”。[12]

此时，索利斯并没有提供C计划或者密码C，来应对密码A和密码B都被破解的情况。但是索利斯这样做并不是因为他对密码A和密码B的安全性特别有信心。索利斯写道：“我认为这样做会降低实验成功的概率，因为要记住第3个关键词对于我的记忆力而言会是一个负担。”[13]

虽然索利斯认为3个关键词太多了，但他似乎又认为至少需要两个关键词。在密码A被破译以后，索利斯又提出了一条新密码来取代密码A，我们将这段密码称为密码C。在密码A被破译以后，索利斯立刻在一篇1948年的论文中提出了密码C。

密码A使用的是普莱费尔加密系统，而密码C使用的是普莱费尔加密系统的升级版，这种加密系统被称为“双重普莱费尔加密系统”（double Playfair cipher）。[14]在双重普莱费尔加密系统下，加密者首先要使用普莱费尔加密方法，根据某个关键词把字母表打乱放入网格中，但是在完成上述步骤以后，双重普莱费尔加密系统还要求加密者继续完成更多的加密步骤。在第一步以后，加密者要把某个特定的字母放到密文的最前面和最后面，然后再次使用普莱费尔加密系统对修改后的密文进行第二次加密，但第二次加密必须使用和第一次加密不同的关键词来打乱字母表。之所以要在进行第二次加密前在密文的最前面插入一个字母，是为了打破之前字母两两配对的方式。而在密文的最后插入一个字母，则是为了让密文的字母总数是偶数。插在密文最前面的字母并不一定要和插在密文最后面的字母相同，但是在这种加密系统的规则中，上述两个字母正好是相同的，这就使得这种加密系统出现了一个弱点！不管怎么说，当索利斯用上述步骤将他的原始信息加密以后，他得到的最终密文如下[15]：

BTYRR OOFLH KCDXK FWPCZ KTADR GFHKA HTYXO ALZUP PYPVF AYMMF SDLR UVUB

这就是索利斯发表的最后一段密码，我将这段密码称为密码C。随着时间的流逝，密码B和密码C都没有被破译出来。索利斯最终于1984年逝世，享年90岁。此时，真正的挑战，或者说真正的实验终于开始了。

所有试图从已逝的索利斯那里获得密钥的努力都以失败告终，我想索利斯的亲朋好友以及和他一起进行心灵学研究的同事应该都很失望吧。又过了许多年，最终在1996年，詹姆斯·J. 吉洛格利和拉里·哈尼施（Larry Harnisch）终于找到了索利斯密码C的密钥。吉洛格利和哈尼施宣布，在这个密码中，两轮普莱费尔加密过程的关键词分别是“BLACK”（黑色）和“BEAUTY”（美丽）。第一轮加密结束后插入密码头尾的字母是T。密码的明文是：

THIS IS A CIPHER WHICH WILL NOT BE READ UNLESS I GIVE THE KEYWORD.

（这是一条没有人能读懂的密码，除非能得到我提供的关键词。）

吉洛格利和哈尼施在他们发表的论文中解释了如何找出关键词的方法。这篇论文的最后一段写道：

> 生存研究基金提供了1 000美元的赏金，奖给第一个解出密码B或者密码C的关键词的人。可惜悬赏期已经于1987年截止，而在

> 1987年时，东芝牌笔记本电脑还没有通灵的本事（甚至还根本没有被制造出来）。我们就把这次成功的电脑通灵当作我们获得的奖赏吧。[16]

假如读者已经忘记了吉洛格利的名字，我有必要提醒你们一下：在《未解之谜（上）》第5章中我们曾经讨论过吉洛格利成功破译IRA密码的故事。他是一名密码学家，而不是灵媒。随着密码C被解开，从1996年开始，索利斯只剩下最后一件工具可以用来证明他死后仍然有知，并且能与这个世界进行通信——他生前留下的密码B。

虽然吉洛格利和哈尼施优雅漂亮地解出了密码C，但他们却无法破译密码B。他们写道：他们“相信要想破译密码B，必须首先找到正确的文本”。[17]吉洛格利和哈尼施也花了不少精力试图破译密码B，他们对数百本书中的文字进行了测试，但是索利斯选择的密钥串显然并不在这几百本书中。也许，研究者可以从一种心理学的角度来试图破译索利斯创造的密码C，如果能够对索利斯的个性有足够深的了解，就有可能猜出他会从什么样的作品选取用于构建密钥的单词串。

1948年，索利斯显然无法猜到未来的计算机会具有如此强大的计算能力，更加不可能猜到未来的密码专家能够利用这种强大的计算能力来破译密码。随着科技的发展，未来又会出现什么样的新工具呢？

索利斯留下的启示

虽然索利斯的实验失败了，但是他的尝试却启发了许多人，让他们也参与到通过密码来测试通灵术的活动中来。英国的一名退休律师托马斯·尤金·伍德（Thomas Eugene Wood）就加入了这一行列。[18]伍德非常明智地选择了索利斯密码B的加密方式来创造他自己的密码。[19]但是他提供的赏金却比索利斯低一些，只有20英镑。1950年，伍德公布了他创造的密码，我将这段密码复制如下[20]：

```
FVAMI NTKFX XWATB OIZVV X
```

伍德写道：

> 破译我的这条加密信息的密钥来自一本常见的书中的一个段落。在我死后，我会试图将这个段落的内容告诉某人，并且我还会试图告诉他这个段落是用哪种外语写成的。如果能够破译我的这段信息的话，你们就会发现这段信息不是用单一语言写成的。我会在死后试图告诉某人这条信息究竟使用了哪几种语言。我之所以要使用外语，是因为索利斯在他的文章第262页的“附加注释”的开头部分提出了要避免使用英语中最常见的单词的想法，因为这样可以增加破译密码的难度。[21]

和索利斯一样，伍德在完成加密过程之后销毁了所有相关文件，并且也没有把他的密钥告诉任何人。[22]但是，伍德在一封信中给出了这个密码的一个解法样例。[23]虽然这个样例并不是真正的正确解法，但是它展示了伍德所能想象的可能性范围。这个样例的密钥是“To be or not to be, that is the question”（生存还是死亡，这是一个问题）。但是，伍德却并没有从莎士比亚的《哈姆雷特》原始作品中直接引用这段话，而是使用了这句话的西班牙语译本。有了这个密钥以后，就可以破译出密码的明文信息是：KUBBA DAJES UISIC IHOLZ。我们应该如何在这段明文中插入空格，得到一段有意义的文字呢？这并不是一件简单的事，但是我们不要忘了，伍德已经告诉我们，这段明文信息并不是用英文写成的。插入空格的正确方式是KUBBADA JE SUIS ICI HOLZ。其中，第一个词是乌尔都语，意思是“注意”；接下来的三个词是法语，意思是“我在这里”；最后一个词是德语，意思是木头，也就是“Wood”（伍德），他的署名。因此，破译出来的明文信息的意思就是：“注意。我在这里。——伍德”

我们不难看出，对于这段密码而言，如果认不出明文中使用的语言，即使知道正确密钥，也很有可能找不到正确的解法。可以设想，如果某人有极强的信心认为自己已经从死去的伍德处获得了正确的密钥，那么此人就一定会花时间在不同的外语词典中查找这几个词，或者拿着这几个词去寻求语言学家的帮助。如果你只有一种可能的解法，并且充分相信这种解法是正确的，那么在这种解法上多花一些时间也是值得的！

伍德出生于1887年，于1972年去世。伍德生前留下的密码至今也未被解开，因此我们不得不怀疑他并不具有死后与世人交流的能力。对于伍德实验的失败，索利斯想必感到十分悲哀吧？但是，可能会令索利斯感到欣慰的是，在他还活着的时候，又有另一位研究者加入到了用密码测试通灵术的行列中。

和伍德一样，弗兰克·C. 特里布（Frank C. Tribbe）也是一名律师，但他执业的地方不在英国，而是在美国。特里布研究的问题不仅仅局限于人死后能否通灵这一项，他著有许多书籍，包括《创造性冥想》（*Creative Meditation*，1975），《耶稣的肖像？都灵裹尸布的配图故事》（*Portrait of Jesus? The Illustrated Story of the Shroud of Turin*，1983），《超自然学习指南》（*Guidebook for the Study of the Paranormal*，与罗伯特·H. 阿什比合著，

1988)，《丹尼和神秘的裹尸布》(*Denny and the Mysterious Shroud*，读者群是青少年和“休闲阅读者”，1998)，《阿瑟·福特选集》(*An Arthur Ford Anthology*，选编，1999)，《我，亚利马太的约瑟》(*I, Joseph of Arimathea*，2000)，《圣杯之谜已被破解》(*The Holy Grail Mystery Solved*，2003)，以及《灵魂图像：第四维度的图片证据》(*Spirit Images: Pictorial Proof of the Fourth Dimension*，2008)。此外，特里布还在这些领域中发表过许多论文，并做过一些演讲。特里布同时还是《灵魂前沿》(*Spiritual Frontiers*)杂志的编辑。

图 7-2　弗兰克·C. 特里布（1914—2006）

1980年，特里布在《宗教与心灵研究会期刊》(*Journal of the Academy of Religion and Psychical Research*)杂志上发表了一篇论文，这篇论文的摘要如下：

> 本文概述了如何通过一些技巧创造出一条加密容易，但解密困难的简单密码，以证明人死后仍然有灵魂存在，并提供了创造这种密码的具体步骤。[24]

然而，在特里布的这篇论文中，所谓“解密困难”的密码只不过是简单的单套字母替代密码（MASC密码）而已。特里布的主要想法是，人们可以在生前通过一个关键词来加密一段信息，然后将密文放在密封的信封里寄给亚利桑那州菲尼克斯市的生存研究基金会。这个密封的信封在寄信人去世之前都不会被打开。如果有人认为自己已经通过与死者的交流获得了密码的密钥，他可以与生存研究基金会联系。

我发现了一个有趣的事实，这些花费大量精力研究死后世界的人都比普通人更加长寿。我们要提到的下一位研究者阿瑟·S. 伯杰（Arthur S. Berger）也不例外。伯杰生于1920年，而在我撰写本书的时候他仍然在世。

图 7-3　阿瑟·S. 伯杰（1920—　）

和我已经介绍过的许多通灵学研究者一样，伯杰也是一名律师。他还是第二次世界大战的退伍老兵。伯杰最新的一本书出版于2012年，在这样的高龄仍然能完成写作，实在是令人印象深刻。伯杰的大部分作品也是讨论与死后世界有关的问题。

1983年，伯杰发现了研究上述问题的另一种方法，这种方法虽然本质上不涉及密码，但至少与数字有关。在他的这种研究方法中，先要通过某个灵媒找到一个能与世人真正交流的灵魂。然后，灵媒会要求这个死去的灵魂提供三个数字，其中一个数字是另外两个数字的和，并对这三个数字保密。然后，需要另外几位灵媒的参与。这个死去的灵魂需要尝试向每一位灵媒分别提供一个数字。伯杰写道："如果第三位灵媒收到的数字正好是前两位灵媒收到的数字的和，那么这项实验就算取得了成功。"[25]

然而，伯杰设计的上述实验存在许多问题。问题之一是有些灵媒之间会互相分享信息。为了理解这个问题的具体机制，我们可以考虑三位灵媒爱丽丝、鲍勃和卡尔。假设一个叫维克托的人先找到了灵媒爱丽丝，但却对爱丽丝的通灵能力持怀疑的态度。因此，维克托没有明确地告诉爱丽丝自己想要联系的已逝亲友究竟是谁。爱丽丝猜维克托想联系的是他的父亲，于是爱丽丝说："我感觉到有一位比你年长的男人想要和你说话。"灵媒这种含糊的说法是故意为之的，因为这样说可以涵盖许多可能性，所谓"一位比你年长的男人"既可以是维克托的父亲，也可以是他的祖父，还可以是他的叔叔，甚至他的哥哥。维克托可能会对爱丽丝的说法嗤之以鼻，说所有比他年长的男性亲戚目前都仍然在世，也可能选择不透露更多信息，只是静静地坐等这次降灵会结束，但他离去的时候却流露出明显不满的神气。不管怎么说，爱丽丝可能会从维克托的反应中知道，自己的猜测是错误的。虽然维克托再也不会来爱丽丝这里，爱丽丝却可以把她了解到的关于维克托的信息分享给另外两位灵媒鲍勃和卡尔。当维克托访问鲍勃的时候，鲍勃可能一上来就会这样说："我感到有一位女性想要与你交流。"如果维克托上次曾经明确告诉爱丽丝，自己想要联系的是已经去世的姐妹的话，鲍勃甚至还可以说得更加具体。如果鲍勃的这次降灵会进行得成功，维克托很可能会再次找鲍勃。而如果鲍勃的某些话错得太离谱，他就有可能失去维克托这名客户。但即便如此，鲍勃仍在降灵会的过程中了解到了关于维克托的进一步信息，他可以把这些信息告诉第三位灵媒卡尔。如果接下来维克托继续访问卡尔，这次降灵会可能会进行得非常成功。鲍勃和爱丽丝从维克托对错误猜测的反应中积累了不少关于维克托的信息，只要卡尔能够获得这些信息，就能够成功地愚弄维克托，并与维克托发展出一种利益关系。

维克托可能会回家对父亲说，卡尔虽然从来没有与他见过面，却"知道一些他不可能知道的事情"。

有些读者可能会问：既然爱丽丝、鲍勃和卡尔之间是一种竞争关系，那么为什么爱丽丝和鲍勃要帮助卡尔呢？这是因为，通常客户并不会像上面的例子那样，按照灵媒名字的英文字母顺序来依次访问。有些客户会先访问爱丽丝，有些客户会先访问鲍勃，而还有一些客户会先访问卡尔。因此，如果这三位灵媒互相分享信息，那么他们每个人就能够从其他两人的失败中获得好处。

灵媒常常会形成有组织的集团，并且在集团内部共享客户信息，尤其是比较富裕的客户的信息。这种情况已经存在很长时间了。在数字时代，上述信息共享的过程变得更加简单和方便。我们可以想象，存在一个有密码保护的灵媒百科，这个网站相当于灵媒界的一个有用的数据库，任何灵媒只要缴纳一定的月费，就能够随时获得这些不断更新的客户信息。

在某些表演中，灵媒会现场为观众席中的某位观众算命。但注意，有时候观众会提前用信用卡买票，并且座位的号码也是确定的。灵媒可以事先通过他信任的某种媒介对观众的信息进行调查和研究，通过这种并不困难的方式，就可以提前知道观众的许多个人信息。

伯杰设计的上述实验存在明显漏洞（他假设灵媒之间是不会互相交流的），在他第一次实验时，遭遇了失败。也许，我们可以乐观地认为，他的实验证明了一项事实，那就是并不是所有灵媒都会联合起来共同欺骗客户！

标准化数字实验

除了上述方法之外，伯杰还在他1987年的书《死者中的精英》（*Aristocracy of the Dead*）中提出了另外一种方法，这种方法确实用到了密码这一工具。伯杰提出，希望死后与世人交流的人可以在生前从一本特定的词典中随机选出一个词。在伯杰给出的例子中，这个人选定的单词是“builder”（建造者）。选定了这个单词以后，此人需要把这本词典该词条下的所有字母一一标上数字，从这个词开始，包括这个词的音标，以及它的所有释义，只可以跳过非字母的字符。伯杰举了一个具体的例子，我把这个例子复制在下面的图7–4中。

1	2	3	4	5	6	7	8	9	10	11	12	13
B	u	i	l	d	e	r	(b	i	l	d	e	r)
14	15	16	17	18	19	20	21	22	23	24	25	26
1. o	n	e	w	h	o	b	u	i	l	d	s	o
27	28	29	30	31	32	33	34	35	36	37	38	39
r	o	v	e	r	s	e	e	s	b	u	i	l
40	41	42	43	44	45	46	47	48	49	50	51	52
d	i	n	g	o	p	e	r	a	t	i	o	n
53	54	55	56	57	58	59	60	61	62	63	64	65
s;	o	n	e	w	h	o	s	e	o	c	c	u
66	67	68	69	70	71	72	73	74	75	76	77	78
p	a	t	i	o	n	i	s	t	o	b	u	i
79	80	81	82	83	84	85	86	87	88	89	90	91
l	d,	a	s	a	c	a	r	p	e	n	t	e
92	93	94	95	96	97	98	99	100	101	102	103	104
r,	a	s	h	i	p	w	r	i	g	h	t,	o
105	106											
r	a											

图7–4　伯杰给出的例子

接下来，就可以对信息进行加密了。加密的方式很简单，只要把每个字母都替换为上图中对应的数字就可以了。如果某个字母在他想要加密的信息中多次出现，加密者就可以用不同的方式来加密这些字母。比如，在词典的“builder”词条中，字母E与以下的多个数字对应，这些数字包括：6，12，16，30，33，34，46，56，61，88和91。伯杰给出了以下明文信息和密文信息作为示例[26]：

明文信息：	W	E	H	A	V	E	S	O	U	L	S
密文信息：	98	6	18	106	29	12	32	26	21	39	53

词典名称和版本信息与加密后的密文信息保存在一起。因此，如果某人能猜出加密系统中的关键词（或者某人宣称死者将这个关键词告诉了他），他就可以利用图7–4中的对应关系来破译这段密码，看看能否获得有意义的明文信息。

虽然通过同音字母的加密方式可以把同一个字母加密成不同的形式，但是这种加密系统的安全性仍然比较弱。伯杰没有意识到这种加密系统的弱点，他认为同音字母的加密方式能让"不知道关键词的人极难破译这段密码"。[27]他写道：

> 由于韦氏词典是一本包含大约600 000个词的足本词典，所以猜到关键词"builder"的概率大约是1/600 000。如此小的概率足以排除破译人恰巧猜到关键词的可能性。[28]

1991年，伯杰提出，可以用这种测试来检验人是否可以轮回重生。这次他给出的密文仍然是98 6 18 106 29 12 32 26 21 39 53。伯杰这次没有给出对应的明文，而是写道："这一串数字当然是毫无意义的，如果没有正确的密钥，就不可能读出这串数字背后的意义。"[29]

此前的这类实验都以失败告终，但伯杰却在这些失败中找到了一些正面的启示。伯杰写道："而且，实际情况显示，此前通过留下测试信息来进行死后实验的所有人都是完全诚实的——比如F. W. H. 迈尔斯、R. H. 索利斯和J. G. 普拉特——因为在他们死后，至今没有人能说出他们设置的密钥。”[30]我同意伯杰的上述判断，并且尊敬这些研究者的诚实。

为了检测人死后灵魂能否继续存在，上述这些研究者使用了一系列密码和数学工具来研究这个问题。但是除了这个方向以外，还有另外一个同样毫无成果的研究方向。接下来，我会向读者介绍这方面的内容。这个研究方向之所以引起了我的兴趣，是因为这方面的研究也与安全问题有关，但这次的安全问题是物理上的问题。

组合锁测试

1968年，弗吉尼亚大学医学院精神病学系的伊恩·史蒂文森（Ian Stevenson）博士提出了一种“组合锁测试”。组合锁测试的主要方法是，让不同的人来设置组合锁的密码（准确地说是萨金特和格林利夫牌8800型号组合锁），然后让灵媒来开锁。如果灵媒能打开生者设置的组合锁，就能验证读心术/透视能力确实存在；而如果灵媒能打开死者生前设置的组合锁，则证明死者的灵魂确实可以与这个世界的人交流。

图7-5　伊恩·史蒂文森（1918—2007）

伊恩·史蒂文森在描述上述实验的时候写道：“因为组合锁的密码有3个数，每个数有50种情况（共6位数），所以通过随机尝试找到正确密码的概率是1/125 000。因为这个概率很低，所以组合锁可以用于我们的实验目的。”[31]

然而，事实上史蒂文森的实验什么也证明不了。人们可以通过许多方法打开组合锁，这些方法既不要求人们会读心术，也不要求人们能和死者交流。[32]史蒂文森也知道一些打开组合锁的方法，他简要地对这个问题进行了以下讨论：

> 参加实验的人必须小心某些廉价的组合锁，因为那些组合锁可以随意重设密码。有时候这种锁的设计有瑕疵，通过强行击打（或者其他一些非正常方式）也能打开。我选用的是萨金特和格林利夫公司制造的组合锁，该锁的生产厂商保证，他们的产品无法通过击打、摇晃、震动等方式打开。此外，也没有办法通过拨动数字、听锁的声音来打开这种锁。[33]

然而，许多密码产品都向客户保证他们的密码绝对安全。所以，这种事情只能靠买家自己小心了！

史蒂文森认为，自己提出的组合锁测试比索利斯的密码测试更适合普通人。虽然伍德选择使用索利斯提出的密码测试方法，但是很多人似乎根本不想学习关于密码学的知识。

事实上，史蒂文森的实验确实吸引了很多参与者。1976年，史蒂文森又发表了一篇论文，为他前一篇论文的读者提供了一些更多的通灵术测试实验。当时，史蒂文森的办公室里一共有11把锁，其中包括他自己的两把锁以及一把由索利斯设置了密码的锁。有趣的是，索利斯为组合锁选择的密码数字是从他的某一段密码中得来的（可能是密码C）。所有这些锁都是希望参加通灵术测试实验的人交给或者寄给史蒂文森的。在史蒂文森的

第二篇论文发表时，上述这些参与者中有两位已经去世，然而他们留下的两把锁在他们死后也依然无人能打开。史蒂文森还提到，英格兰地区还有另外两把类似的组合锁，那两把锁的主人也想参加测试通灵术的实验，但是那两把锁并不在史蒂文森的手上。在那两把锁的主人去世以后，那两把锁也同样没有被打开过。

1989年，史蒂文森又与两名合作者联名发表了一篇论文。这篇论文记述了通灵术测试实验的进展。此时，索利斯已经去世，另一个给史蒂文森留下组合锁的人——超心理学家J. G. 普拉特也已辞世。虽然人们尝试用各种方法开锁，但不管是索利斯留下的组合锁，还是普拉特留下的，都和其他人留下的锁一样打不开。在1989年的那篇论文发表时，已经有42人次尝试打开帕特的锁，然而这42次尝试全部以失败告终。

在这篇1989年的论文中，史蒂文森还向读者提供了关于索利斯的密码B和密码C的最新情况。密码B和密码C分别被142人次和87人次尝试破译，所有这些尝试全部失败。当然，后来吉洛格利和哈尼施终于利用计算机解开了密码C。

史蒂文森本人于2007年去世，他留下的组合锁在他死后也无人能够打开。

以上这些专业人员（主要是律师）试图证明通灵术存在的努力似乎都不太成功。然而，接下来，另一位来自纽约州罗切斯特市的律师又设计出了一种比以上所有方法更不可能获得结果的方法。在阅读下一部分内容时，读者应该始终记住，专业学术期刊中的内容（以上介绍的所有内容都刊登在专业学术期刊上）和作者自行出版的书籍中的内容在可信度上是有很大区别的。

雷达之爱

通过密码与死者交流的方式千奇百怪，然而在这些方式中，最奇怪的绝对是约瑟夫·马丁·菲利的方法。在《未解之谜（上）》的第1章中我们曾经提到，约瑟夫·马丁·菲利是众多声称自己已经破译了伏尼契手稿的人之一，然而他并未就此停止在密码破译方面的脚步。在声称破译了伏尼契手稿以后，菲利又试图用换位重排密码与一些已经逝世的名人取得联系，这些名人包括达·芬奇、托马斯·爱迪生，以及富兰克林·德拉诺·罗斯福。当其他研究者还在进行实验或者测试的时候，菲利却认为科学早已进入下一阶段。菲利写道：

> 在科学圈中，科学家们已经认同，死后的交流有相当大的概率是可能发生的。目前唯一的问题是，这种死后的交流究竟是以何种方式发生的，以及如何有效地确认号称发送这些信息的人的真实身份。[34]

菲利在1954年出版的著作《来自极乐世界的电描记图》（*Electrograms from Elysium*）中声称自己不仅找到了死后交流的方式，还发现了一种确认发信人身份的有效方法。菲利在这本书中写道：

> 在讨论这本书时，我提出这样一种理论：死去的人可以在另一个世界发送出用他们自己的名字换位重排的密码。对死者的名字进行换位重排的方式是非常多的，为了让生者能够确认死者的身份，死者会从多种方式中挑选出一个字串或者句子，透露死者生前某种隐藏的、

难解的事实，而除了死者本人以外，几乎没有人知道这些隐藏的事实。在选好了上述字串或者句子以后，死者会通过他的电波脑，或者意识波，不断将这些信息直接辐射到仍然在世的信息接收者内耳中的电波神经终端上。因此，活着的人就可以通过心灵感应的方式收到死者的信息。在生者接收到这些信息以后，他们会立刻通过透视术看到死者名字中的这些字母。这些字母会逐个地按照顺序排好，最终显现出死者想要发送的句子。在收到死者传来的这些信息以后，生者可以从外界寻找相关证据，证明这些隐藏的事实属实，这样就能确认死者的身份。在确认这些隐藏的事实属实以后，接收到信息的生者就能够判断出：这些电信号一定是拥有这个名字的死者发来的密码。在某些情况下，只有通过上下文，或者对特定问题的问答，或者交叉通信，才能确认死者的身份。由于现代电学的发现，特异功能感知变得更有说服力了。而这种人名密码能够让与特异功能感知领域并重的另一个领域——超心理学（Parapsychology）变得更有说服力。[35]

菲利相信，这种方法能让死者和生者进行双向交流：

不言自明的是……人脑确实可以发射电波。而在某些情况下，这种电波可以用无线的方式从一个世界传播到另一个世界。人脑也具有接收这种电波的能力，因此上述这种电波信号可以被接收者感知。对接收者而言，通过这种方式获得的信息不仅是可解读的，而且是亲密和有意义的。这种信息中的很大一部分可能是死者特意设计的，目的是作为外部证据，帮助生者确认他们的身份。[36]

菲利从托马斯·阿尔瓦·爱迪生（Thomas Alva Edison）处获得的信息

包括："TO LIVE ON AS HAD MAD"（像疯了一样继续生活），"STAMINA SHOVE LOAD"（毅力能帮我们推动重负），以及"HO! LOVE ADS STAMINA"（哦！爱能增强毅力）。[37]

此外，菲利还号称他从阿瑟·柯南·道尔处获得了一些信息。在《未解之谜（上）》的第3章中，我已经把其中一条信息作为一种重排字母顺序的挑战密码介绍给读者。其他信息包括："ON, SIR, CLEAR THY AROUND"（开始了，先生，清理干净你的附近），"HAND, SIR, ANY TRUE COLOR"（手，先生，任何真正的颜色），"AIRS HOLD COUNTRY NEAR"（空气让乡村靠近），"AIRS CHAT: ONLY ROUNDER"（空气说：只有酒鬼），"AIRS RECOUNT LADY HONOR"（空气细数女士荣誉），"AIR RECOUNT LADYS' HONOR"（空气细数女士们的荣誉），"LADY AIR CURETH NO SON"（女士空气没有治好儿子）以及"AIRS CON HEART ROUNDLY"（空气充分欺骗心灵）。[38]

读者不难看出，如果说这些句子确实有意义的话，那也是需要牵强的解释才能有一点儿意义。事实上，我挑出的还远不是这些信息中最令人费解的。以上这几句话实际上已经是菲利"收到"的信息中最值得展示的成果了！菲利"收到"的大部分电波信息包含一些可怕的拼写错误（因为某些名字里缺少字母），还经常出现一些外文词汇。此外，菲利还常常无法用上名字中的所有字母，或者必须添补一些名字里没有的字母。

奇怪的是，阿瑟·柯南·道尔爵士的名字实际上可以重排成"HORRENDOUS CARNALITY"（可怕的肉欲），但不知道为什么，菲利并没有给出过这种组合。如果菲利能给出"可怕的肉欲"一词，还真能为死者在另一个世界的生活增添一丝非传统（但并非不受欢迎）的色彩呢。

菲利声称自己一共收到来自38人的1 500条这样的电波信息。

菲利相信自己收到了一条来自已故美国总统富兰克林·D. 罗斯福的信

息。菲利称，这条信息中提到，罗斯福曾摔过一跤并一直对此事保密。为了确认这一事实，菲利给罗斯福的遗孀埃莉诺（Eleanor）写信求证。令人惊讶的是，埃莉诺居然给菲利回了信。埃莉诺在回信中写道："我不知道我丈夫指的是什么，我也不知道你引用的这句话的出处是哪里。"[39]

埃莉诺的回信并没有浇灭菲利打破砂锅问到底的勇气。菲利再次写信给埃莉诺，要求她提供更多信息。菲利声称自己又收到了一条来自罗斯福的新的电波信息，信息中称：罗斯福乘坐的一架电梯曾经突然下降了7层，这起事故导致罗斯福的左脚踝受伤。这一次，菲利居然又收到了埃莉诺的回信。埃莉诺在信中写道："我非常确定你描述的这起事故从来没有发生过。"[40]

菲利在一本书中描述了以上事件，在几页以后，菲利说他最终从罗斯福的侄子处确认罗斯福确实发生过一起这样的事故（罗斯福的这位侄子也已经过世，据说，这位侄子通过电波信息的方式给菲利发来了对这次事故的确认）。

用现代密码学证明灵魂的存在

1994年，一位研究者尝试用现代密码学研究死者灵魂是否存在。这名研究者的名字是迈克尔·莱文（Michael Levin），他供职于哈佛医学院的遗传学系。迈克尔·莱文写道：

这种方法依赖于一种活板门算法，或称单向算法。这种算法可

> 以加密一段信息，但这个过程却不能反向进行。这种方法是这样进行的：一个人在计算机上利用上述算法将他选择的信息加密成一串符号，并且绝不留下任何关于原始信息的记录和文件。接着，此人将加密后的信息公之于众，并且同时公布用于加密的算法。在此人去世后，除非能通过超自然的方式从死者处获得信息原文，否则任何人都不可能破解这段密码。如果有人认为自己从死者处收到了信息原文，他只要用死者生前公布的算法再次处理这条信息，并将算法给出的密文信息与死者生前发布的密文进行比较就可以了。如果算法给出的结果和死者生前公布的密文信息完全一致，就说明此人收到的确实是完全正确的原始信息。[41]

然而，事实上并没有任何已知的算法确实能满足上述单向算法的要求。我们认为某些算法可能是单向的，却不能证明它们一定是单向的。索利斯万万没有想到，几十年后会有人利用计算机破解他生前设定的密码C，同样，我们也无法预测今天所谓的单向算法在未来是否会被逆向破解。在莱文的这篇论文中，还有这样一个让人有些迷惑的段落：

> 若干算法都可能符合上述要求，其中最常见的一种是RSA算法。找到一个大质数是比较容易的，但是对一个大数做分解质因数处理在计算上则是非常困难的，RSA算法利用的正是上述原理。除了RSA算法以外，还有DES密码系统和布卢姆–布卢姆–舒布密码系统。关于以上这些密码系统的详细描述在一些文献中也有所记载。[42]

在本书的第11章中，我会向读者进一步介绍RSA密码。而另外两种密码——DES密码系统和布卢姆–布卢姆–舒布密码系统则不会出现在本书中。但是莱文提到的这3种密码都是20世纪70年代以后才出现的，因此我不清楚为什么莱文要引用1944年和1968年发表的论文，并称这些论文里有“关于以上这些密码系统的详细描述”。莱文还写道：

> 这种测试方法的最后一个缺点是，在未来的某一天，数学知识和计算机的计算能力可能会发展到一个更高的水平，到那时，这些今天所谓的单向算法就有可能被逆向破解，用这些算法加密的信息也就被破译了。此外，到那时，也许用穷举法尝试所有可能的字符串也不再是不可能的事情。这些算法中的一部分已经被数学证明是不可逆的。另外，大部分普通加密算法都在不断地受到破译的测试，如果某种算法被破译了，这一消息很快就会在计算机安全界和密码学界成为公开的信息。如果有人声称他破解了某条密码，只要向相关的专业组织确认，就能很容易地判断他解开的密码是不是已经有了可逆的算法。如果答案是肯定的，则这段密码的破译并不能证明超自然现象的存在。尽管如此，加密算法仍然为超自然现象的验证提供了一种强有力的工具。[43]

虽然莱文声称“这些算法中有一部分已经在数学上证明是不可逆的”，但实际上这种说法是没有根据的，目前没有任何算法经数学证明是不可逆的。我在上文中已经说过，事实上并不存在已知的单向算法。

莱文的这篇论文引用了索利斯和史蒂文森之前发表的论文。因此，虽然莱文的方法在细节上有错误，但是他很可能是受到了索利斯和史蒂文森的启发，才在20世纪90年代对通灵术测试实验做出了这些更新和提高。然而，他的努力似乎也没有取得什么成果。莱文并没有像史蒂文森一样继续发表论文不断更新这项研究，在“谷歌学术”网站上也找不到任何其他作者对他的这篇论文的引用。[44]如果莱文真的留下了某个密码，并希望人们在他死后尝试破译这段密码，至少这段密码目前还没有被公布出来。

然而，测试通灵术的基本想法并没有因此消亡！

一败涂地

在上述所有测试通灵术的人中，除了菲利声称自己成功了，其他所有人都失败了。我想我们完全可以无视菲利取得的所谓“成果”，因为那不过是没有事实支持的自我幻想而已。但是，前人的失败并没有浇灭后人的热情，仍有各种各样的研究者在不断修改和重复测试通灵术的实验。距今较近的一次尝试是苏茜·史密斯（Susy Smith）的实验。史密斯是非营利性组织“生存研究基金会”的创始人，她生前创造了两项密码测试，希望在她死后人们能通过与她交流解开这些密码。2001年2月11日，苏茜·史密斯逝世。

当史密斯尚在人世的时候，她认为自己联系上了她已故的母亲，以及美国著名哲学家和心理学家威廉·詹姆斯（William James）。威廉·詹姆斯也对超心理学有很强的兴趣，他还是美国心灵学研究协会的创始人之一。

因此，史密斯可能对自己死后能与世人交流有着很高的期望。在一本名为《死后的密码》（*The Afterlife Codes*）的书中，史密斯不仅分享了她在这个世界中的故事，还写了自己对死后交流的希望，以及进行这种交流的具体计划。[45]她甚至还为能收到她死后传来的密钥并破解密码的人提供了10 000美元的赏金。

史密斯的网站www.afterlifecodes.com现在已经不存在了，但是“互联网档案馆时光机”（the Internet Archive Wayback Machine）[46]保存了这个网站在不同时间的样子。最后一个能为我们提供有用信息的存档时间点是2001年5月17日。[47]在上述网页的一个链接中，我找到了关于史密斯的实验方法的以下解释。[48]

对于密码以及相关问题的解释

在生存研究基金会成立后不久，我请董事会成员弗兰克·特里布和克拉丽莎·马尔德斯（Clarissa Mulders）创造了一种人人都会使用的简单密码。这样，希望死后与世人交流的人就可以留下一条这样的密码，并在死后发来破解密码的密钥。弗兰克是美国政府的一名律师；而克拉丽莎则是一名教师，她同时还是美国大学优等生协会的成员，拥有该协会的钥匙。①弗兰克和克拉丽莎用他们的聪明头脑发明了一套加密系统，密码方面的权威专家认为，这套系统设计得准确、成功且科学。[49]

特里布–马尔德斯密码需要对好几套字母表进行操作，因此对于普通人而言，要手动用这种方法给信息加密是极其困难的。我曾尝试过手动用该系统加密信息，因此我非常清楚这些操作的复杂程度。但是对于今天的计算机而言，用这套系统加密信息是一件非常简单的事情。该系统的加密步骤如下：加密者先要选定一条简短的秘密信息（特里布和马尔德斯把这种信息称为“密钥词组”）。加密者应该选择

① 美国大学优等生协会（Phi Beta Kappa Society）是由美国大学优秀学生组成的全国性荣誉组织，其标志是一把钥匙。——译者注

对他们而言足够重要的内容，因为这样才能保证他们死后仍然能记得这些内容。

此外，密钥词组应该是公众熟知的内容，这样当你死后把密钥词组发送过来时，接收者才能从空气中（或者通过其他方式）接收到这个词组。比如，你可以使用“正面思维的力量”作为密钥词组，或者你也可以从一首诗或者一首摇篮曲中选取一个词组作为密钥词组，你还可以选择一本书的题目作为你的密钥词组，比如《生命是永恒的》。注意，这个词组并不是你的密码，而是密码的密钥，有了这个密钥，就可以破译你留下的密码。在你选定密钥词组，并且确认自己不会忘记这个词组以后，你应该保证不将这个密钥词组告诉任何人。完成了上述步骤以后，你就可以开始填写调查问卷了。填写完调查问卷以后，就成功注册了。

在注册的过程中，计算机的秘密内核会把你提供的密钥词组和标准字母表相结合，得到一种特定的字母表。然后，计算机会随机从一本书中挑出一句话（这一内容是保密的），并用上一步中的字母表来加密这句话。上述加密步骤完成以后，就会得到一串看起来很奇怪的字母，这串字母就是你的密码。没有人知道这段密码是什么，也没有人知道这段密码的含义。只有当计算机获得了你提供的原始密钥词组时，才能够再次破译这段密码。

为了防止作弊，在注册的过程中你必须回答46个问题，计算机会将你对这46个问题的答案和你的密码一起秘密储存起来。在未来的某个时刻，当你过世以后，你的某位朋友或亲戚，或者某个好心的陌生人，或者灵媒，或者心灵学研究者可能会宣称自己收到了你发来的秘密信息。当此人带着这条秘密信息来找我们时，他们也必须填写上述问卷，并完成注册的步骤。然后，计算机会自动比较此人带来的词组是否与你事先设定的秘密词组一致。如果（我们热切盼望着这样的“如果”）某人带来的词组与你留下的密钥词组完全一致，这个词组就能够破译你的密码。如果我们能够得到这样的实验结果，我们将会非常欢欣鼓舞。在此，我并不想提出太多虚假的希望，但我认为，上述情况是完全有可能发生的。在某个时候，某个人也许就会经历这样一个奇异而愉快的事件。

在我们这里登记了密码的人还应该采取一些措施来保护自己。如果在你死后某人相信自己收到了你的信息，你一定希望我们可以展开必要的调查。所以，你应该在你的遗嘱中或者其他重要文件中留下我们的地址，我们的地址是：图森市亚利桑那大学心理学系，人类能源系统实验室，苏茜·史密斯计划。

你不能向收到信息的人提供任何奖金。我已经说过，设置奖金并不是一个好主意。[50]但是有一些事是你可以去做的——你可以设置一种控制机制，这种机制从科学角度来讲是很有价值的。告诉你周围的亲戚朋友，你头脑中有一个秘密的词组，你希望能在自己死后把这个词组告诉仍然在世的人。如果当你还在世时，你的某位朋友或熟人就猜出了这个词组，或者通过读心术得到了这个词组，又或者一名灵媒或者心灵科学研究者获得了你头脑中的这个词组，那都没有关系。你只需要给苏茜·史密斯计划写邮件，要求重置密钥就可以了。（如果真的发生了这样的事情，请一定告诉我们此人是如何得到你脑中的密钥词组的。我们之后会对统计数据进行标准化处理，所以你提供的这方面信息对我们很有价值。）如果你在世的时候没有人能够获得这个密钥词组，而在你去世以后的某天，某人联系我们声称他取得了你的密钥词组，我们就会把此人提供的密钥词组和你留下的词组进行比对。如果比对成功，我们就有了充足的证据来证明这条信息确实来自已经去世的你，这也可以很好地证明你的灵魂在死后仍然存在。上述控制机制能给你提供多一层的保护。

请记住，当你在我们这里注册你的密码时，你在告诉整个世界：虽然你会先某些人而去，你对他们的爱却是永远存在的。但请不要太着急去另一个世界。

现在，请点击这里进入**调查问卷/注册**的页面。

虽然史密斯在《死后的密码》一书中大力宣传了这种方式，并且提供了10 000美元的奖金，但她的努力也和所有前人的努力一样以失败告终。以上内容引自史密斯创建的网站，这个网站现在早已不存在了，我也不知道在这个网站注册的那些密码现在究竟怎么样了。据说大约有1 000人在这个网站注册了自己的密码。

一个怀疑论者的挑战

克劳斯·施梅在他2012年出版的书《无法破解》(*Nicht zu Knacken*，这本书目前只有德文版)[51]中讨论了一些我们在本章中提到的密码。虽然施梅对通灵术持怀疑态度，但他也设计了自己的挑战密码，他的密码使用的是当今最先进的加密系统之一——高级加密标准系统（Advanced Encryption Standard，简称AES系统）。

毫无疑问，以后一定还会有人继续尝试用密码来证明人死后灵魂仍然存在。但对于我们来说，现在是转向下一个挑战的时候了。

第8章

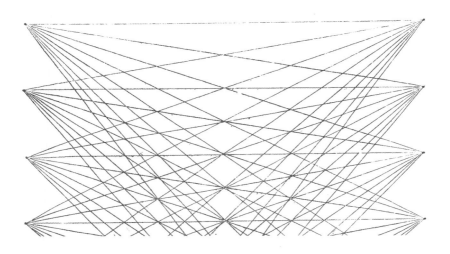

是否有绝对安全的密码？

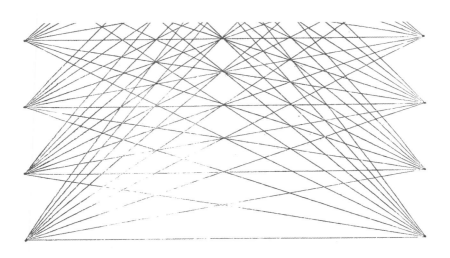

世界上是否存在一种具有绝对安全的密码，不管破译者多聪明，不管用于破译的计算机的计算能力多强，不管花多长时间，都绝对没有可能成功地破译这种密码？是的，世界上真的存在这种绝对安全的密码。1917—1918年，约瑟夫·O. 莫博涅（Joseph O. Mauborgne）和吉尔伯特·韦尔纳姆（Gilbert Vernam）就发现了一种这样的密码系统。[1]这种系统叫作"一次性密码本"（one-time pad）系统，有时简写为OTP系统。下面，我就来向读者介绍一次性密码本系统的工作原理。

假设有一串随机数字，其中每个数字都在0~25之间。比如以下这串随机数：`6 3 6 25 16 6 10 6 18 1 3 7 24 25 18 21 8 10 12 1 13 0 2 17 0 13 22 16 4 24 4 10 4 13`。

接下来，我们就用上面的这串随机数字来加密我们选中的明文信息，这段明文信息是："YOU MET ME AT A VERY STRANGE TIME IN MY LIFE."（你在我人生中一个非常奇怪的时间点遇到了我。）[2]

加密的方法很简单：我们先把明文信息中的字母逐一和这串随机数字对齐，然后再把每个明文字母向后移动几位，移动的位数等于和这个字母对齐的随机数字的值。如果字母移动后超过了Z的位置，就重新绕回字母表的开头，从A开始继续移动。具体的加密过程如下：

```
YOU M ET ME ATAV E R Y ST R AN GET IM
6 3 6 25 16 6 10 6 18 1 3 7 24 25 18 21 8 10 12 1 13 0 2 17 0
ERA L UZ WK SUDC C Q Q NB B MO TEV ZM
```

E I NM YL IF E.

13 22 16 4 24 4 10 4 13

R E DQ WP SJ R

我们最终得到的密文信息是："ERALU ZWKSU DCCQQ NBBMO TEVZM REDQW PSJR. "

如果有人能猜到密钥的话，他就可以把密文中的字母一一移回原位，得到原始的明文信息。既然如此，为什么说这种加密系统是无法破解的呢？请读者想一下，一个人如何能猜到上面的一串数字密钥？如果此人猜出的不是正确的密钥，而是以下的错误密钥：23 3 4 25 0 23 15 9 7 6 15 25 6 8 5 2 3 13 18 22 12 0 18 6 24 25 11 3 18 22 4 10 14 13，他就会得到这样一条明文信息：

HOW MUCH BLOOD WILL YOU SHED TO STAY ALIVE? [3]

（为了继续活着你要流多少血？）

又或者，如果某人猜出的是以下这个错误密钥：6 3 6 9 20 12 3 3 24 3 10 16 24 3 2 20 5 19 19 7 7 6 19 18 8 13 12 25 9 18 4 6 5 24，他又会得到另一条不一样的明文信息，这条明文信息是：

YOU CAN'T HURT ME, NOT WITH MY CHEESE HELMET! [4]

（你无法伤害我，我有我的奶酪头盔保护我！）

为什么一条密码可以被破译为这么多种完全不同的信息呢？要回答这个问题，我们不妨自己看看密文的第一个字母 E。如果我们猜测密钥的第一个数字是0，那么E破译后还是E。如果我们猜测密钥的第一个数字是1，那么E就被破译为D。类似地，如果我们猜测密钥的第一个数字是2、3、4、5、6、7、8、9、10，那么同一个密文字母E就会分别被破译为C、B、A、Z、Y、X、W、V、U。在此，我将不再列出所有可能性。读者很容易理解，如果密钥的第一个数字分别是26个不同的数值，明文信息的第一个字母就可以被分别破译为26个字母中的任何一个。也就是说，没有密钥的话，破译者完全无法知道明文的第一个字母究竟是什么，因为这个字母可能是26个字母中的任何一个。上述情况也同样适用于密文中的第二个字母、第三个字母等。也就是说，只要选取合适的密钥，密文中的每一个字母都可以被破译为26个字母中的任何一个。换句话说，只要找到合适的密钥，我们就可以破译出任何我们能想象出来的信息，唯一的限制条件是：破译出来的明文信息的长度必须和原始密文信息的长度一致。

对于一个给定的长度而言，可能的信息有许多种，破译者无法以任何方式从数量极大的可能信息中选出符合加密者原意的原始明文信息。因此，我们说上述加密系统是不可破解的。要证明这种系统是无法破解的，需要用到一些数学工具。利用这些工具，克劳德·香农不仅证明了上述系统确实无法被破解，而且还证明了任何不可破解的加密系统都一定与上述系统等价。也就是说，如果存在另一种无法被破解的加密系统，即使那种系统表面看起来和一次性密码本系统不同，从更本质的层次上看，那种系统也必须等同于一次性密码本系统。

间谍身上常常会携带一种非常小的密码本，这种密码本上写满了数字或者字母。有了这种密码本，就可以用上述方式来加密信息了。有的密码本上写的不是数字，而是字母，这其实只是用一种更精简的方式来表示0~25之间的数字而已。数字和字母的对应规律是：A=0，B=1，C=2，…Z=25。不管这类系统用的是数字还是字母，我们都把它统称为一次性密码本加密系统。这是因为，如果加密者用以前用过的一串密钥来加密一条新信息，那么新旧两条信息都可能会被破译（这种事情确实发生过）。但是，每次更换密码本是很麻烦的，尤其是当要加密的信息很长的时候（想象一下，如果我们要加密的内容是音频文件或者视频文件）。因此，这种一次性密码本系统并不适用于所有需要加密的内容。

此外，保证密钥的随机性非常重要！如果密钥具有一定的规律，这种一次性密码本系统就可能被破解，这种情况在现实中也发生过。

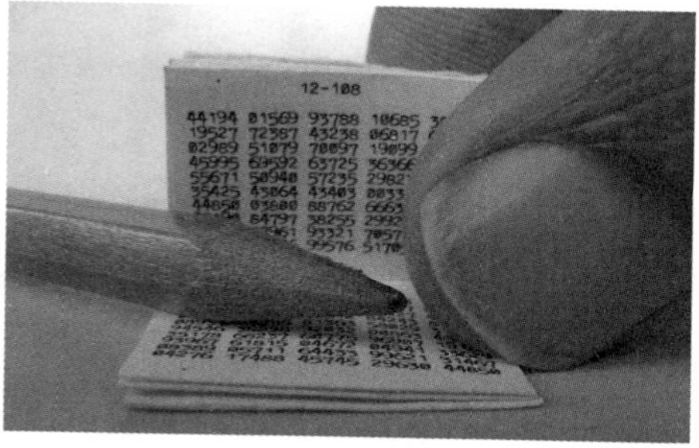

图 8-1　一次性密码本

近年来，大量新的密码系统涌现出来。要对这些系统一一进行分析和研究，搞清楚这些系统的安全性是不是足够高、有没有一些微妙的弱点（因此不适合用来加密）需要花费大量时间。一般来说，专家不愿意花时间研究无名的加密系统，只有已经在该领域中取得了较高声誉的加密系统才会获得专家的关注。这听起来有些像《第22条军规》中的情况。①

① 《第22条军规》是一本黑色幽默风格的小说，这本小说里有这样的情节："第22条军规"规定，疯子可以免于飞行，但同时又规定，任何人都必须由本人提出申请才能免于飞行。然而，如果本人提出申请不参加飞行，便证明此人并未变疯，因为"对自身安全表示关注，乃是头脑理性活动的结果"。因此，根本没有人可以根据第22条军规不飞行。——译者注

如果专家不愿意检视你的作品，你究竟要怎样在业内获得较高的声誉呢？这个问题的答案是：专家看的不是加密系统本身，而是攻击这些加密系统的方式。当某个人针对已有的加密系统提出一种新的攻击方式时，专家就可以相对比较容易地评估被攻击的加密系统能否抵抗这种攻击，这比直接评估一种新的加密系统要高效得多。当有人提出一种好的攻击方法时，专家会更愿意对号称能够抵御这种攻击的加密系统进行研究和分析。简单来说，有时候研究系统安全性的最佳方式就是看看这个系统能否经受住外界的攻击。

1914年，莫博涅证明普莱费尔密码是可以被破解的，这一结论为他赢得了很高的声誉。在上一章中，我们已经讨论过普莱费尔密码的加密原理，但是，莫博涅的工作首次公开发表了攻击该密码的途径。此外，莫博涅还破译了爱德华·埃尔加曾经破译过的挑战密码，这则挑战密码的作者是斯库林。1939年，斯库林的这段挑战密码被再次刊登在了当年的某期《通信兵团公报》（*The Signal Corps Bulletin*）上。[5] 莫博涅在该杂志上看到斯库林的密码以后，很快就成功地破译了这则挑战密码。他的破译结果被刊登在了接下来的一期《通信兵团公报》上。

图 8-2　约瑟夫·O. 莫博涅（1881—1971）

前文已经说过，莫博涅在1917—1918年间发明了一次性密码本加密系统。在破解普莱费尔密码以后，莫博涅就想要找出一种不可攻破的加密系统。在找到最终解答（一次性密码本加密系统）之前，他还测试过另外一种加密系统。他测试这种加密系统的方式是向军方提出一项挑战，看看军方能否破译这种密码。结果这个密码至今未被破解。虽然有些未被解开的密码很可能是没有任何意义的恶作剧，但是莫博涅似乎不太可能与军方开这样的玩笑。他在密码界有着崇高的声誉和地位，因此一旦他要讨论与密码相关的问题，所有人都会非常严肃地对待他提出的挑战。

莫博涅提出上述挑战密码是在1915年。1991年（也有可能稍早于1991年），路易斯·克鲁（Louis Kruh）在纽约公共图书馆的珍稀藏书部中再次发现了莫博涅的这段密码。这段密码是在陆军通信学校（Army Signal School）的一些老资料里被发现的。我将这段密码以及包含这段密码的文件的上下文全部摘录如下[6]：

> 以下密码及解释这则密码的信件是由第8步兵团的约瑟夫·O. 莫博涅中尉提交给通信班的。我们邀请所有对密码学有兴趣的军官尝试破译这段密码，并将破译结果交给校长办公室。
>
> 菲律宾，马尼拉，西班牙军营
>
> 1915年10月23日

> 寄信人：约瑟夫·O. 莫博涅中尉，第8步兵团
>
> 收信人：陆军通信学校校长
>
> 主题：一种新型的军事密码
>
> 1. 我发明了一种密码。我认为这种密码非常适合用作军事用途。它从许多方面来看都比普通的密码盘或者普莱费尔密码更有优势，因为它非常适宜士兵使用。在使用得当的前提下，这种密码不仅操作超级简单，并且完全无法被敌方破解。我请陆军通信学校将我发明的

这种密码交给学校的教师和学生军官进行破译，以测试该密码的抗破译能力，也是为了方便大家讨论这种新的加密系统是否可以被用作军队内部的标准加密系统。

2. 为了达到上述目的，我建议校长批准将这封信中所附的密码信息贴在学校的公告板上，这样教师和学生军官就可以把这段密码抄下来，并利用他们的业余时间尝试破译这段密码了。我建议给学生们3个月的时间，因为我们都知道陆军通信学校的学生们的业余时间是十分有限的。
3. 在上述时间段结束以后，如果校长愿意的话，可以打开随信附上的标有“1号密码，解答”的密闭信封。校长可以把这段密码的解答贴在学校的公告牌上，这样学员们就可以看出为什么破译替代式密码的标准方法对我的密码不起效果了。此外，学校的军官们还可以尝试开发其他方法。我相信此举可以激发学员们的创造性，也许还能产生某种对军方有益的成果。
4. 再过一个月以后，校长就可以打开随信附上的另一个信封了，那个信封上标有“1号密码，加密方法及密钥”的字样。如果校长同意的话，可以将我的密码提交学校讨论，讨论的目的我已经在第一段中写明了。
5. 我听说通信学校培养了一些密码方面的人才，他们对破译从敌方截获的密码有着很高的热情。我相信校长会认为我的请求有一定道理，并允许这些军官以我描述的方式对我的密码展开破译工作。
6. 如果任何人能成功破译我的密码，我将非常高兴听到他们的破译结果。

J. O. 莫博涅中尉，第8步兵团

1 号密码

这是一种简单的单字母替代密码，专门为军事用途设计。我相信，只要用我创造以下密码的方式来加密信息，这种密码在实际中将是无法破译的。只要经过5分钟的培训，士兵就能学会使用这套加密系统。在精确度和方便性两个方面，我的这套加密系统都比目前军队中使用的普莱费尔密码或者密码盘更加优越。如果你能够破译以下密码，请通知我你花费了多久，并告知使用了哪种方法。我相信我的密码能够躲过目前我们已知的所有破译密码的技巧。

PMVEB DWXZA XKKHQ RNFMJ VATAD YRJON FGRKD TSVWF TCRWC
RLKRW ZCNBC FCONW FNOEZ QLEJB HUVLY OPFIN ZMHWC RZULG
BGXLA GLZCZ GWXAH RITNW ZCQYR KFWVL CYGZE NQRNI JFEPS
RWCZV TIZAQ LVEYI QVZMO RWQHL CBWZL HBPEF PROVE ZFWGZ
RWLJG RANKZ ECVAW TRLBW URVSP KXWFR DOHAR RSRJJ NFJRT
AXIJU RCRCP EVPGR ORAXA EFIQV QNIRV CNMTE LKHDC RXISG
RGNLE RAFXO VBOBU CUXGT UEVBR ZSZSO RZIHE FVWCN OBPED
ZGRAN IFIZD MFZEZ OVCJS DPRJH HVCRG IPCIF WHUKB NHKTV
IVONS TNADX UNQDY PERRB PNSOR ZCLRE MLZKR YZNMN PJMQB
RMJZL IKEFV CDRRN RHENC TKAXZ ESKDR GZCXD SQFGD CXSTE
ZCZNI GFHGN ESUNR LYKDA AVAVX QYVEQ FMWET ZODJY RMLZJ
QOBQ-

J. O. 莫博涅中尉，第8步兵团

菲律宾，马尼拉

1915年10月21日

陆军通信学校，堪萨斯，利文沃思堡，1915年12月13日克鲁把莫博涅的这段密码公布在了《密码》（*The Cryptogram*）杂志和《密码学》（*Cryptologia*）杂志上。到目前为止，这两本杂志的读者都未能交出该密码的解法。在《密码学》杂志上，克鲁还写道：“我找到这份文件时，文档中并没有收录解法。但是这个密码很可能被破译了，或者因为其他原因被认为不适合做军事用途，因为没有任何信息显示当时军方采用了这种新型密码。”[7]

如果莫博涅的这段密码真的从来未被破解过，那么莫博涅不仅发明了世界上唯一一种不可能被破解的加密系统（一次性密码本加密系统），还创造了另一种加密系统——这种系统虽然在理论上是可以被破解的，却一直没有人能够真正破解它。如果莫博涅今天还活着的话，我想他会这样说：“我并不经常创造密码。但一旦我创造出来，这些密码都是不可破解的。”

莫博涅的职业生涯又继续了许多年，直到他1941年以少将军衔退役。除了在密码方面的贡献，莫博涅在职业生涯中还有许多其他成就：他研发了用于飞机的无线电话，这项发明使飞机和地面的通信首次成为可能。他从1937年10月开始担任美国陆军通信兵的总负责人，直到1941年到达退役年龄，他才卸任。由于在第二次世界大战期间莫博涅已经退役，所以他并没有在这次战争中为国家效力。但是在莫博涅退役之前，他曾支持过雷达的研发工作，而雷达在第二次世界大战中发挥了至关重要的作用。由于以上所有这些贡献，1988年，莫博涅进入军事情报兵团的名人堂。

对答案的追求

为了破译莫博涅留下的这则挑战密码，我研究了针对这则密码进行的各种统计结果。但是，这些结果并没能让我判断出莫博涅使用的究竟是何种加密系统。然而，莫博涅在上述信件中对这则密码的一些描述引起了我的注意。莫博涅称“这是一种简单的单字母替代密码，专门为军事用途设计”，他还称这种加密系统“非常适宜士兵使用”、“操作超级简单”，因此优于其他系统。莫博涅还说道：“只要经过5分钟的培训，士兵就能学会使用这套加密系统。在精确度和方便性两个方面，我的这套加密系统都比目前军队中使用的普莱费尔密码或者密码盘更加优越。”

根据以上这些描述，我判断莫博涅在这套加密系统中使用了一种类似于图8–3中展示的密码轮。密码轮使用起来很简单，但它的加密安全性却比许多更复杂的加密系统还要高。

图 8-3　美国军方使用的一个密码轮

在人类的密码史上，密码轮这种工具曾数次被发明、使用、遗忘，然后又再次被发明。目前我们知道，密码轮这一工具可以追溯到15世纪。后来，美国总统托马斯·杰斐逊于18世纪再次发明了这种工具。而到了19世纪末，这种工具又被法国密码分析师艾蒂安·巴泽里（Étienne Bazeries）再次发明。

不同的密码轮上配有的密码盘数量不一，但是所有密码轮都具有以下这个共同点：它们由一系列可以移动的密码盘组成，每个密码盘上都写着完整的字母表。如果每个密码盘上的字母以不同顺序排列，那么这种密码轮的安全性就会更高。在加密信息的时候，加密者只需要转动密码轮上的密码盘，使得这条待加密的信息出现在密码轮的同一横行中即可。然后，加密者再挑选密码轮的另一行并抄下这行所有字母，这就是经过加密后的密文信息。

加密信息的接收者也有一个与加密者完全一样的密码轮，在收到密文信息以后，接收者转动密码轮上的密码盘，让密文信息出现在密码盘的同一横行中。然后，接收者会逐一查看密码轮上的其他25行内容，直到他发现其中一行的内容是有意义的文字为止。这行有意义的文字就是加密者想要传达的原始明文信息。

克鲁称，1915年，美国军方没有采用任何新的加密系统，因此他判断莫博涅的密码很可能被别人破译了。但是，军方从什么时候开始能够快速地采纳建议了？又有多少军官会询问他们的下级：在工作中是否有什么需要改进的地方？事实上，我描述的这种密码轮系统后来确实被军方采用了，只不过这是一段时间之后才发生的事情。

美国陆军在第一次世界大战之后采用了这种密码轮系统。陆军将这种密码轮命名为“密码装置M–94”，并于1922年2月发布了这种密码装置的使用说明。美国海军也于1926年10月采用了这种系统，海军将这种密码装置命名为“CSP–488”。

另一种可能性

下面这段话的作者是赫伯特 · O. 亚德利（Herbert O. Yardley）。这段话为我们打开了莫博涅留下的未解密码的另一种可能性，而我要做的是试图排除这种可能性！亚德利写道：

> 为了解释我对莫博涅上校的看法，我将引述一段我与他的短暂接触经历。在战前的2~3年中，莫博涅上校和（帕克·）希特在通信兵团中推广了一种不容易破解的加密系统，这种加密系统用到了美国陆军的密码盘，也就是所谓的“滚动密钥”。滚动密钥是这样一种密钥：它不包含一串字母或一个词，而是包含某本书中的一个段落或一页内容，而这些内容的字母长度必须与待加密信息的长度相等。有了这种滚动密钥，我们就可以做到密钥永不重复。莫博涅和希特认为，这样一种加密系统是不可能被破解的。[8]

那么，莫博涅的这段1915年的密码是否可能是用滚动密钥加密的呢？我们会考虑这种可能性。但是，由于在上面这个含糊的段落中，亚德利对滚动密钥系统的描述并不清楚，所以在分析这种可能性之前，让我们首先来看一个滚动密钥的例子。

滚动密钥加密系统的第一步是：密码的加密方和接收方必须选择同一本书，或者同一段较长的文字。这一步骤的重点是：被选中的这段文字至少要和待加密的信息一样长。假设我们选择理查德·莱蒙（Richard Laymon）的《寂寞十月的一个夜晚》（*Night in the Lonesome October*）（这是一本很棒的书！）。这篇小说的开头是这样写的：

> I was twenty years old and heartbroken the night it started.
>
> My name is Ed Logan.
>
> Yes, guys can be heartbroken, too. It isn’t an affliction reserved for

women only.

（在一切开始的那个夜晚，我20岁，正处于心碎的状态之中。

我的名字是埃德·洛根。

是的，男人也是会心碎的。心碎并不是一种只属于女人的痛苦。）

好吧，我承认这本书的开头节奏有点儿慢，但这篇小说真的很好看。

接下来，假设我们想要加密的信息是："When I was a little kid my mother told me not to stare into the sun. So, when I was six I did.（在我还是小孩的时候，妈妈告诉我不要盯着太阳看。于是，我等到6岁时，才去盯着太阳看。）"[9]

首先，我们把这段待加密的信息和《寂寞十月的一个夜晚》开头的那段文字一一对齐：

```
WHENIWASALITTLEKIDMYMOTHERTOLDMENOTTOSTAREINTOTHESUNSO
IWASTWENTYYEARSOLDANDHEARTBROKENTHENIGHTITSTARTEDMYNAM

WHENIWASSIXIDID
EISEDLOGANYESGU
```

然后，我们把上一行中的每个字母向字母表的后方移动几位，这个位数等于第二行字母对应的数字。在这个加密步骤中，我们事实上相当于使用了一次性密码本。只不过在这种系统中，密钥不是一串数字，而是一串字母。为了方便读者参考，我把字母表和数字的对应关系复制如下：

A	B	C	D	E	F	G	H	I	J	K	L	M	N	O	P	Q	R	S	T
0	1	2	3	4	5	6	7	8	9	10	11	12	13	14	15	16	17	18	19

U V W X Y Z

20 21 22 23 24 25

在我们要加密的这段信息中，第一个字母是W，字母W下方是字母I，因此我们要把字母W向字母表后移8位。这时，我们已经超过了字母表的最后一个字母Z，但没有关系，我们只要再绕回字母表的开头，从字母A开始继续向后数就可以了。后移8位以后，字母W变成了字母E。待加密信息中的第二个字母是字母H，向后移动22位就变成了字母D。接着，我们要处理第三个字母E，向后移动0位后仍然是字母E。对接下来的所有字母进行以上操作，我们就得到了这样一段密文信息：

EDEFBSEFTJGXTCWYTGMLPVXHVKUFZNQRGVXGWYATZXAGTFMLHE
SASAAPWRLHOYSVVMVOX

下面，让我们把原始的明文信息和加密后的密文信息一一对齐进行比较：

WHEN I WAS A LITTLE KID MY MOTHER TOLD ME NOT TO STARE
EDEF B SEF T JGXTCW YTG ML PVXHVK UFZN QR GVX GW YATZX

INTO THE SUN SO WHEN I WAS SIX I DID
AGTF MLH ESA SA APWR L HOY SVV M VOX

注意，一开始的3个单词“WHEN I WAS”变成了“EDEF B SEF”。但是，在明文信息中，“WHEN I WAS”这3个单词后来又出现了一遍。第二遍出现时，它们被加密成了“APWR L HOY”。这是因为，同样的3个单词两次出现时下方对齐的密钥单词（即理查德·莱蒙小说中的文字）并不相同。

这种变化会给想要破译密码的人造成一定的麻烦，使他们以为这是一种比较高级的密码，而这实际上是一种简单密码。但是，这样的加密系统也有一些弱点，我们很快就会谈到。

由于以下这几点原因，我认为莫博涅1915年的挑战密码并没有使用这种滚动密钥的加密系统。

1. 虽然这种滚动密钥加密系统并不是非常复杂，但它也不像莫博涅在1915年的那封信中写的那么简单。我曾经向不同的学生解释过滚动密钥加密系统的工作原理，但我从来没有能够在5分钟内教会学生使用这种加密系统。也许你是第一次听说滚动密钥加密系统，并且只花了不到5分钟的时间就读完并理解了以上所有内容。但是，请记住，我在本书中首先介绍了一次性密码本加密系统，然后在介绍滚动密钥加密系统时，我通过引述这一系统与一次性密码本加密系统的相似之处而简化了说明文字。如果你从来没有学习过一次性密码本加密系统的原理，我想你是很难在5分钟内搞清楚滚动密钥加密系统的原理的。
2. 在莫博涅的信中，他认为即使把密文信息和原始的明文信息并排放在一起，学员们也没有办法看出这段密码的加密方式。如果这段挑战密码是一段滚动密钥密码的话，只要把明文信息和密文信息对应起来，就很容易判断出加密所用的方法了。
3. 这段挑战密码中的字母频率与滚动密钥加密系统应该产生的字母概率不符。比如，在英文中，最常见的一个单词是“THE”，由于在滚动密钥加密系统中，明文信息和密钥都是有意义的英文文字，所以有时候明文信息中的单词“THE”会和密钥中的单词“THE”正好对齐。在这种情况下，我们就应该看到密文信息中出现“MOI”这个单词。但是在莫博涅的密码中，从来没有出现过“MOI”这三个字母连在一起的情况 。当然，这一现象本身并不足以证明这段密码使用的不是滚动密钥加密系统，也许在这段密码中只是由于巧合而没有出现明文中的“THE”和密钥中的“THE”正好对齐的情况。

在莫博涅的这段挑战密码中，经常出现的一个字母是字母R。字母R一共在密文中出现了47次，这个字母出现的概率比其他所有字母都高出许多。那么，在一段用滚动密钥加密系统加密的密文中，字母R究竟应该以多高的概率出现呢？如果明文中的字母是A，而对应的密钥字母是R，我们就会得到密文字母R，我们把这种情况简写为A+R=R。除了这种情况以外，以下这些可能性也都会让密文字母为R，这些可能性包括：B+Q，C+P，D+O，E+N，F+M，G+L，H+K，I+J，S+Z，T+Y，U+X，以及V+W。

现在，我们只研究其中一种可能性，即A+R=R。明文信息中的字母A和密钥中的字母R正好对齐的可能性有多大呢？要计算出这个概率很简单，只要用字母A出现的概率乘以字母R出现的概率就可以了。根据《未解之谜（上）》第1章中的字母概率表，我们可以计算出这个概率是$0.081\,67 \times 0.059\,87 \approx 0.004\,889$。但是，如果明文信息中的字母是R，而密钥中的字母是A，我们也同样会得到密文字母R，也就是R+A=R。因此，上面的这个概率应该乘以2，才是我们真正想要的概率。也就是说，通过A+R或者R+A的方式得到密文字母R的概率总共是$2 \times 0.081\,67 \times 0.059\,87 \approx 0.009\,779$。接下来，我们对其他能够得到密文字母R的组合（B+Q，C+P，D+O，E+N，F+M，G+L，H+K，I+J，S+Z，T+Y，U+X，以及V+W）

分别进行上述计算，并将所有这些概率相加，得到的总和是0.042 47。也就是说，如果一则密码使用的是滚动密钥加密系统，那么密文中应该有4.247%的字母是字母R。莫博涅留下的这段挑战密码中共有499个字母，499×4.247%≈21。因此，如果这段密码用的是滚动密钥加密系统，我们大约应该在这段密码中看到21个字母R，但实际上在这段密码中字母R一共出现了47次！这种预期值和实际值之间的巨大差别提示我们，莫博涅的挑战密码并不是用滚动密钥加密系统加密的。我们可以对密码中的每一种字母都进行上述比较分析。某些字母出现的概率与预期值比较接近，但这很可能只是因为巧合。也有相当多的字母的出现概率与预期值相差很远，这种情况足以让我们排除滚动密钥加密系统的可能性了。

4. 用滚动密钥加密系统加密的信息并不难破译。事实上，上面的第3点原因就是这种加密系统的弱点。在滚动密钥密码中，密文中的每一个字母都可能由频率较高的明文字母和密钥字母组合产生。通过研究一组密文字母，并且尝试各种频率较高的明文与密钥的组合形式，破译者就有可能同时猜出明文信息和密钥中的单词，这样这段密码也就被破译了。在我与几位本科生合作撰写的论文中，我提出了两种攻击滚动密钥加密系统的方法。此外，另一位学生（虽然不是我的学生）以我的一篇论文为基础，提出了一种破译能力非常强的攻击算法。如果莫博涅留下的挑战密码使用的是滚动密钥加密系统，这段密码就不会至今无人能破。但是，如果如我所说，这段密码并不是用滚动密钥加密系统加密的，我们究竟应该如何破译这段密码呢？我在上文中描述的所有破译技巧都没有办法破译用**密码轮**加密的信息。而要证明一段信息确实使用了密码轮，唯一的方法可能就是用同样的密码轮来解开这段信息了。

如何破解密码轮

首先，读者应该理解，密码轮加密系统的安全性并不依赖于对每个密码盘上的字母顺序进行保密。当然，我们希望尽量不让敌方知道各个密码盘上的字母顺序，但是在现实中，我们必须假设敌方可能获知密码盘上的字母顺序。

使用密码轮进行加密的人会定期把密码轮上的各个密码盘取下来，然后再以不同的顺序把这些密码盘重新装配好。也就是说，真正的密钥是这些密码盘的排列顺序。即使敌方俘获了一个我方的密码轮，或者通过叛徒或间谍知道了我方密码轮上各个密码盘的字母排列顺序，他们仍然无法立刻破译我方密码，这是因为密码轮上的密码盘可以用不同的顺序排列，这种排列顺序的可能性非常多。如果一个密码轮上有25个密码盘，那么这些密码盘的排列顺序就一共有25! = 15 511 210 043 330 985 984 000 000 种可能性。

敌方绝不可能对如此大数目的可能性逐一进行检查，即使计算能力再强的计算机也无法完成这一任务。但是，一种更加聪明的现代攻击方法却可以解决这个问题。

如果某种攻击方法对多种加密系统都有效的话，这种破译方法就具有很高的价值。一种名为“爬山”的攻击方法就具有这种特点。在下文中，我会向读者解释如何用“爬山法”破解密码轮加密系统。除了密码轮系统以外，爬山法还可以破解MASC密码、普莱费尔密码，以及许多其他种类的密码。

用爬山法进行攻击的第一步是把密码轮上的密码盘以随机的方式排

列。这种我们随意选定的顺序不太可能是正确的，但是这并不要紧！接着，我们在密码轮上选定一行特定的文字，我们评估这段文字与正常英语的匹配度有多高，然后根据这种匹配度给这段文字打个分。评估一行文字与正常英语的匹配度高低的方式有很多，其中一种比较好的方法是比较每3个连续字母（三字母）出现的概率。比如，假如我们选定的这段文字是："INEVERTAKEMYSKATESOFF"（我从不会脱下我的冰鞋），我们就要先把这段文字中的每3个相邻单词都分出来一组，即INE NEV EVE VER ERT RTA TAK AKE KEM EMY MYS YSK SKA KAT ATE TES ESO SOF OFF。然后，我们要把上述每组字母在正常英语中出现的概率加起来。

如果这个密码轮上的密码盘是以正确顺序排列的，并且我们选择的这行文字也正好正确，我们就应该得到一个非常高的值。当然，在我们进行第一次尝试时，这个值通常都非常低。由于我们对这个较低的值不满意，我们就会在密码轮上随机选取两个密码盘，并调换它们的位置。然后，我们再次计算密码盘换位之后的新值。如果新值比旧值高，我们就保持交换顺序后的状态。如果新值比旧值更低，我们就把这两个密码盘换回原来的位置。接下来，我们再次从密码轮上随机选取两个密码盘，并调换它们的位置，如果换位后值有所提高，我们就保留交换顺序后的状态。

现在，想象我们已经把上述操作重复了几万次，甚至几十万次。在我们不断调换密码盘顺序的过程中，我们得到的文字（从统计上来看）越来越像正常的英文文字了。通过一系列这种微小的提高，最后我们会得到一个最高值。如果我们把这些值的变化画在一张图表上，就会呈现一条像山一样的曲线。我们沿着这条曲线向上爬，只保留能够提高分数的位置变化，那么我们最终就会爬到这座山的山顶。在山顶处，任何新变化都只能降低分数。当密码轮上的密码盘以这种顺序排列时，我们就应该已经得到了正确解法。

手动完成上述操作是极其烦琐和困难的。因此，在上面的文字中我一直说“如果”我们调换了密码盘的位置，因为这只是一个思想实验！现在，你需要做的是想象用一台计算机来帮助我们调换密码盘的位置，然后重新计算值。有了这台计算机，上述思想实验就变成了现实中的实验。对于一台计算机来说，进行几十万次顺序调换并不是一件难事，它们可以很快地完成这项任务。所以，这种攻击方法在实际中是具有可操作性的。

当然，上述攻击方法存在若干可能出现的问题，在这里我有必要向读者介绍一下。第一，在这种攻击算法中，有可能到达一个极大值，在这个极大值处，任何进一步的变化都只能降低分数，但这个极大值却不是正确解法。如果我们继续把这种算法比作爬山的话，这种情况就相当于我们爬到了一个小山峰的顶点。在这个小山峰处，我们可以看到更高的山峰，但要想爬到更高的山峰，就必须先走一些下坡路。我们可以对这种算法进行一些简单的修改，来避免上述情况的发生。

首先，在最初调换密码盘位置时，我们不要调换两个密码盘的位置，而应该同时调换多个密码盘的位置。在早期阶段，不太可能只有少数密码盘处于错误的位置上，因此，只进行微小的位置变化是不合理的。更合理的做法是每次移动多个密码盘，比如每次移动6个。在进行了几千次，甚至几万次上述循环以后，我们可以开始进行较小的位置变换。再经过很多次循环以后，我们又可以再改用更小的位置变化。最终，我们可以每次只调换两个密码盘的位置。如果用爬山作为类比的话，这就相当于：在山脚下刚开始爬的时候，我们应该大步地向上爬，而当我们比较接近山顶的时候，则应该以较小的步伐向上爬。如果我们这时候还以很大的步伐来爬山的话，就可能会错过山顶而走到山的另一侧去。在初始阶段采取较大的步伐，能够帮助我们避开较小的山峰，因为我们的目标是一直向着最高峰攀爬。

此外，我们还应该对这种算法进行另外一项修正，那就是：我们不能总是保留能提高值的变化。也就是说，有时虽然一个变化能够提高分数，但我们却不应该保留这种变化，而有时我们则需要保留某种会降低值的变化！听上去这是一种很不合理的做法，然而实际上这种做法是非常合理的。还是以爬山作为类比：如果我们发现自己正在向一个错误的山顶进发，我们就必须先走一段下坡路，然后再朝着正确的最高峰向上爬。那么，在这种攻击算法中，如何让计算机走下坡路呢？我们可以设定一个表示向上走的概率的百分比。比如，一开始，我们希望计算机只在60%的情况下保持能提高值的变化。在经过许多次循环以后，我们可以把上述百分比调高到65%或者70%。再经过许多次循环以后，我们又可以再把这个百分比提高到80%。最终，我们会把这个百分比提高到接近100%。

在上述这种修正方法中，这种表示向上走的概率的百分比被称作算法的“温度”（temperature）。在不断调整密码盘位置的过程中，我们也会逐渐调高算法的温度。“温度”这个词来自铸剑的过程：在锻造一把剑的时候，我们需要把剑反复加热到不同的温度，再反复冷却下来。这个过程被称为“淬火”（annealing，也称“退火”）。当然，在我们的算法中，我们并不是真的在铸剑，而是在用一种和铸剑类似的方法来破译密码。因此，上述过程被称为“模拟退火”过程。如果你想用一些专业词语来吓唬别人的话，你可以说用“有模拟退火过程的爬山算法”（hill climbing with simulated annealing）来破译密码。我觉得以上这串术语太复杂了，应该用一种缩写来指代这一长串文字。就像我把“单套字母替代密码”简写为MASC 密码一样，我也可以把“有模拟退火过程的爬山算法”简写为HCWSA算法，可惜这种缩写念起来一点儿也不顺口。

对爬山算法进行的最后一项修正是三种修正中最简单的一种。在前面的段落中，我曾经说过，在破解密码轮的第一步中，我们应该在密码轮上选定一行特定的文字。但是，如果我们选择了错误的文字，该怎么办？如果明文信息实际上在密文信息下的第15行，那么我们研究密文信息下的第2行是不会有什么作用的。而且，在调换密码盘位置的时候，有可能调换位置后某一行文字的值会提高，而另一行文字的值会下降，我们应该如何处理这种情况呢？这个问题的答案是：因为我们是用计算机来完成这项任务，所以我们只要对每一行文字都使用上述算法就可以了。除了密文信息本身以外，密码轮上共有25行文字，把上述整个过程重复25遍对于计算机来说并没有什么困难。最后，电脑会分别找出25座山的山顶，然后我们可以对这25个山顶进行比较，其中海拔最高的山顶就是我们要找的最终结果。

密码轮的种类

目前美国军方使用的密码轮已经不再对外保密，这些工具已经成了一种昂贵的收藏品。有时我们可以在网络拍卖中买到这种军方的密码轮，有时这种密码轮会出现在一些博物馆中，比如国家密码博物馆，这个博物馆就位于美国国家安全局附近。因此，我们完全可以用上述HCWSA算法来破译1915年的密码。

问题是，不同的军方密码轮上的字母会以不同的顺序排列。下面，我将向读者展示我找到的几种1915年前后不同版本的军方密码轮，如果在这几种密码轮中有一种与莫博涅使用的密码轮完全一样，那么用这个密码轮来破译莫博涅留下的挑战密码就应该能够成功。

我们先来看一段由威廉·F. 弗里德曼提供的记录，其中记载了一组帕克·希特使用过的字母表。帕克·希特是莫博涅的朋友，他自己也拥有十分辉煌的职业生涯。希特的著作《军事密码解法手册》（*Manual for the Solution of Military Ciphers*，1916）是第一本由美国人写成的密码学著作。此外，希特还担任过美国陆军第一军团的首席通信官。

弗里德曼这样写道：

> 这似乎是帕克·希特上校的条状密码装置的原始模型，这种装置上的字母表就是所谓的“星辰密码”字母表。希特上校曾说这个系统是他于1915年独立设计发明出来的，早于莫博涅发明的密码装置。莫博涅一开始把它称为QDEYAUB密码装置，因为在他的字母表中，QD、EY、AU和UB是唯一几种重复出现的字母组合。

事实证明，弗里德曼的记性似乎并不是特别好——莫博涅其实把这种密码装置称为DEYAUB密码装置（没有一开头的字母Q）。但是对于这种密码装置得名的根本原因，弗里德曼的复述是正确的。在莫博涅的字母表中，唯一几种重复出现的字母组合是DE、YA和UB。

图8–4展示的就是弗里德曼在上述段落中提到的希特的字母表。

从图8–4中我们可以看出，最下面的两行是数字，这两行起到对字母表进行编号的作用。在某些版本中，希特用一行星辰的符号来隔开字母表和数字，以避免阅读表格的人将数字0和1误读作字母O和I。也许正是因为这一行星辰符号的存在，所以后来人们才把希特的这套字母表称为“星辰密码”字母表。

当然，在加密的过程中，可以以任何一种顺序来使用图8–4中的这些字母表。字母表的排列顺序就是密码的密钥。但是，为什么在这套字母表

星辰密码字母表

A	X	Q	J	X	H	F	C	G	D	B	W	M	Y	P	L	O	E	U	I	X	S	N	T	R	0
E	U	V	K	Z	N	T	D	B	F	C	G	P	L	W	M	Y	I	A	O	Z	V	R	H	S	1
I	A	O	Q	J	R	H	S	C	G	D	B	F	M	Y	P	L	W	E	U	J	X	Q	N	T	2
O	E	U	I	K	S	N	T	R	B	F	C	G	D	L	W	M	Y	P	A	K	Z	V	K	H	3
U	I	A	O	E	T	R	H	S	N	G	D	B	F	C	Y	P	L	W	M	Q	J	X	Q	J	4
H	O	E	U	I	L	S	N	T	R	A	F	C	G	D	V	W	M	Y	P	V	K	Z	V	K	5
N	T	I	A	O	M	Y	R	H	S	E	U	D	B	F	X	Q	P	L	W	B	Q	J	X	Q	6
R	H	S	E	U	P	L	W	N	T	I	A	O	C	G	Z	V	K	M	Y	C	G	K	Z	V	7
S	N	T	R	A	W	M	Y	P	H	O	E	U	I	B	J	X	Q	J	L	D	B	F	J	X	8
T	R	H	S	N	Y	P	L	W	M	U	I	A	O	E	K	Z	V	K	Z	F	C	G	D	Z	9
B	S	N	T	R	A	W	M	Y	P	Q	O	E	U	I	Q	J	X	Q	J	G	D	B	F	C	10
C	G	R	H	S	E	U	P	L	W	V	K	I	A	O	H	K	Z	V	K	L	F	C	G	D	11
D	B	F	N	T	I	A	O	M	Y	X	Q	J	E	U	N	T	J	X	Q	M	Y	D	B	F	12
F	C	G	D	H	O	E	U	I	L	Z	V	K	Z	A	R	H	S	Z	V	P	L	W	C	G	13
G	D	B	F	C	U	I	A	O	E	J	X	Q	J	X	S	N	T	R	X	W	M	Y	P	B	14
L	F	C	G	D	K	O	E	U	I	K	Z	V	K	Z	T	R	H	S	N	Y	P	L	W	M	15
M	Y	D	B	F	Q	J	I	A	O	H	J	X	Q	J	B	S	N	T	R	A	W	M	Y	P	16
P	L	W	C	G	V	K	Z	E	U	N	T	Z	V	K	C	G	R	H	S	E	U	P	L	W	17
W	M	Y	P	B	X	Q	J	X	A	R	H	S	X	Q	D	B	F	N	T	I	A	O	M	Y	18
Y	P	L	W	M	Z	V	K	Z	V	S	N	T	R	V	F	C	G	D	H	O	E	U	I	L	19
J	W	M	Y	P	J	X	Q	J	X	T	R	H	S	N	G	D	B	F	C	U	I	A	O	E	20
K	Z	P	L	W	B	Z	V	K	Z	L	S	N	T	R	A	F	C	G	D	H	O	E	U	I	21
Q	J	X	M	Y	C	G	X	Q	J	M	Y	R	H	S	E	U	D	B	F	N	T	I	A	O	22
V	K	Z	V	L	D	B	F	V	K	P	L	W	N	T	I	A	O	C	G	R	H	S	E	U	23
X	Q	J	X	Q	F	C	G	D	Q	W	M	Y	P	H	O	E	U	I	B	S	N	T	R	A	24
Z	V	K	Z	V	G	D	B	F	C	Y	P	L	W	M	U	I	A	O	E	T	R	H	S	N	25
A	X	Q	J	X	H	F	C	G	D	B	W	M	Y	P	L	O	E	U	I	X	S	N	T	R	0
E	U	V	K	Z	N	T	D	B	F	C	G	P	L	W	M	Y	I	A	O	Z	V	R	H	S	1
I	A	O	Q	J	R	H	S	C	G	D	B	F	M	Y	P	L	W	E	U	J	X	Q	N	T	2
O	E	U	I	K	S	N	T	R	B	F	C	G	D	L	W	M	Y	P	A	K	Z	V	K	H	3
U	I	A	O	E	T	R	H	S	N	G	D	B	F	C	Y	P	L	W	M	Q	J	X	Q	J	4
H	O	E	U	I	L	S	N	T	R	A	F	C	G	D	V	W	M	Y	P	V	K	Z	V	K	5
N	T	I	A	O	M	Y	R	H	S	E	U	D	B	F	X	Q	P	L	W	B	Q	J	X	Q	6
R	H	S	E	U	P	L	W	N	T	I	A	O	C	G	Z	V	K	M	Y	C	G	K	Z	V	7
S	N	T	R	A	W	M	Y	P	H	O	E	U	I	B	J	X	Q	J	L	D	B	F	J	X	8
T	R	H	S	N	Y	P	L	W	M	U	I	A	O	E	K	Z	V	K	Z	F	C	G	D	Z	9
B	S	N	T	R	A	W	M	Y	P	Q	O	E	U	I	Q	J	X	Q	J	G	D	B	F	C	10
C	G	R	H	S	E	U	P	L	W	V	K	I	A	O	H	K	Z	V	K	L	F	C	G	D	11
D	B	F	N	T	I	A	O	M	Y	X	Q	J	E	U	N	T	J	X	Q	M	Y	D	B	F	12
F	C	G	D	H	O	E	U	I	L	Z	V	K	Z	A	R	H	S	Z	V	P	L	W	C	G	13
G	D	B	F	C	U	I	A	O	E	J	X	Q	J	X	S	N	T	R	X	W	M	Y	P	B	14
L	F	C	G	D	K	O	E	U	I	K	Z	V	K	Z	T	R	H	S	N	Y	P	L	W	M	15
M	Y	D	B	F	Q	J	I	A	O	H	J	X	Q	J	B	S	N	T	R	A	W	M	Y	P	16
P	L	W	C	G	V	K	Z	E	U	N	T	Z	V	K	C	G	R	H	S	E	U	P	L	W	17
W	M	Y	P	B	X	Q	J	X	A	R	H	S	X	Q	D	B	F	N	T	I	A	O	M	Y	18
Y	P	L	W	M	Z	V	K	Z	V	S	N	T	R	V	F	C	G	D	H	O	E	U	I	L	19
J	W	M	Y	P	J	X	Q	J	X	T	R	H	S	N	G	D	B	F	C	U	I	A	O	E	20
K	Z	P	L	W	B	Z	V	K	Z	L	S	N	T	R	A	F	C	G	D	H	O	E	U	I	21
Q	J	X	M	Y	C	G	X	Q	J	M	Y	R	H	S	E	U	D	B	F	N	T	I	A	O	22
V	K	Z	V	L	D	B	F	V	K	P	L	W	N	T	I	A	O	C	G	R	H	S	E	U	23
X	Q	J	X	Q	F	C	G	D	Q	W	M	Y	P	H	O	E	U	I	B	S	N	T	R	A	24
Z	V	K	Z	V	G	D	B	F	C	Y	P	L	W	M	U	I	A	O	E	T	R	H	S	N	25
★	★	★	★	★	★	★	★	★	★	★	★	★	★	★	★	★	★	★	★	★	★	★	★	★	
0	0	0	0	0	0	0	0	0	1	1	1	1	1	1	1	1	1	1	2	2	2	2	2	2	
1	2	3	4	5	6	7	8	9	0	1	2	3	4	5	6	7	8	9	0	1	2	3	4	5	

图 8-4　希特的 25 个字母表

中，每张纸条上都把26个字母写了两次呢？这是因为，希特在设计这套加密系统时，打算用一种和我之前描述的密码轮不同的方法来使用这种密码装置，但是，其最终效果与密码轮的加密方式是一样的。在密码轮中，每个字母表都是以环状形式出现在一个密码盘的圆周上的，而希特则是把每一个字母表写到一张长条形的纸上。在图8–5中，读者可以看到一部1916年的希特密码装置的样子。

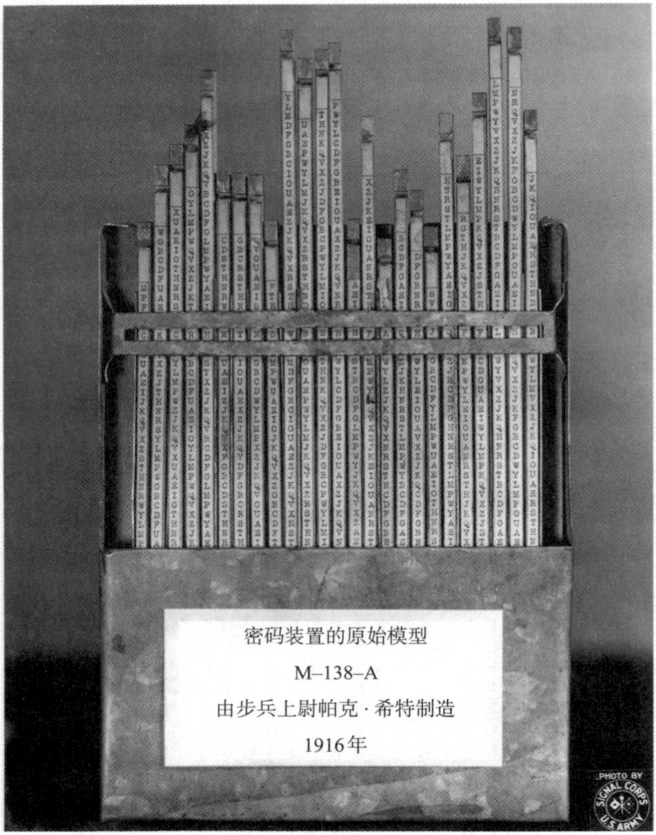

图8–5　和密码轮作用等同的一个装置

从图8–5中可以看出，在这个密码装置中有一个水平的窗口，从窗口中能够读到一行文字。在使用这种装置时，加密者需要将纸条向上或者向下拉，直到这个水平的窗口中显示的文字是他想要的为止。在完成上述步骤以后，加密者可以选择加密装置上的另一行字母作为密码的密文。

在这幅摄于1916年的照片中，我们还可以看到这种加密装置是如何解密信息的。水平窗口中出现的是密码的密文，而最后一个横行中出现的就是原始的明文信息。之所以要在每一张纸条上把字母表重复两次，是为了保证还存在其他每一个位置上都有字母的横行。如果在每张纸条上把26个英文字母只写一遍，让我们想象一下遇到以下这种情况时会出现什么问题：假设加密者需要让第一张纸条的最后一个字母和第二张纸条的第一个字母作为信息最开始的两个字母，那么，就没有其他横行同时贯穿这两张纸条了，于是加密的过程也就无法进行。希特的这种加密装置是一种条状密码装置，这种装置的正式名称是M–138–A。虽然这是一种很简单的装置，但是却被美国军方频繁使用。一直到第二次世界大战，美国军方仍在继续使用希特的这种加密装置。

我将希特的这套字母表改写如下。在改写时，我只把每个字母表抄了一遍，因为我并不是在希特设计的条状密码装置中使用这套字母表，而是在密码轮上使用这套字母表。密码轮和希特设计的条状密码装置是完全等效的。

```
A A A A A A A A A A A A A A A A A A A A A A A A A
E E E E E E E E E V E E E E X E E E E M E E E E N
I I I R R I I I X X I I I Z Z I I I P P I I I R R
O O S S S O O Z Z Z O O J J J O O W W W O O S S S
U T T T T U J J J J U K K K K U Y Y Y Y U T T T T
```

H	H	H	H	H	K	K	K	K	K	Q	Q	Q	Q	Q	L	L	L	L	L	H	H	H	H	H
N	N	N	N	C	Q	Q	Q	Q	Q	V	V	V	V	V	M	M	M	M	Z	N	N	N	N	J
R	R	R	D	D	V	V	V	V	C	X	X	X	X	N	P	P	P	J	J	R	R	R	K	K
S	S	F	F	F	X	X	X	D	D	Z	Z	Z	R	R	W	W	K	K	K	S	S	Q	Q	Q
T	G	G	G	G	Z	Z	F	F	F	J	J	S	S	S	Y	Q	Q	Q	Q	T	V	V	V	V
B	B	B	B	B	J	G	G	G	G	K	T	T	T	T	V	V	V	V	V	X	X	X	X	X
C	C	C	C	M	B	B	B	B	B	H	H	H	H	H	X	X	X	X	X	Z	Z	Z	Z	Z
D	D	D	P	P	C	C	C	C	N	N	N	N	N	M	Z	Z	Z	Z	N	J	J	J	J	C
F	F	W	W	W	D	D	D	R	R	R	R	R	P	P	J	J	J	R	R	K	K	K	D	D
G	Y	Y	Y	Y	F	F	S	S	S	S	S	W	W	W	K	K	S	S	S	Q	Q	F	F	F
L	L	L	L	L	G	T	T	T	T	T	Y	Y	Y	Y	Q	T	T	T	T	V	G	G	G	G
M	M	M	M	Q	H	H	H	H	H	L	L	L	L	L	H	H	H	H	H	B	B	B	B	B
P	P	P	V	V	N	N	N	N	M	M	M	M	M	C	N	N	N	N	C	C	C	C	C	M
W	W	X	X	X	R	R	R	P	P	P	P	P	D	D	R	R	R	D	D	D	D	D	P	P
Y	Z	Z	Z	Z	S	S	W	W	W	W	W	F	F	F	S	S	F	F	F	F	F	W	W	W
J	J	J	J	J	T	Y	Y	Y	Y	Y	G	G	G	G	T	G	G	G	G	G	Y	Y	Y	Y
K	K	K	K	K	L	L	L	L	L	B	B	B	B	B	B	B	B	B	B	L	L	L	L	L
Q	Q	Q	Q	E	M	M	M	M	E	C	C	C	C	E	C	C	C	C	E	M	M	M	M	E
V	V	V	I	I	P	P	P	I	I	D	D	D	I	I	D	D	D	I	I	P	P	P	I	I
X	X	O	O	O	W	W	O	O	O	F	F	O	O	O	F	F	O	O	O	W	W	O	O	O
Z	U	U	U	U	Y	U	U	U	U	G	U	U	U	U	G	U	U	U	U	Y	U	U	U	U

在希特的这套字母表中会出现重复的字母，我在前文中已经说过，莫博涅认为这是希特加密系统的一个弱点。在莫博涅设计字母表的时候，他希望不同字母表之间的区别越大越好。因此，在莫博涅设计的字母表中，

只有以下几种字母组合会多次连续出现在不止一个密码盘上，这几种字母组合是DE、YA和UB。莫博涅设计出的这套密码就是我们之前提到的DEYAUB密码。

我将莫博涅设计的25列字母表复制如下。我们需要搞清楚的问题是，莫博涅是否在1915年时就已经开始以某种形式（条状或轮状）使用这套字母表了？如果答案是肯定的，那么莫博涅1915年留下的这条挑战密码就有可能是用这套字母表加密而成的。资料显示，1917年年初，莫博涅在陆军通信学校的商店里定制了两个印有这套字母表的密码轮。然而，在1917年之前，莫博涅会不会已经制造出了不太完美的雏形了呢？

A	A	A	A	A	A	A	A	A	A	A	A	A	A	A	A	A	A	A	A	A	A	A	A	A
B	C	D	E	F	G	H	I	J	K	L	M	N	O	P	Q	R	S	T	U	V	W	X	Y	Z
C	D	K	D	N	P	X	H	D	E	T	N	C	D	B	J	M	D	O	T	N	V	K	J	D
E	E	O	C	Q	O	J	P	S	L	M	F	J	W	V	N	Y	M	J	R	K	S	W	P	N
I	H	M	B	U	C	E	J	K	B	S	L	I	P	H	U	O	C	Y	Z	H	F	R	X	B
G	F	J	I	K	I	Z	O	Q	D	X	H	L	K	I	B	F	N	L	X	R	D	E	M	U
D	I	U	F	D	X	B	B	O	F	V	Q	D	J	Y	T	T	E	F	Q	G	L	V	V	H
J	J	B	G	O	L	N	W	I	J	Q	G	H	V	K	G	H	Q	X	L	O	I	D	K	Y
F	K	G	J	P	U	I	K	V	G	P	C	B	I	S	I	E	B	N	Y	X	E	T	B	F
V	T	E	H	I	R	K	C	T	H	N	U	M	U	G	M	U	O	G	I	E	B	U	Q	W
U	L	P	L	T	N	P	V	Z	O	O	J	K	Q	U	W	S	Z	W	O	Y	H	F	W	J
Y	M	H	K	J	D	V	F	E	N	H	T	G	H	E	Z	Z	P	H	V	B	K	O	U	L
M	O	S	M	B	Y	R	Z	F	M	U	B	X	Z	N	R	J	L	V	B	F	N	Y	G	V
H	U	C	R	R	Z	O	L	H	T	W	Y	U	C	T	V	X	G	C	P	S	R	H	L	G
T	V	Z	U	H	H	G	Q	G	P	D	P	Z	T	C	L	D	V	M	E	J	J	M	O	T

Q Y I O C W S E Y R I Z T X X X P J I S M Q L S C
K G N Q Y B Y R U Q Z K S B O C C R R N U Z S T Q
Z Z X V S J D Y N S Y X W L W S W K B H D G I E M
O N F P L S U N L V C I Q E F H G Y S J Q M Q C P
L P Y T W Q L S P Z G S Y G Q D Q T E W C X N H S
R Q Q N E F C U M U K R V N D E I F K M L P J N O
X X R W M K F M B X R D O Y R O B U U D Z U C Z E
S R T Y Z V M G X Y F V R R L K K I P G W C P F X
P W V X V M Q T W W B E P S J F L W D F T O G R T
W S W Z X E T D C I E W F M Z P N X Z C I T B I K
N B L S G T W X R C J O E F M Y V H Q K P Y Z D I
1 2 3 4 5 6 7 8 9 25

莫博涅的密码盘也用字母进行了编号，但是因为一共只有25个密码盘，而英文有26个字母，所以必须要去掉一个字母。莫博涅去掉了字母A，也就是说，第一个密码盘是B盘，第二个密码盘是C盘，以此类推，最后一个密码盘是Z盘。这种编号方式是合理的，因为每一个密码盘的字母编号正好是该密码表上A下面的那个字母。

不知道读者有没有注意到这样一个细节：在莫博涅设计的这套字母表中，第17个字母表的开头是“ARMY OF THE US…（美国陆军……）”？当美国海军采用莫博涅设计的这种加密装置时，他们给这种装置起了一个新名字，但是他们却不能改动密码盘上的字母表。于是，第17个密码盘上的字母发挥了一种有趣的作用：这些字母将不断提醒美国海军的人员，他们使用的这个装置是由一名陆军军官（再次）发明的。

另外一个值得注意的现象是，莫博涅设计的密码轮共有25个密码盘，而他1915年提出的挑战密码共有499个字符——只差1个字符就是25的倍数！如果这段1915年的挑战密码真的是用这种密码轮加密的，那么莫博涅需要每次在密码轮上调出25个字母，然后记下另一行显示出的密文，并且需要把这个操作重复20次。

要想破译莫博涅1915年留下的这段挑战密码，不仅需要找到正确的字母表，还需要克服另一重困难。在下文中，我将向读者详细解释这重额外的困难。

新的挑战密码

莫博涅1915年留下的挑战密码是路易斯·克鲁于1991年发现的，在大约10年前，克鲁还发表过莫博涅留下的另一则挑战密码，这则密码是莫博涅1918年创造的。资料显示，这段1918年的密码是莫博涅用上文中介绍的密码轮创造出来的。这段密码共有25条，都是用同一个密码轮加密的，并且这个密码轮上的密码盘顺序也没有发生改变。

这25条密码如下所示：

1 VFDJL QMMJB HSYVJ KCJTJ WDKNI

2 CGNJM ZVKQC JPRJR CGOXG UCZVC

3 CSTDT SSDJN JDKKT IXVEX VHDVK

4 OZBGF VTUEC UGTZD KYWJR VZSDG

5 QIRMB FTKBY CGAQV DQCVQ AHZGY

6 VQWRM IHDHB RQBWU LKJCS KEYUU

7 SSEIQ DWHNH QHGIK HAADN GNFBY

8 VXDVX NIGJO PCOTN GKWAX YTNWL

9 QJRLH AWTWU CYXVM BGJCR SBHWF

10 DULPK UXMVL XFUPS ULRZK PDALY

11 DCAIY LUPMB NACQE OPTLH KKRGT

12 MGODT VGUYX NHKBE WPOUR VTQOE

13 TBVEB QDXGP LCPUY AVVBK ZEOZY

14 FIJDW WBKTY GBSMB PZWYP RRZCW

15 DYVPJ CLNXE SCMF0 YPIZF PEBHM

16 MYYTJ RFMEP PHDXP ODFZO WLGLA

17 EYKKD XHTEV TRXWK CJPSG MASCY

18 LGQLV HTUIP YAUGJ PGDLH UZTKV

19 BRKTJ RGGTB HMLXX FRHOA AZVWU

20 CDUDV DBZUA ELRPO SPUJD XRZWA

21 EUFBT TWNIY HHTNW QNFVE NYGBY

22 TUTVY NGLPG TYOLI HXZQT XSGOJ

23 PBTJC CJONJ UNIXB UAQBI WNIHL

24 VHNKR XVZMD KFHUY XRNDD KXXVM

25 NNHBF VQH0B LXCYM AKFLS SSJXG

这25条密码一开始是为了挑战赫伯特·O. 亚德利和威廉·F. 弗里德曼而设计的，然而这两位密码专家都没有能破解出来。于是，莫博涅为他们提供了一条小提示："ARE YOU"在其中一条密码中紧挨着出现。然而，即使有了莫博涅的这条提示，亚德利和弗里德曼还是没能取得任何进展。但正是因为用密码轮加密的这25条密码如此难以破解，美国陆军才于1921年采用了这种密码轮作为军方的加密设备。

弗里德曼最终于1941年找到了这段密码的解答，但他却并不是通过密码分析解开的！在莫博涅退休以后，弗里德曼检查了莫博涅办公室里的文件，并且在这些文件中找到了这25条密码的明文。

克鲁在重新公布这段挑战密码时，还为第一位将25条密码全部破译出来，并解释自己是如何破译的人提供了一件奖品——一本最新版的《美国黑箱》（*The American Black Chamber*），这是一本非常生动的书，作者就是亚德利。

克鲁的奖品很快就找到了主人。第一个解开全部25条密码的是翠欧［TRIO，伦敦的芬威克·韦森克拉夫特（Finwick Wesencraft）在美国密码协会的笔名[10]］。翠欧很快就提交了答案，他的解法发表在当年（1982年）11—12月号的《密码》杂志上。此外瑞伊［REEY，同样来自伦敦的马克·E. R. 艾尔（Mark E. R. Eyre）］和布鲁因［THE BRUIN，来自耶路撒冷的摩西·鲁宾（Moshe Rubin）］也解出了这25条密码，但他们没有翠欧动作快。翠欧在发表这段解法的文章说的第一句话是："莫博涅浪费了许多时间来构想一些荒谬的测试语句。"

这些荒谬的语句如下：

1 14 Chlorine and oxygen have not b（氯和氧没有……）

2 8 Where did you meet each other（你们俩是在哪里遇见的）

3 1 Drink this potion quickly for（快喝下这服药，为了……）

4 20 Well, make me the same shape but（好吧，让我变成同样的形状，但是）

5 13 Cyanogen is a colorless gas in(在……里面氰是一种无色气体）

6 25 Phenols are benzene derivati（苯酚是苯的衍生……）

7 2 Xylonite and artificial ivor（赛璐珞和人造象……）

8 12 I went to a new theatre, the Pala（我去了一家新戏院，叫……）

9 24 Picric acid is explosive and i（苦味酸易爆炸，且……）

10 18 Llangollen is a town in Wales a（兰戈伦是威尔士的一个城市……）

11 6 Yvette, are you going shopping（伊薇特，你要去买东西吗）

12 23 Orthophosphoric is the compo（正磷酸是化合……）

13 16 Caoutchouc is closely allied（生橡胶和……联系很紧密）

14 17 Olefiant gas, ethene or ethyle（乙烯气、乙烯或者乙基……）

15 3 See the terrible tank tackle a（看那可怕的坦克抓住……）

16 7 It is a thin limpid liquid that（它是一种稀薄清澈的液体）

17 22 If it is insoluble in water it i（如果它不溶于水，它……）

18 11 Silver has been known from rem（人们知道银……）

19 21 Hot concentrated sulphuric a（热的浓硫……）

20 4 Small coefficient of expansi（小的膨胀系数……）

21 9 Palladium possesses a power o（钯有一种能……）

22 19 Absorbing and condensing thi（吸收和冷凝……）

23 15 Compounds of platinum form tw（铂的化合物形成……）

24 10 Gold occurs widely dristribut（金广泛分布……）

25 5 Oxidation caused by it probab（它造成的氧化很可能……）

这篇文章的作者还说："这些语句读起来真是枯燥极了。"

每行的第二个数字（在密码编号和明文之间的数字）标示的是明文信息和密文信息在密码轮上隔了多少行。

密码轮上各密码盘的顺序是：4，14，5，6，2，13，3，24，21，11，19，8，7，16，15，17，10，9，1，12，25，23，22，20，18。

解出这25段密码确实是相当了不起的成果，但这并不能说明这些破译者就比亚德利和弗里德曼更加优秀，因为当年这两名政府密码专家并不知道莫博涅使用的字母表是什么样的。而克鲁在公布这段挑战密码时，也同时公布了莫博涅的字母表。但是，克鲁犯了一个错误，在一个应该写L的地方错误地写上了K。读者很快就会知道我为什么要特别谈到这个细节。

如果读者有兴趣了解破译上述密码的细节，可以阅读破译者发表在《密码》杂志上的原文，该文对上述问题做了详细的解释。但是，我必须提醒读者注意，破译者说他使用了一种“简单却极其烦琐”的破译方法。[11]在破译这段密码的过程中，这名破译者并没有使用计算机！

再信任也得小心查证

克鲁在1993年4月的《密码学》杂志上公布了莫博涅1915年留下的密码。在这篇文章中，克鲁写道：“去年早些时候，这则密码被发表在了美国密码协会的官方出版物《密码》杂志上。”然而，这则密码并不是1992年发表在《密码》杂志上的，实际上是刊登在1991年3—4月号的《密码》杂志的第7页上，文章的署名是克鲁在美国密码协会的笔名“米罗克”。

也就是说，克鲁说错了出处。我估计他很可能以为自己的后一篇文章1992年就可以发表在《密码学》杂志上，但是事实上该文被推迟到了1993年才刊登出来，而克鲁却忘记修改这篇文章中的“去年早些时候”的说法。当然，这没什么大不了的。虽然引用参考文献时犯了小错误，但对于更重要的密码抄录问题，克鲁一定会非常小心的。

作为一名密码学家，我总是充满了怀疑精神，因此我没有理所当然地接受克鲁公布的莫博涅密码，而是想要看到载有这段密码的原始文献。也许原始文献里还有一些克鲁漏掉的细节能为破译这段密码提供更多的线索，也许克鲁在抄写密码字母时犯了一些错误。不幸的是，后一种情况是有先例的。

1990年，克鲁与另外两名作者合作发表过一篇文章，讨论了另一则挑战密码（准确地说是一组3条挑战密码）。这段挑战密码使用的加密系统是由约翰·F. 伯恩（John F. Byrne）于1918年发明的，叫作“混沌密码”（Chaocipher）。[12]伯恩一直希望美国政府能对他发明的这种加密系统感兴趣，但是他反复尝试了近40年，还是没有能够获得美国政府的青睐。约翰·F. 伯恩于1960年去世，但是他的儿子约翰·伯恩（John Byrne）知道混沌密码系统的工作原理，并且愿意继续推广父亲发明的这种加密系统。在这3条挑战密码中，虽然第3条是正确的，但前两条却都有一些错误。

最严重的错误发生在第2条密码中，第2条密码如下所示：

```
ENWSC EAQGI VIDEM WUMSN ZMNTV UFDLB JKKMR HHSNB
KTJBH VPTWH FMQQJ PGRWF FVJMD HFUZO XEOZT MKZSA
MJYRL SQSXU ZYEKR JBFRE SGGFX FEGXL PWTWL ZAVIM
TBDTQ BLVRZ VEMMT LXITZ
```

破译后的明文信息应该是：

```
    Our own memories are mysterious enough even to
```

ourselves and we should recognize that they do not exist in space only in time and that they are in all respects immaterial.

（我们的记忆对自己而言也已经很神秘了。我们应该认识到，记忆并不存在于空间中，而是仅仅存在于时间中，并且不管从哪个方面来看，记忆都是一种非物质的存在。）

但是，上面的密文根本无法破译出这段明文。第三组密文字母“VIDEM”是错误的，正确的密文字母应该是“UIOEM”，由于作者把U误写成了V，又把O误写成了D，因此即使一位破译者知道正确的破译算法，并且拥有正确的密钥（破译算法和密钥都是保密的），他也只能得到以下结果：

OUROW NMEMO OIPXA AEJZG GRYTF NYMRS DJOIM USHNV
LLFIN WOJWK JAGTA MUQVR CBGGS HJAMW CCRGY JKAEY
SPFAP PMHPZ VXWFQ DGHXA HKHRJ EWLST ALBPT JSZAQ
HQXVI WBLIC QTHQJ MZVWU[13]

然而，这篇包含严重错误的论文并不是克鲁一人写成的。他还有另外两名合作作者，其中一人还是混沌密码系统的发明人的儿子约翰·伯恩！既然我们知道克鲁有过抄错密码的先例，还知道克鲁在复制莫博涅挑战密码（后一则挑战密码）的字母表时又一次犯了错误，我们就不得不怀疑莫博涅1915年留下的这则挑战密码也有被抄错的可能。

于是，我联系了纽约公共图书馆珍本书部门的一位图书馆员，请他帮我寻找相关文献，只是为了保证我向读者展示的莫博涅在1915年留下的挑战密码是准确无误的。以下是我收到的回复：

亲爱的克雷格，

感谢你发电子邮件询问关于密码的问题。我搜索了我们的图书编目，书架上也都找过了，但是找不到符合你描述的文献。我能明白你所描述的意图，但是就像你指出的那样，你既没有查询号码，也没有确切的标题，这种情况使得找到这份文献变得有些困难。如果你能找到更多线索，请一定要联系我，我会继续为你搜索你想要的文献。在我们的普通藏书部分中，有一份1912年的作品的作者是莫博涅，但是这份文献的日期早于你想查询的日期。如果我能为你提供其他帮助，请再联系我。

祝你好运。

凯尔·R.特里普利特，图书管理员

纽约公共图书馆

档案、手稿及珍本书部

布鲁克·拉塞尔·阿斯特珍本图书及手稿阅览室

斯蒂芬·A.施瓦茨曼楼328室

纽约州，纽约市，第五大道476号，邮编：10016

manuscripts@nypl.org

看来，在我们找到这份文献（或者在别处找到它的复制本）之前，我们的这一重疑问恐怕暂时无法解开了！

第9章

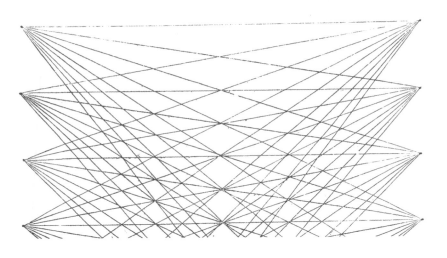

欲言又止的挑战密码

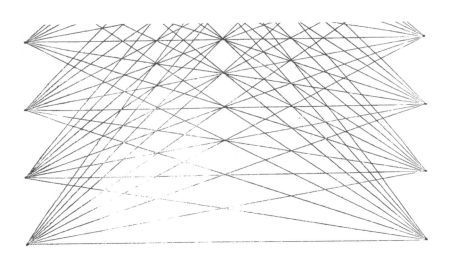

在前一章中，我向读者介绍过莫博涅留下的挑战密码。莫博涅创造这条挑战密码是为了证明自己的观点——他所发明的加密系统比当时正在使用的其他加密系统更加先进。然而，还有许多人为了各种五花八门的原因留下了许多其他挑战密码。有些挑战密码的创作者只是想发明一些能够娱乐他人的挑战游戏，就像填字游戏和数独一样。而另一些挑战密码则似乎是为了筛选才智过人的人才而设计的。还有一些挑战密码是为了欺骗、愚弄他人而产生的。在本章中，我将从以上各类挑战密码中挑选一些例子详细介绍。本章中介绍的所有密码至今仍是一种挑战，因为目前尚没有人能找到这些密码的解答。

被遗忘的密码

我们的第一个例子是亚历山大·达加佩耶夫（Alexander d’Agapeyeff）密码。这则密码出现于1939年，当时整个世界正在走向战争的深渊。从密码学的角度来看，1939年之后的一段时间十分奇怪。一方面，战争提高了人们对暗号和密码的兴趣，但是另一方面，政府却在试图采取一些措施来抑制人们对密码的兴趣，至少在美国是这样。为了满足公众对密码知识的需求，一些以密码学为主题的新书涌现了出来。但是，战争部部长亨利·L. 斯廷森（Henry L. Stimson）认为这些知识过于危险，不宜广泛传播。因此，斯廷森发出了以下的这封信件，要求秘密禁止这类书籍的传播。[1]

***

保密文件

战争部

华盛顿

1942 年 8 月 13 日

WD 461 (8−10−42) MS

美国图书馆协会

密歇根大道 750 号

芝加哥，伊利诺伊州

先生们：

战争部注意到以下事实：隶属于美国图书馆协会的各成员图书馆收到了大量图书借阅申请，要求借阅有关爆炸物、秘密墨水，以及密码的书籍。

在此，战争部要求这些图书馆停止向公众出借以上种类的书籍。同时，各图书馆还应向当地联邦调查局办公室提供要求借阅这些书籍的人的名单。

我们感谢你们在这方面的支持和配合。

您忠诚的，

亨利 · L. 斯廷森（签名）

战争部部长

本文件包含影响美国国防的机密信息。根据间谍法案（Espionage Act,《美国法典》第 50 卷第 31 和 32 章)，以任何形式向未经授权的人员泄露或传播本文件内容的行为都是违法行为。

保密文件

***

一般来说，政府的任何审查行为只会进一步提高民众对违禁材料的好奇和需求。然而，在上述例子中，美国的普通民众根本不知道联邦政府对这方面的书籍颁布了禁令，因为斯廷森签名之后的这一小段文字要求对这封信件严格保密。

因为斯廷森颁布的上述禁令而被禁止流通的密码学书籍有很多，这些书籍的作者的背景也五花八门。在这些作者中，亚历山大·达加佩耶夫（1902—1955）在密码学方面的资历恐怕称得上数一数二。在他的著作的防尘套上，印有这样一段作者简介：

> 亚历山大·达加佩耶夫先生有足够的资格来撰写一部关于暗号和密码的书籍。他16岁加入英国陆军，成为一名二等兵，17岁升任军官。之后，他成了北俄罗斯远征军的情报官员，并担任波罗的海地区的英国军事代表团成员直至1921年。这些经历使得他非常了解德国和苏俄之间的各种间谍和反间谍活动。战后，达加佩耶夫在剑桥大学短暂地停留了一段时间，然后前往非洲进行探险测绘工作。在非洲，达加佩耶夫不仅对前往乍得湖的路径进行了粗略的测绘工作，还成为某个非洲野生部落的“结盟兄弟”。在未知世界的广袤丛林中，达加佩耶夫以地图测绘者和管理人员的身份学习了当地习俗、神话，以及宗教信仰方面的知识。[2]

达加佩耶夫出生于俄国的圣彼得堡，这一点在上述书的作者简介部分中并未被提及。虽然我们非常有兴趣了解达加佩耶夫在非洲的经历，但是这本书仅仅提到他“在非洲丛林中度过了数年的时间”，并了解到当地的原住民如何通过击鼓的方式来传递信息。他在当地使用的名字是“恩科伊”（Nkoy）。[3]虽然达加佩耶夫的名字广为流传，但是出乎意料的是，关

图 9-1 《暗号与密码》一书的第一版

于他生平信息的进一步资料却难以找到。但是我们有理由相信，他的活动范围严重超出了他的妻子能够接受的范围。他的妻子于1929年与他离婚，理由是达加佩耶夫“频繁通奸”。法庭采纳了妻子的上述指控。[4]

1940年，英国成立了一个名为“特别行动执行委员会”（Special Operations Executive）的新组织。该组织的目标是秘密渗透被德军占领的欧洲地区，并组织发动各种破坏活动。按照温斯顿·丘吉尔的说法，这个组织的目标是“在欧洲燃起反抗的火焰”。从某些方面来看，达加佩耶夫可能完全有资格加入特别行动执行委员会，然而该组织却拒绝吸收达加佩耶夫。委员会的这一决策可能与达加佩耶夫背叛妻子有关。[5]如果一个人曾经背叛自己最亲密的人，谁还敢信任他呢？但是，即便未能加入该委员会，达加佩耶夫仍然以自己的方式在第二次世界大战中为国家做出了贡献——战时，他曾是英国皇家空军的中校。

达加佩耶夫留下的这段至今未被解开的挑战密码出现在《暗号与密码》一书的第158页上。在这本书中，他对这段挑战密码做了这样的介绍："这里有一段密码，我邀请读者试着破译这段密码，以测试自己的能力。"在此，我将达加佩耶夫的这段挑战密码完整地复制如下：

75628 28591 62916 48164 91748 58464 74748 28483
81638 18174 74826 26475 83828 49175 74658 37575
75936 36565 81638 17585 75756 46282 92857 46382
75748 38165 81848 56485 64858 56382 72628 36281
81728 16463 75828 16483 63828 58163 63630 47481
91918 46385 84656 48565 62946 26285 91859 17491
72756 46575 71658 36264 74818 28462 82649 18193
65626 48484 91838 57491 81657 27483 83858 28364
62726 26562 83759 27263 82827 27283 82858 47582
81837 28462 82837 58164 75748 58162 92000

既然这段密码是用来测试读者的能力的，我们有理由相信，达加佩耶夫在这段密码中所使用的加密方法应该与他在《暗号与密码》一书中讨论过的加密方法比较接近。因此，我们可以排除这本书中未提及的加密系统，以节省时间。尤其值得我们注意的是，虽然在《暗号与密码》一书出版时，机器密码已经被广泛使用，但是达加佩耶夫在这本书中几乎没有谈到机器密码。在这本书中，机器密码的讨论只占了一页多的篇幅。而且在这段讨论的结尾，达加佩耶夫还写下了这样一句话："然而，所有这些机器（即使是有打字键盘的机器）都无法提供政府密码所需要的保密程度。因此，这些机器并未被大规模使用。"从这句话中我们可以看出，达加佩耶夫对机器密码的认识已经落后于当时的时代了。[6]

虽然《暗号与密码》一书没有谈到当时最先进的加密系统，并且整本书只有160页，但是达加佩耶夫在这本书中仍然讨论了各种不同的加密系统，包括（但不局限于）以下这些：恺撒的移位加密系统、斯基塔里密码棒、波利比奥斯加密系统、简写加密法、共济会石匠标记（mason mark）、欧甘文字（Ogham）、阿特巴什密码（Athbash）、特里特海姆密码（Tritheim's code，用14个单词替代每个字母，以创造另一段假文字）、格子密码［我在《未解之谜（上）》第1章中将这种格子称为"格栅"］、维吉尼亚密码、所罗门字母表密码（一种MASC密码）、弗朗西斯·培根的双字母密码、"黑话"、多重密钥的换位重排密码、路易十四的"伟大密码"（grand chiffre）、拿破仑的"小密码"（petit chiffre）、双列换位重排密码、信号密码（信号旗、信号火把、旗语等）、莫尔斯电码、"丛林电报"(即非洲原住民用的鼓点密码，通过敲鼓来传递信息）、流浪汉密码、标记卡密码、商店密码（这种密码是用来标记店主的进货价的）、布莱叶盲文（Braille）、手语字母（hand alphabets）、童子军信号密码、商业密码、圣西尔密码带（St. Cyr slide）、MASC密码、纵列换位重排密码、密码盘、普莱费尔密码、替代—换位组合密码、路径换位重排密码、书本密码、加密密码、使用字典的书本密码、密码轮以及音乐密码（密码被伪装成乐谱）。

在以上这些加密系统中，有若干种系统都用到了插入无意义字符的加密方法。

我们可以快速地排除以上的大部分加密系统，因为这些加密系统不会把文字信息转化为数字。波利比奥斯加密系统会把文字信息加密为数字，因此它是一种我们可以考虑的可能性。但是，波利比奥斯加密系统产生的是MASC密码，这种密码是很容易破译的，如果达加佩耶夫使用的是

MASC密码，那么这则挑战密码肯定早就被破解出来了。

在《暗号与密码》一书的第127页上，达加佩耶夫给出了一种与波利比奥斯加密系统类似的加密系统，我将这种加密系统复制于图9–2中。

	1	2	3	4	5	6
1	AB	AH	AN	AG	AM	AT
2	EC	EK	EG	EF	EM	ES
3	ID	IE	IR	ID	IL	IW
4	OF	OM	OS	OC	OK	OG
5	UG	UN	UB	UT	UF	UP

图9–2 一种与波利比奥斯加密系统类似的加密系统

利用图9–2中的网格，我们可以把音节（而不是单个字母）替换为数字。比如，如果需要加密的明文信息是“a huge cat ate Sam”（一只巨大的猫吃掉了萨姆），那么我们首先需要把这条明文信息中的每两个字母分成一组，即“AH UG EC AT AT ES AM”。然后，我们再将这串音节加密为“12 51 21 16 16 26 15”。

我花费了很大精力才构思出以上这个不太恰当的例子。因为图9–2的网格给出的音节组合选择太少，所以大部分信息根本无法用这个网格进行加密。此外，我还想提醒读者注意，在图9–2的网格中，“ID”这个字母组合出现了两次。这很可能是作者的一个笔误，而不是为了设置某种同音字母密码。那么，达加佩耶夫留下的这则挑战密码是否可能是用某种比图9–2中的网格规模更大，且性质相同的网格进行加密的呢？

在《暗号与密码》一书中，达加佩耶夫还提到了另一种工作原理与波利比奥斯加密系统类似的加密系统，并解释了这种加密系统如何将明文转化为字母。我将这种加密系统的工作原理列于图9–3中。[7]

德国陆军密码

第2个字符

第1个字符

	3	6	0	7	4	8	1	9	5	2
2	a	ä	ai	au	aü	b	c	ch	ck	d
6	e	ei	eu	f	ff	g	h	i	ie	j
3	k	l	ll	m, mm	n, nn	o	ö	p	pp	r
7	s	sch	sp	spr	ss	st	str	t	tt	u
4	ü	v	w	x	y	z	0	1	2	3
0	4	5	6	7	8	9	.	,	;	?
8	section	army	artillery	Battalion	Battery	Brigade	bridge	Division	railway	squadron
1	field	flight	flying corps	engineers	rifles	trench mortars	group	infantry	scouts	guns
9	cavalry	company	command	corps	men	morse	muni-tions	officer	horse	pioneer
5	regiment	Red Cross	snipers	sappers	staff	tanks	train	troop	watch	balloons

MILITARY CODES AND CIPHERS

Example:

Message: Three Companies to attack bridge-head at once.

Plain text: 3 Companies t o a tt a ck bridge h e a d a t o n c e .

Cipher: 42, 96, 79, 38, 23, 75, 23, 25, 81, 61, 63, 23, 22, 23, 79, 38, 34, 21, 63, 01.

Cryptogram: 42967 93823 75232 58161 63232 22379 38342 16301

图 9-3　受到波利比奥斯加密系统启发而诞生的另一种加密系统。标题意为“德国陆军密码”，表格下方是将一句明文加密为密文的示例，从上到下依次为信息、明文、密文、密码

但是，如果达加佩耶夫以这样的方式加密挑战密码的话，显然对读者太不公平了。对于一段如此短的加密信息，读者怎么可能判断出加密网格中出现的究竟是哪些字母或者单词呢？

在《暗号与密码》一书的第102页和第104页，达加佩耶夫还写下了另一段密码，这段密码使用的密钥如下：

A	B	C	D	E	F	G	H	I	J	K	L	M	N	O	P	Q	R
21	27	12	29	35	19	26	36	11	28	37	18	25	31	13	38	17	24

S	T	U	V	W	X	Y	Z
32	14	23	33	16	34	22	15

然而，这段密码是MASC密码。由于这类密码太容易被破解了，我们

可以排除挑战密码使用这种加密系统的可能性。

于是，《暗号与密码》一书中剩下的唯一可能性就在该书的第106页和第107页了。在这种加密系统中，加密的第一步是为每一个字母分配一个数值，字母和数值的具体对应关系如下：

A	B	C	D	E	F	G	H	I	J	K	L	M	N	O	P	Q	R	S	T
1	2	3	4	5	6	7	8	9	10	11	12	13	14	15	16	17	18	19	20

U	V	W	X	Y	Z
21	22	23	24	25	26

根据上表中的对应关系，我们可以把明文信息“ENEMY MOVES EAST”（敌人向东方移动）加密为以下的这串数字：5 14 5 13 25 13 15 22 5 19 5 1 19 20。

然而，在这个加密系统中，以上步骤只是对信息进行加密的第一步。我们还需要用另一套字母表来对一个关键词进行加密，这套字母表是：A=101，B=102,C=103，以此类推。在我们的例子中，假设关键词是“HONOUR”（荣誉），那么它经过加密后就变成了：108 115 114 115 121 118。

接下来，我们把加密后的关键词与第一步加密后得到的信息一一对齐（如果关键词的长度不足，则可多次重复这个关键词）。接下来，我们用第一行的数字减去第二行的数字，这一步骤如下表所示：

108	115	114	115	121	118	108	115	114	115	121	118	108	115
- 5	14	5	13	25	13	15	22	5	19	5	1	19	20
103	101	109	102	96	105	93	93	109	96	116	117	89	95

将完成以上步骤后得到的数字分组，每组包含5个数字，我们就得到以下密文：

10310 11091 02961 05939 31099 61161 17899 5。

这种加密方式和我们在《未解之谜（上）》第3章中提到的“无政府主义密码”十分类似。无政府主义密码是由斯库林提出的一则挑战密码，这个密码后来被爱德华·埃尔加破译。因此，我们可以借助埃尔加使用的方法来破译这条密码。

考虑到《暗号与密码》一书中介绍的密码学知识已经落后于当时密码学的发展，因此达加佩耶夫留下的这段挑战密码应该不是一段特别难破译的密码。然而，却没有任何人能够解开它。甚至连达加佩耶夫本人也无法帮助我们破译这段密码，因为当一筹莫展的破译者联系他时，他居然已经忘记了自己创造这则密码时所使用的加密方法！也许是为了避免更多的询问，《暗号与密码》一书再版时，达加佩耶夫从书中删除了这则挑战密码。这则挑战密码仅出现于该书的第一版中。

然而，虽然达加佩耶夫删除了这则挑战密码，它却并没有被世人遗忘。美国密码协会的一位成员戴维·舒尔曼（David Shulman）将这段密码发表在了该组织的出版物《密码》杂志的1952年4—5月号上。舒尔曼还承诺向第一个提交正确解答的读者提供一项“适当的奖励”。此前，舒尔曼曾经通过这种方法成功征集到未解密码的答案。在纽约密码协会的一次会议上，舒尔曼向与会者公布了一则未解密码，这段密码来自W. E. 伍德沃德（W. E. Woodward）的一篇小说。和达加佩耶夫一样，这段密码的创造者伍德沃德事后也忘记了自己当初使用的是哪种加密方式。在舒尔曼提出挑战以后，美国密码协会的一位笔名为“雅姆雅姆”（Yum Yum）的会员成功破译了这段密码，并领取了舒尔曼提供的奖品——一本作者亲笔签名的小说。

事实上，在公布达加佩耶夫的这则挑战密码之前，舒尔曼已经对它进行了初步分析，希望能帮助读者完成破译的任务。舒尔曼认为，这段密码结尾处的3个0是无意义字符，因此他删除了这3个字符，并且将剩下的数字每2个分为一组。在完成上述两项处理以后，舒尔曼向读者提供了这段密码的词频统计数据：

04＝1	72＝9	84＝11
61＝0	73＝0	85＝17
62＝17	74＝14	91＝12
63＝12	75＝17	92＝3
64＝16	81＝20	93＝2
65＝11	82＝17	94＝1
71＝1	83＝15	95＝0

注意，在这段密码中，每组数字的第一个数总是6，7，8，9，0中的一个，而第二个数总是1，2，3，4，5中的一个。这种现象说明，这则密码使用了一种波利比奥斯网格，网格的形式如下：

	1	**2**	**3**	**4**	**5**
6	A	B	C	D	E
7	F	G	H	I	K
8	L	M	N	O	P
9	Q	R	S	T	U
0	V	W	X	Y	Z

虽然上述网格中的字母是按顺序排列的，但在实际情况中，加密者可以把字母以任意顺序放入上述形式的网格中。不管怎么说，这种密码都是比较简单的，很难想象达加佩耶夫的所有读者都无法破译这样一则简单的密码。事实上，舒尔曼指出，上述词频统计数据不符合MASC密码的特征。比如，在上述词频统计中，常见字母的频率和少见字母的频率差距过大。

在舒尔曼的上述论文发表几年之后，破译达加佩耶夫密码的工作却没有任何进展。最终，在1959年，舒尔曼向读者提供了一些关于达加佩耶夫密码破译工作的新情况。舒尔曼称，一名陆军上尉写信给他，提供了关于破译达加佩耶夫密码的一些进一步的想法。根据舒尔曼的说法，这名陆军上尉将达加佩耶夫密码中的成对字母重排（对最后三个0忽略不计）成了一个14×14的网格。我将这个网格复制如下。这些成对的字母是沿竖直方向被逐列地放入网格中的。在《未解之谜（上）》的第5章中，我们讨论过纵列换位重排密码，当时我们也进行过这样的操作。

75	67	64	63	85	64	63	74	62	65	65	83	72	72
62	74	75	65	74	85	75	81	62	83	62	83	63	84
82	74	83	65	63	85	82	91	85	62	64	85	82	62
85	82	82	81	82	63	81	91	91	64	84	82	82	82
91	84	84	63	75	82	64	84	85	74	84	83	72	83
62	83	91	81	74	72	83	63	91	81	91	64	72	75
91	81	75	75	83	62	63	85	74	82	83	62	83	81
64	63	74	85	81	83	82	84	91	84	85	72	82	64
81	81	65	75	65	62	85	65	72	62	74	62	85	75
64	81	83	75	81	81	81	64	75	82	91	65	84	74

91 74 75 64 84 81 63 85 64 64 81 62 75 85

74 74 75 62 85 72 63 65 65 91 65 83 82 81

85 82 75 82 64 81 63 62 75 81 72 75 81 62

84 62 93 92 85 64 04 94 71 93 74 92 83 92

这位陆军上尉注意到，所有低频数字（92，93，94，以及04）都仅出现在最后一行中。因此，他判断这些低频数字可能是用来填满某些列的无意义字符。在去掉了这些无意义的字符以后，剩下的密码中只有13种不同的成对数字。

但是，接下来我们应该怎么做呢?

舒尔曼对能成功破译密码的人提供了奖励，这种方式在过去曾经获得了成功，而在破译达加佩耶夫密码的过程中，舒尔曼的这种做法也使得破译工作取得了一定的进展。然而，这种方法并没能帮我们找到最终的解法。在接下来的数十年中，其他密码研究者仍在继续努力试图破译达加佩耶夫密码。

1978年，韦恩·G. 巴克（Wayne G. Barker）发表了他对达加佩耶夫密码的分析结果。贝克很可能并不知道舒尔曼提到的那名陆军上尉的分析结果，因为在贝克发表的内容中，他通过一种稍微不同的方式得到了和那名陆军上尉相同的结果。

关于达加佩耶夫密码的最新成果来自戈登·鲁格［在《未解之谜（上）》的第1章中我们已经提到过他的名字，在之后的第10章中，读者还会再次见到这个名字］和加文·泰勒（Gavin Taylor）。鲁格和泰勒不仅对达加佩耶夫密码进行了研究，还仔细检查了阿曼德·范赞特（Armand Van Zandt）提出的一种解法。他们在网上发表了一系列论文，阐述了他们对达加佩耶夫密码的研究成果。[8]

范赞特认为，达加佩耶夫密码的明文只是把英文字母表按顺序抄了 7 遍而已，也就是：

ABCDEFGHIJKLMNOPQRSTUVWXYZ
ABCDEFGHIJKLMNOPQRSTUVWXYZ
ABCDEFGHIJKLMNOPQRSTUVWXYZ
ABCDEFGHIJKLMNOPQRSTUVWXYZ
ABCDEFGHIJKLMNOPQRSTUVWXYZ
ABCDEFGHIJKLMNOPQRSTUVWXYZ
ABCDEFGHIJKLMNOPQRSTUVWXYZ

但是范赞特并未非常详细地解释他究竟是如何得到上述解法的。或许在解法的某一步中，范赞特在做替换时过于主观，而靠这种方式得到的解法一般都是不正确的。然而，包括鲁格和泰勒在内的许多研究者（其中也包括我自己）都认为，达加佩耶夫在加密这段密码时可能犯了一些错误。因此，鲁格和泰勒正在试图找出一些达加佩耶夫在加密时可能会犯的简单错误，他们希望，这些错误能够解释为什么范赞特在破译的过程中不得不进行一些前后不一致的替换操作。

鲁格和泰勒希望能与范赞特讨论他提出的解法，然而他们却找不到范赞特的联系方式。在该系列论文的第6篇的草稿中，鲁格和泰勒写道："阿曼德——如果你在读我们的文章，我们希望告诉你，你的结果给我们留下了深刻的印象，我们非常希望和你取得联系。你可以很容易地找到戈登的联系方式，只要用谷歌引擎搜索他在基尔大学的电子邮件地址就可以了。"[9]

鲁格和泰勒表示，他们会继续发表这一系列的论文。我非常期待看到他们的更新！

物理学家费曼的密码

公众一向偏爱一些“怪人式”的物理学家，而理查德·费曼（Richard Feynman）恰好填补了阿尔伯特·爱因斯坦和斯蒂芬·霍金之间的空白。在专业方面，费曼因量子电动力学方面的工作获得过诺贝尔奖，此外他还曾在洛斯阿拉莫斯实验室和其他科学家一起成功研发出了原子弹。除了专业方面的成就以外，费曼还出版过好几本畅销书，其内容主要是费曼生活中的一些有趣的故事。电影《情深我心》（*Infinity*）就是根据费曼的一本畅销书改编的，在该片中扮演费曼的是马修·布罗德里克（Mathew Broderick）。

在洛斯阿拉莫斯实验室工作期间，费曼非常喜欢去破解实验室的各种安保措施，其中之一就是学习开保险箱的技术。费曼读了好几本关于如何开保险箱的书，后来他写下了下面这段话（这是完全真实的）：

> 我心想：“现在我也能写一本如何开保险箱的书了，这本书肯定写得比任何人都好，因为在这本书的一开头，我就打算告诉读者我是如何打开一些超级保险箱的，里面的物品比任何开保险箱的盗贼能偷到的东西价值都更高（当然，除了人命以外）。一般的保险箱里存着的可能是皮草或者金条，但我能开的保险箱里的东西价值远远超过了皮草和金条：里面有关于原子弹的所有秘密——生产钚的程序、净化的步骤、需要多少原料、原子弹如何工作、如何产生中子、原子弹的设计细节、各部件的尺寸——在洛斯阿拉莫斯实验室中我们知道的所有信息，关于原子弹的一切！”[10]

图 9-4　费曼（1918—1933）的形象出现在一枚美国邮票上

除了开保险箱以外，费曼还喜欢用密码和妻子通信，此举令洛斯阿拉莫斯实验室的邮件审查部门十分恼火。但是，后来的一份报告显示（如果这份报告的内容可信的话），费曼也有被别人难倒的时候。1987 年 12 月 10 日，克里斯·科尔（Chirs Cole）在 sci.crypt 论坛①上发表了如下内容（该内容发表两个多月以后，费曼就去世了）：

> 当我在加州理工学院读研究生时，费曼教授向我展示了 3 段密码样本。费曼教授称，这 3 段密码是他在洛斯阿拉莫斯实验室的一名科学家同事给他的，目的是为了挑战他的密码破译能力。费曼教授称，他一直无法破译这 3 段密码。我也没有能够成功地破译这 3 段密码。现在，我将这 3 段密码发表出来，网友们可以尝试破译一下。[11]

第一段密码被标记为"较简单"的密码，内容如下：

① sci.crypt 论坛是一个以密码技术问题为主题的论坛。——译者注

```
MEOTAIHSIBRTEWDGLGKNLANEAINOEEPEYST
NPEUOOEHRONLTIROSDHEOTNPHGAAETOHSZO
TTENTKEPADLYPHEODOWCFORRRNLCUEEEEOP
GMRLHNNDFTOENEALKEHHEATTHNMESCNSHIR
AETDAHLHEMTETRFSWEDOEOENEGFHETAEDGH
RLNNGOAAEOCMTURRSLTDIDOREHNHEHNAYVT
IERHEENECTRNVIOUOEHOTRNWSAYIFSNSHOE
MRTRREUAUUHOHOOHCDCHTEEISEVRLSKLIHI
IAPCHRHSIHPSNWTOIISISHHNWEMTIEYAFEL
NRENLEERYIPHBEROTEVPHNTYATIERTIHEEA
WTWVHTASETHHSDNGEIEAYNHHHNNHTW
```

在这段密码中，各字母出现的概率与正常英语中的字母概率基本一致，这说明这是一段换位重排密码。在美国国家航空航天局喷气推动实验室工作的杰克·莫里森（Jack Morrison）注意到了这段密码的特点。1987年12月21日，莫里森在sci.crypt论坛上发表了以下内容：

> 我对第一段密码的看法如下：这是一段非常典型的换位重排密码。我们首先把这些文字分成几个部分，每部分包含5列。然后，我们从右下角开始向上读。明文的内容有些奇怪，我猜想这段话可能是来自某部我不熟悉的文学作品。也许其他网友可以帮我修正一下标点符号以及断字的处理……[12]

下面就是莫里森破译出来的明文。这段文字确实摘自一部莫里森不熟悉的文学作品，即乔叟的《坎特伯雷故事集》。只要看一下破译后的原始明文，我们就会知道给这段话断字以及加标点有多困难了。换位完毕后，得到的原始明文如下：

WHANTHATAPRILLEWITHHISSHOURESSOOTET
HEDROGHTEOFMARCHHATHPERCEDTOTHEROOT
EANDBATHEDEVERYVEYNEINSWICHLICOUROF
WHICHVERTUENGENDREDISTHEFLOURWHANZE
PHIRUSEEKWITHHISSWEETEBREFTHINSPIRE
DHATHINEVERYHOLTANDHEETHTHETENDRECR
OPPESANDTHEYONGESONNEHATHINTHERAMHI
SHALVECOURSYRONNEANDSMALEFOWELESMAK
ENMELODYETHATSLEPENALTHENYGHTWITHOP
ENYESOPRIKETHHEMNATUREINHIRCORAGEST
HANNELONGENFOLKTOGOONONPILGRIM

因为自乔叟的时代之后，英文已经发生了大幅改变，所以即使在完成断字和加标点的工作之后，这段话也仍然相当难懂：

Whan that Aprille, with his shoures soote
The droghte of March hath perced to the roote,
And bathed every veyne in swich licour
Of which vertu engendred is the flour;
Whan Zephirus eek with his sweete breeth
Inspired hath in every holt and heeth
The tendre croppes, and the yonge sonne

```
Hath in the Ram his half coursy ronne,
And smale foweles maken melodye,
That slepen al the nyght with open ye
(So priketh hem nature in hir corages),
Thanne longen folk to goon on pilgrim[ages,]
```
[13]

（中文译文：当四月的甘霖渗透了三月枯竭的根须，沐灌了丝丝茎络，触动了生机，使枝头涌现出花蕾，当和风吹香，使得山林莽原遍吐着嫩条新芽，青春的太阳已转过半边白羊宫座，小鸟唱起曲调，通宵睁开睡眼，是自然拨弄着它们的心弦：这时，人们渴想着朝拜四方名坛。）

虽然莫里森很快就破译了第一段“较简单”的密码，但是他却完全没有办法破译科尔贴出的第二段密码。第二段密码被标记为“较难”的密码，其内容如下：

```
XUKEXWSLZJUAXUNKIGWFSOZRAWURORKXAOS
LHROBXBTKCMUWDVPTFBLMKEFVWMUXTVTWUI
DDJVZKBRMCWOIWYDXMLUFPVSHAGSVWUFWOR
CWUIDUJCNVTTBERTUNOJUZHVTWKORSVRZSV
VFSQXOCMUWPYTRLGBMCYPOJCLRIYTVFCCMU
WUFPOXCNMCIWMSKPXEDLYIQKDJWIWCJUMVR
CJUMVRKXWURKPSEEIWZVXULEIOETOOFWKBI
UXPXUGOWLFPWUSCH
```

这段密码不仅难倒了费曼、科尔，以及莫里森，也难倒了世界上的其他人。直到今天，仍没有人能够成功地破译这段密码。由于这段密码中出现概率最高的两个字母是U和W，所以这显然不是一段换位重排密码。

如果我们研究这段密码中出现的三字母和多字母组合，就会发现一些字母组合在这段密码中出现了不止一次。下表中列出了一些这样的字母组合：

字母	两次出现的间隔
WUR	193
RKX	186
VTW	63
UFP	87
WUF	76
MVR	6
CMUW	102
CMUW	26
WUID	39
CJUMV	6
JUMVR	6

破译者常常可以利用这些重复出现的字母组合来破译整段密码。比如，在维吉尼亚密码中就出现了重复的字母组合，这使得破译维吉尼亚密码的工作变得特别简单（在本章接下来的内容中，我会向读者介绍维吉尼亚密码）。然而，费曼的第二段密码看起来不太可能是维吉尼亚密码，因为上述这些重复出现的字母组合并没有给破译工作带来什么帮助。

费曼的第三段密码被标记为“新信息”，然而，这一标签并没有告诉我们这段密码的难度有多高。鉴于至今仍没有人能够成功解开第三段密码，恐怕这也是一段很难解的密码。第三段密码的内容如下：

WURVFXGJYTHEIZXSQXOBGSVRUDOOJXATBKT

ARVIXPYTMYABMVUFXPXKUJVPLSDVTGNGOSI

GLWURPKFCVGELLRNNGLPYTFVTPXAJOSCWRO

DORWNWSICLFKEMOTGJYCRRAOJVNTODVMNSQ

IVICRBICRUDCSKXYPDMDROJUZICRVFWXIFP

XIVVIEPYTDOIAVRBOOXWRAKPSZXTZKVROSW

CRCFVEESOLWKTOBXAUXVB

在第三段密码中，出现概率最高的字母是字母V，因此，我们可以排除这段密码是简单换位重排密码的可能性。在第三段密码中，不止一次出现的字母组合以及它们之间相隔字符数如下表所示：

字母	两次相隔字符数
WUR	72
RVF	165
GJY	115
RUD	125
PYT	49
PYT	92
ICR	4
ICR	19

和第二段密码的情况一样，第三段密码中重复出现的这些字母组合似乎也并没有给破译者提供有用的线索。

于是，对于费曼留下的这3段挑战密码，目前密码破译者的战绩是一胜两负。接下来，我要向读者介绍一段距今更近的密码。在破译下面这段密码时，破译者们取得了比破译费曼挑战密码时更好的成绩。

中央情报局的密码

为了创造“更加令人愉悦的工作环境”，20世纪80年代末期，美国中央情报局决定在新的总部大楼中放置一些原创的艺术品。为此，他们专门拨出了一笔资金，用于向艺术家支付佣金。为了完成上述任务，美国中央情报局成立了一个委员会，专门负责征集美国艺术家的作品。在这个委员会的一次会议上，委员们总共审阅了超过700张幻灯片，这些幻灯片上展示的艺术作品来自275名不同的艺术家。[14]

在漫长的审阅工作结束后，最终美国中央情报局与詹姆斯·桑伯恩（James Sanborn）签订了一份合同。合同规定，桑伯恩负责向中央情报局提出一份艺术作品的提案。当然，在开始制作之前，该提案还必须获得中央情报局的一些领导（包括局长）的最终批准。最后，桑伯恩获得了250 000美元的佣金，而中央情报局则获得了一件十分独特的雕塑作品——这件作品上有一段密码。读者可以在图9–5中看到这座雕塑上的部分密码。这件雕塑作品被命名为“克里普托斯”（Kryptos）。

“克里普托斯”雕塑于1990年10月5日正式揭幕启用。在当天的庆祝仪式上，中央情报局局长威廉·H. 韦伯斯特（William H. Webster）做了揭幕致辞。韦伯斯特在演讲中讲述了以下内容：

> 我们一直希望达成的目标是创造一个理想的环境。在我们的总部大楼，以及赋予我们总部大楼生命的艺术作品中，我们都一直追求着这样的目标。对我们来说，所谓理想的环境，就是一个既适合沉思，又富有挑战性的环境。[15]

图 9-5 “克里普托斯”雕塑

此外，韦伯斯特还在致辞中对桑伯恩说了如下的话：

> 你在这个庭院中向我们展示了一段密码，我知道所有专业情报人员都会享受破译这段密码的机会。[16]

在中央情报局的所有情报人员中，只有韦伯斯特本人可以不做任何密码分析就读出“克里普托斯”雕塑上的信息。在“克里普托斯”雕塑的揭幕仪式上，桑伯恩将两个信封交给了韦伯斯特。其中一个信封中写着可用于破译这段密码的关键词，而另一个信封中则是这段密码的明文。在将这两个信封交给韦伯斯特的时候，[17]桑伯恩对韦伯斯特说：“我希望你能保守秘密。”桑伯恩只把“克里普托斯”雕塑的秘密告诉了韦伯斯特一个人。

“克里普托斯”雕塑包含四个面板。左边的两个面板上是加密后的密文，而右边的两个面板上有一份字母表和另一份用关键词KRYPTOS（克里普托斯）打乱后的字母表，还列出了 KRYPTOS 一词出现在各种起始位置的可能性。由于与中央情报局总部没有业务往来的人无法看到总部庭院中的这座雕塑，因此中央情报局在自己的网站上贴出了“克里普托斯”雕塑的照片，并且列出了卷曲面板上呈现的文字。[18]我将中央情报局网站上的这些图片复制于图 9–6 和图 9–7 中。

面板 1

E M U F P H Z L R F A X Y U S D J K Z L D K R N S H G N F I V J

Y Q T Q U X Q B Q V Y U V L L T R E V J Y Q T M K Y R D M F D

V F P J U D E E H Z W E T Z Y V G W H K K Q E T G F Q J N C E

G G W H K K ? D Q M C P F Q Z D Q M M I A G P F X H Q R L G

T I M V M Z J A N Q L V K Q E D A G D V F R P J U N G E U N A

Q Z G Z L E C G Y U X U E E N J T B J L B Q C R T B J D F H R R

Y I Z E T K Z E M V D U F K S J H K F W H K U W Q L S Z F T I

H H D D D U V H ? D W K B F U F P W N T D F I Y C U Q Z E R E

E V L D K F E Z M O Q Q J L T T U G S Y Q P F E U N L A V I D X

F L G G T E Z ? F K Z B S F D Q V G O G I P U F X H H D R K F

F H Q N T G P U A E C N U V P D J M Q C L Q U M U N E D F Q

E L Z Z V R R G K F F V O E E X B D M V P N F Q X E Z L G R E

D N Q F M P N Z G L F L P M R J Q Y A L M G N U V P D X V K P

D Q U M E B E D M H D A F M J G Z N U P L G E W J L L A E T G

面板 2

E N D Y A H R O H N L S R H E O C P T E O I B I D Y S H N A I A

C H T N R E Y U L D S L L S L L N O H S N O S M R W X M N E

T P R N G A T I H N R A R P E S L N N E L E B L P I I A C A E

W M T W N D I T E E N R A H C T E N E U D R E T N H A E O E

T F O L S E D T I W E N H A E I O Y T E Y Q H E E N C T A Y C R

E I F T B R S P A M H N E W E N A T A M A T E G Y E E R L B

T E E F O A S F I O T U E T U A E O T O A R M A E E R T N R T I

B S E D D N I A A H T T M S T E W P I E R O A G R I E W F E B

A E C T D D H I L C E I H S I T E G O E A O S D D R Y D L O R I T

R K L M L E H A G T D H A R D P N E O H M G F M F E U H E

E C D M R I P F E I M E H N L S S T T R T V D O H W ? O B K R

U O X O G H U L B S O L I F B B W F L R V Q Q P R N G K S S O

T W T Q S J Q S S E K Z Z W A T J K L U D I A W I N F B N Y P

V T T M Z F P K W G D K Z X T J C D I G K U H U A U E K C A R

图 9–6　“克里普托斯”雕塑左侧的两个面板，来自中央情报局的网站

面板3

A B C D E F G H I J K L M N O P Q R S T U V W X Y Z A B C D

A K R Y P T O S A B C D E F G H I J L M N Q U V W X Z K R Y P

B R Y P T O S A B C D E F G H I J L M N Q U V W X Z K R Y P T

C Y P T O S A B C D E F G H I J L M N Q U V W X Z K R Y P T O

D P T O S A B C D E F G H I J L M N Q U V W X Z K R Y P T O S

E T O S A B C D E F G H I J L M N Q U V W X Z K R Y P T O S A

F O S A B C D E F G H I J L M N Q U V W X Z K R Y P T O S A B

G S A B C D E F G H I J L M N Q U V W X Z K R Y P T O S A B C

H A B C D E F G H I J L M N Q U V W X Z K R Y P T O S A B C D

I B C D E F G H I J L M N Q U V W X Z K R Y P T O S A B C D E

J C D E F G H I J L M N Q U V W X Z K R Y P T O S A B C D E F

K D E F G H I J L M N Q U V W X Z K R Y P T O S A B C D E F G

L E F G H I J L M N Q U V W X Z K R Y P T O S A B C D E F G H

M F G H I J L M N Q U V W X Z K R Y P T O S A B C D E F G H I

面板4

N G H I J L M N Q U V W X Z K R Y P T O S A B C D E F G H I J

O H I J L M N Q U V W X Z K R Y P T O S A B C D E F G H I J L

P I J L M N Q U V W X Z K R Y P T O S A B C D E F G H I J L M

Q J L M N Q U V W X Z K R Y P T O S A B C D E F G H I J L M N

R L M N Q U V W X Z K R Y P T O S A B C D E F G H I J L M N Q

S M N Q U V W X Z K R Y P T O S A B C D E F G H I J L M N Q U

T N Q U V W X Z K R Y P T O S A B C D E F G H I J L M N Q U V

U Q U V W X Z K R Y P T O S A B C D E F G H I J L M N Q U V W

V U V W X Z K R Y P T O S A B C D E F G H I J L M N Q U V W X

W V W X Z K R Y P T O S A B C D E F G H I J L M N Q U V W X Z

X W X Z K R Y P T O S A B C D E F G H I J L M N Q U V W X Z K

Y X Z K R Y P T O S A B C D E F G H I J L M N Q U V W X Z K R

Z Z K R Y P T O S A B C D E F G H I J L M N Q U V W X Z K R Y

A B C D E F G H I J K L M N O P Q R S T U V W X Y Z A B C D

图 9-7 “克里普托斯”雕塑右侧的两个面板，来自中央情报局的网站

在我之前出版的第一本书中，我也复制了上述两张图片，当时我认为中央情报局的官方网站应该是最可靠的来源了。然而，后来我发现，这个网站上的文字实际上存在一些错误。其中一个错误对于该密码的破译过程有着非常重要的影响，我将在接下来的篇幅中详细解释这一点。

“克里普托斯”雕塑因其神秘的气息而获得了极大的关注。我相信它一定是同时代的雕塑中知名度最高的作品之一。不幸的是，在这座雕塑还没完工的时候，它就受到了来自中央情报局内部的一些负面关注。16名中央情报局的雇员联名请愿，对“克里普托斯”雕塑提出强烈抗议。我根据信息自由法案请求公开这份请愿书，中央情报局去掉了签名后公开的内容如下：

> 我们——在这份请愿书末尾签名的所有人——对中央情报局总部进行的这项所谓的“艺术工程”表示强烈抗议。在这个时代，我们面临着大规模的预算赤字和无家可归的人们，国防预算也有可能被削减。面对这样的情况，滥用资金是极不合理的。考虑到目前的环境，将我们的资金花费在任何种类的艺术作品上都是不合适的。而此处涉及的艺术作品尤其应该受到谴责和蔑视，因为它们根本毫无意义。中央情报局的任务是以宪法的形式捍卫国家利益，因此，在这样的机构中设立这种无意义的艺术作品是非常不合适的。这些令人不悦的艺术作品不仅没有提升我们的工作环境，反而让我们的工作环境变得更加糟糕，我们有比如何设置这些昂贵的物件更加重要的事情需要讨论。
>
> 如果这些雕塑真的有任何实际价值的话，我们建议将它们出售，然后把纳税人的钱花在更合适的地方。点缀在走廊周围的各种几何图形也应该做同样的处理。在各种各样的展览中，有许多由中央情报局雇员创作的艺术作品，用这些作品可以更好地装饰我们的办公大楼，同时也能更道德、更明智地使用我们有限的资金。

究竟什么是艺术？关于这个问题当然可以有各种各样的争论，但是上述请愿书中显然犯了一个错误：“克里普托斯”并不是一件“无意义”的雕塑作品。

虽然吉姆·吉洛格利并不是中央情报局的雇员，但他却是第一个向公众揭示“克里普托斯”雕塑的部分意义的人。1999年6月16日，吉洛格利公布了他对“克里普托斯”密码的解法，此时距离“克里普托斯”揭幕10周年还有不到5个月的时间。由于这是一个公众等待已久的答案，媒体对吉洛格利发表的结果进行了大量报道，《纽约时报》也专门刊登了一篇文章。

然而，问题比大部分人想象的更加复杂。吉洛格利发现，光找到一条密钥并不足以破译整段密码。事实上，吉洛格利认为，“克里普托斯”雕塑左侧的面板中包含了4段不同的密码。

“克里普托斯”雕塑上的第一段密码现在也被称为K1密码，这段密码很短，具体内容如下：

EMUFPHZLRFAXYUSDJKZLDKRNSHGNFIVJ

YQTQUXQBQVYUVLLTREVJYQTMKYRDMFD

为了让读者理解这种密码的工作原理，我必须向读者介绍另一种加密系统——维吉尼亚密码。在本书的前文中，我们曾经提到维吉尼亚密码——比如，在讨论费曼留下的挑战密码时，我就曾经提到这个名词。不过，现在我们要更加详细地了解一下这一加密系统的细节。

MASC密码是一种很弱的密码，因为在这种加密系统下，加密者总是用相同的字符来替换同一个字母。而在维吉尼亚密码中，加密者可以通过一种方式系统地用不同的密文字母去替代同样的明文字母。为了向读者清楚地说明维吉尼亚密码的工作原理，下面我会举一个具体例子。假设我们想要加密的明文信息是：

If you're going to threaten me with a knife, you may as well cut me a little.

（如果你要拿刀威胁我，那你还不如干脆割我几刀算了。）

为了用维吉尼亚加密系统来加密以上信息，我们先要选择一个关键词。在这个例子中，我会用“JIGSAW”（拼图）作为关键词。由于“JIGSAW”一词中包含6个字母，我们可以得到6种不同的替换字母表。这6种字母表是：

ABCDEFGHIJKLMNOPQRSTUVWXYZ（明文）

JKLMNOPQRSTUVWXYZABCDEFGHI（密码字母表1）

IJKLMNOPQRSTUVWXYZABCDEFGH（密码字母表2）

GHIJKLMNOPQRSTUVWXYZABCDEF（密码字母表3）

STUVWXYZABCDEFGHIJKLMNOPQR（密码字母表4）

ABCDEFGHIJKLMNOPQRSTUVWXYZ（密码字母表5）

WXYZABCDEFGHIJKLMNOPQRSTUV（密码字母表6）

要得到这6种不同的字母表很容易，只要将“JIGSAW”一词放入左边第一列中，然后在水平方向上依次写完剩下的字母表就可以了。如果在写字母表的过程中写到了字母Z，我们就回到字母表的开头，从字母A开始继续，直到26个字母都被列出为止。

有了这6个不同的字母表以后，我们就可以开始对明文信息进行加密了。我们用字母表1来加密明文中的第一个字母，用字母表2来加密第二个字母，以此类推。在对第六个字母进行加密以后，所有字母表都用完了。因此，我们就要重新回到字母表1，用它来加密明文中的第7个字母，然后再用字母表2来加密明文中的第8个字母，以此类推。

为了保证明文中的每个字母都能用正确的字母表进行加密，在加密过程开始之前，加密者通常会在明文信息下重复书写关键词，也就是：

```
IFYOUREGOINGTOTHREATENMEWITHAKNIFEYOUMAYASWELLCUT
JIGSAWJIGSAWJIGSAWJIGSAWJIGSAWJIGSAWJIGSAWJIGSAWJ

MEALITTLE
IGSAWJIGS
```

接下来，只要仔细完成整个加密过程，我们就可以得到以下结果：

```
IFYOUREGOINGTOTHREATENMEWITHAKNIFEYOUMAYASWELLCUT
JIGSAWJIGSAWJIGSAWJIGSAWJIGSAWJIGSAWJIGSAWJIGSAWJ
RNEGUNNOUANCCWZZRAJBKFMAFQZZAGWQLWYKDUGQAOFMRDCQC

MEALITTLE
IGSAWJIGS
UKSLECBRW
```

下面，我们去掉关键词所在的一行，将明文和密文以一对一的形式写好，并且在单词与单词之间插入空格。这样做以后，我们就会发现维吉尼亚密码具有一些非常好的特点：

```
IF YOU'RE GOING TO THREATEN ME WITH A KNIFE, YOU MAY AS
RN EGU NN OUANC CW ZZRAJBKF MA FQZZ A GWQLW  YKD UGQ AO

WELL CUT ME A LITTLE.
FMRD CQC UK S LECBRW
```

注意，明文中的第一个“YOU”（短语YOU'RE的一部分）被加密为EGU，而当同样的单词在明文中再次出现时，却被加密成了YKD。这是因为，当“YOU”第二次出现时，它对应的关键词字母已经和第一次不同了。因此，当“YOU”第二次出现时，它的每一个字母都被加密成了和第一次完全不同的密文字母。

当然，如果同一个单词两次出现时恰好与关键词中相同的字母对齐，那么这个单词就会被加密为同样的密文字母。此时，重复出现的两个单词之间的距离应该是关键词长度的整数倍。这是维吉尼亚密码的一个弱点：如果破译者知道了关键词的长度，他就可以把以相同字母表加密的单词单独挑出来，然后通过统计字母频率来破译这些字母。

举例来说，假设现在破译者知道上面这段密码的关键词由6个字母组成，那么他就知道相隔6个字母的字母之间（第1个字母、第7个字母、第13个字母、第19个字母、第25个字母等）都是用同一个字母表（即字母表1）加密的。因此，破译者可以把上述这些字母看成一组，找出这组字母中出现频率最高的字母，并暂时假设这个字母对应的明文字母是字母E。接下来，破译者需要利用这一加密系统的以下性质：虽然每个字母表以不同的字母开头，但是后面的字母都是按顺序排列的。因此，假设破译者判断在这组字母中密文字母M对应明文字母E，那么密文字母N就对应明文字母F，密文字母O对应明文字母G，以此类推。也就是说，只要破译了一个字母，整个字母表就被完全破译了！同样，第2、8、14、20、26等位置上的字母都是用字母表2进行加密的，破译者可以用以上方法继续破译这组字母。这样，破译者很快就可以把整段密码完全破译出来。

为了防御上述这种攻击（或者至少稍微降低这种攻击成功的概率），加密者可以打乱字母表中的字母顺序。在这种情况下，即使破译者已经知道了关键词的长度，他也无法通过一个字母知道字母表中的所有字母，因为破译者不知道加密者究竟以何种方式打乱了字母表中的字母顺序。

“克里普托斯”雕塑上的第一段密码就是按照以上方式进行加密的。加密者首先用关键词“KRYPTOS”打乱字母表（如“克里普托斯”雕塑右侧面板上的字母所示），然后再用关键词“PALIMPSEST”（重写本）来选择字母表。

因此，这段密码使用了如下字母表：

ABCDEFGHIJKLMNOPQRSTUVWXYZ（明文）

PTOSABCDEFGHIJLMNQUVWXZKRY（密码字母表1）

ABCDEFGHIJLMNQUVWXZKRYPTOS（密码字母表2）

LMNQUVWXZKRYPTOSABCDEFGHIJ（密码字母表3）

IJLMNQUVWXZKRYPTOSABCDEFGH（密码字母表4）

MNQUVWXZKRYPTOSABCDEFGHIJL（密码字母表5）

PTOSABCDEFGHIJLMNQUVWXZKRY（密码字母表6）

SABCDEFGHIJLMNQUVWXZKRYPTO（密码字母表7）

EFGHIJLMNQUVWXZKRYPTOSABCD（密码字母表8）

SABCDEFGHIJLMNQUVWXZKRYPTO（密码字母表9）

TOSABCDEFGHIJLMNQUVWXZKRYP（密码字母表10）

事实上，吉洛格利早就能够熟练地破译这种密码了。他是美国密码协会的会员，因此他经常在该协会的官方出版物《密码》上看到这类密码。在《密码》杂志上，这类密码被称为“泥沼3”密码（Quagmire Ⅲ）。事实上，了解“泥沼3”密码的破译者只要看看“克里普托斯”雕塑右侧面板上的内容，就可能猜到这是“泥沼3”密码的字母表，接下来他们自然会尝试用面板上的字母表来破译密码。但是，在猜到密码的性质以后，吉洛格利还需要找到这段密码的另一个关键词，即“PALIMPSEST”（重写本）一词。最终，他破译出的明文信息如下：

BETWEEN SUBTLE SHADING AND THE ABSENCE OF LIGHT LIES THE NUANCE OF IQLUSION.

（在微妙的阴影和光线的缺失之间，就有了幻影的微妙区别。）

在破译出来的明文信息中，“IQLUSION”一词中的字母Q是桑伯恩故意犯的一个拼写错误（“幻影”一词应该为ILLUSION）。这个问题我暂时放在一边，等会再做详细解释。

接下来，“克里普托斯”密码的第二部分，即K2的密文如下：

VFPJUDEEHZWETZYVGWHKKQETGFQJNCE

GGWHKK?DQMCPFQZDQMMIAGPFXHQRLG

TIMVMZJANQLVKQEDAGDVFRPJUNGEUNA

QZGZLECGYUXUEENJTBJLBQCRTBJDFHRR

YIZETKZEMVDUFKSJHKFWHKUWQLSZFTI

HHDDDUVH?DWKBFUFPWNTDFIYCUQZERE

EVLDKFEZMOQQJLTTUGSYQPFEUNLAVIDX

FLGGTEZ?FKZBSFDQVGOGIPUFXHHDRKF

FHQNTGPUAECNUVPDJMQCLQUMUNEDFQ

ELZZVRRGKFFVOEEXBDMVPNFQXEZLGRE

DNQFMPNZGLFLPMRJQYALMGNUVPDXVKP

DQUMEBEDMHDAFMJGZNUPLGEWJLLAETG

吉洛格利发现，K2也是一段“泥沼3”密码，这段密码的关键词是“KRYPTOS”和“ABSCISSA”（意为“横坐标”），前者用于打乱字母表，后者用于选择字母表并决定字母表的顺序。最终，吉洛格利破译出了以下明文信息：

IT WAS TOTALLY INVISIBLE. HOW'S THAT POSSIBLE? THEY USED THE EARTH'S MAGNETIC FIELD. X THE INFORMATION WAS GATHERED AND TRANSMITTED UNDERGRUUND TO AN UNKNOWN LOCATION. X DOES LANGLEY KNOW ABOUT THIS? THEY SHOULD: IT'S BURIED OUT THERE SOMEWHERE. X WHO KNOWS THE EXACT LOCATION? ONLY WW. THIS WAS HIS LAST MESSAGE. X THIRTY EIGHT DEGREES FIFTY SEVEN MINUTES SIX POINT FIVE SECONDS NORTH, SEVENTY SEVEN DEGREES EIGHT MINUTES FORTY FOUR SECONDS WEST. ID BY ROWS.

（中文译文：它完全是隐形的。这怎么可能呢？原来是利用了地球的磁场。X信息被收集之后，从地下传输到一个未知地点。X中央情报局知道这件事吗？他们应该知道：那东西就埋在那里的某处。X谁知道确切的地点？只有WW。这是他的最后一条信息。X北纬38度57分6.5秒；西经77度8分44秒。由罗斯确认）

在密码的第二部分中，也有一个加密者故意为之的拼写错误："UNDER- GRUUND"（地下）一词中原本的字母O错写成了字母U。然而，这段密码中还包含一个桑伯恩无意中写错的地方。2006年，桑伯恩承认了这个他无意中犯下的错误，就出现在密码的结尾处。密文最后几行中的"EWJLLAETG"应作"ESWJLLAETG"。对错误的密文进行破译后得到的明文是"ID BY ROWS"，如果在密文中加上一个字母S，明文的信息就完全改变了。正确的密文对应的明文信息为"X LAYER TWO"（X第二层）。一般来说，如果加密者在加密信息的过程中出现了错误，或者在抄写最终的密文信息时犯了错误，这部分密码就会被译成毫无意义的明文。而在"克里普托斯"密码的第二部分中，虽然桑伯恩漏写了一个字母S，这段密码却仍然能被翻译成有意义的明文信息，这事实上是一个非常惊人的巧合。当然，由错误的密文翻译出来的明文信息和加密者本来想要加密的明文信息是完全不同的。

在破译了"克里普托斯"密码的第一部分和第二部分以后，吉洛格利将注意力转向密码的第三部分，即K3。我将K3的内容复制如下。在破译第三部分时，吉洛格利注意到了一个变化。首先，这段密码的结尾处有一个问号，提供了一个非常明显的潜在停顿点。然而，问号之前的字母频率与正常英语的字母概率相匹配，因此第三部分的密码不可能是"泥沼3"密码。桑伯恩一定是在这段密码中改变了他使用的加密系统，这部分密码是用换位重排的方法进行加密的。K3密码的密文如下：

ENDYAHROHNLSRHEOCPTEOIBIDYSHNAIA

CHTNREYULDSLLSLLNOHSNOSMRWXMNE

TPRNGATIHNRARPESLNNELEBLPIIACAE

WMTWNDITEENRAHCTENEUDRETNHAEOE

TFOLSEDTIWENHAEIOYTEYQHEENCTAYCR

EIFTBRSPAMHHEWENATAMATEGYEERLB

TEEFOASFIOTUETUAEOTOARMAEERTNRTI

BSEDDNIAAHTTMSTEWPIEROAGRIEWFEB

AECTDDHILCEIHSITEGOEAOSDDRYDLORIT

RKLMLEHAGTDHARDPNEOHMGFMFEUHE

ECDMRIPFEIMEHNLSSTTRTVDOHW?

根据吉洛格利发表的解答，我发现至少可以通过两种不同的方式破译出K3的明文。在接下来的段落中，我只向读者展示其中比较容易理解的一种方式。[19]桑伯恩并没有公布他究竟使用的是哪种方式。

首先，我们忽略K3密码末尾的问号，然后将剩下的字母重新排列为14行，其中每一行中有24个字母：

ENDYAHROHNLSRHEOCPTEOIBI

DYSHNAIACHTNREYULDSLLSLL

NOHSNOSMRWXMNETPRNGATIHN

RARPESLNNELEBLPIIACAEWMT

WNDITEENRAHCTENEUDRETNHA

EOETFOLSEDTIWENHAEIOYTEY

QHEENCTAYCREIFTBRSPAMHHE

WENATAMATEGYEERLBTEEFOAS

FIOTUETUAEOTOARMAEERTNRT

IBSEDDNIAAHTTMSTEWPIEROA

GRIEWFEBAECTDDHILCEIHSIT

EGOEAOSDDRYDLORITRKLMLEH

AGTDHARDPNEOHMGFMFEUHEEC

DMRIPFEIMEHNLSSTTRTVDOHW

接着，我们将上述长方形字母矩阵顺时针旋转90度，这样就得到24行字母，其中每一行中有14个字母：

DAEGIFWQEWRNDE

MGGRBIEHONAOYN

RTOISONEEDRHSD

IDEEETAETIPSHY

PHAWDUTNFTENNA

FAOFDEACOESOAH

ERSENTMTLELSIR

IDDBIUAASNNMAO

MPDAAATYERNRCH

ENREAEECDAEWHN

HEYCHOGRTHLXTL

NODTTTYEICEMNS

LHLDTOEIWTBNRR

SMODMAEFEELEEH

SGRHSRRTNNPTYE

TFIITMLBHEIPUO

TMTLEABRAUIRLC

RFRCWETSEDANDP

TEKEPEEPIRCGST

VULIIREAOEAALE

DHMHETFMYTETLO

OELSRNOHTNWISI

HEEIORAHEHMHLB

WCHTATSEYATNLI

接着，我们改变字母的分行方式，得到42行字母，其中每一行中有8个字母：

DAEGIFWQ

EWRNDEMG

GRBIEHON

AOYNRTOI

SONEEDRH

SDIDEEET

AETIPSHY

PHAWDUTN

FTENNAFA

OFDEACOE

SOAHERSE

NTMTLELS

IRIDDBIU

AASNNMAO

MPDAAATY

ERNRCHEN

REAEECDA

EWHNHEYC

HOGRTHLX

TLNODTTT

YEICEMNS

LHLDTOEI

WTBNRRSM

ODMAEFEE

LEEHSGRH

SRRTNNPT

YETFIITM

LBHEIPUO

TMTLEABR

AUIRLCRF

RCWETSED

ANDPTEKE

PEEPIRCG

STVULIIR

EAOEAALE

DHMHETFM

YTETLOOE

LSRNOHTN

WISIHEEI

ORAHEHMH

LBWCHTAT

SEYATNLI

接着，我们再次把这个字母矩阵顺时针旋转90度，得到以下结果：

SLOWLYDESPARATLYSLOWLYTHEREMAINSOFPASSAGED
EBRISTHATENCUMBEREDTHELOWERPARTOFTHEDOORWA
YWASREMOVEDWITHTREMBLINGHANDSIMADEATINYBRE
ACHINTHEUPPERLEFTHANDCORNERANDTHENWIDENING
THEHOLEALITTLEIINSERTEDTHECANDLEANDPEEREDI
NTHEHOTAIRESCAPINGFROMTHECHAMBERCAUSEDTHEF
LAMETOFLICKERBUTPRESENTLYDETAILSOFTHEROOMW
ITHINEMERGEDFROMTHEMISTXCANYOUSEEANYTHINGQ

最后，我们对上述这段文字进行断字，并加入标点符号，同时把最后一个字母Q改为问号，这样就得到了最终的明文信息：

SLOWLY, DESPARATLY SLOWLY, THE REMAINS OF PASSAGE DEBRIS THAT ENCUMBERED THE LOWER PART OF THE DOORWAY WAS REMOVED. WITH TREMBLING HANDS I MADE A TINY BREACH IN THE UPPER LEFT-HAND CORNER. AND THEN, WIDENING THE HOLE A LITTLE, I INSERTED THE CANDLE AND PEERED IN. THE HOT AIR ESCAPING FROM THE CHAMBER CAUSED THE FLAME TO FLICKER, BUT PRESENTLY DETAILS OF THE ROOM WITHIN EMERGED FROM

THE MIST. X CAN YOU SEE ANYTHING?

（慢慢地，之前堵住门廊下半部分的走廊残骸被移除了，这个过程缓慢得令人绝望。我用颤抖的手在左上角挖出了一个小洞，然后将这个小洞稍微扩大了一点儿。我插入蜡烛，并向里面窥看。从房间中溢出来的热空气使得火焰跳动不已，但是不一会儿，雾气散去，眼前便清楚了。X你能看到什么吗？）

在这段密码中，加密者又故意拼错了一个词，这个词是“DESPARATLY”（“绝望”这个词正确拼法应为DESPERATELY）。

至此，“克里普托斯”密码的前三部分都被吉洛格利成功破译了。接下来只剩下密码的第四部分——K4了。我将K4的内容抄录如下，这段密码至今仍无人能解。

OBKR

UOXOGHULBSOLIFBBWFLRVQQPRNGKSSO

TWTQSJQSSEKZZWATJKLUDIAWINFBNYP

VTTMZFPKWGDKZXTJCDIGKUHUAUEKCAR

至此，在破译“克里普托斯”密码的竞赛中，吉洛格利的战绩是3胜1负。考虑到其他破译者的成绩都是0胜4负，他的成绩已经非常优异了，他差点儿就解开了所有“克里普托斯”密码……

在吉洛格利公布了他对“克里普托斯”密码的解法以后，人们发现其实中央情报局的一名物理学家兼高级分析师戴维·斯坦（David Stein）早在1999年2月就得出了结果。也就是说，他是比吉洛格利更早解开这道谜题的人。他的破译结果被发表在中央情报局的保密期刊《情报研究》

（*Studies in Intelligence*）上。目前这些内容已经解密，读者可以在网上查到这篇文章的内容。在破译“克里普托斯”密码的过程中，斯坦并没有使用计算机，他是完全靠纸和笔完成破译工作的。

约翰·马尔科夫（John Markoff）在《纽约时报》上这样写道：

> 当斯坦得知吉洛格利是用计算机解开“克里普托斯”密码时，他似乎有些生气。
>
> 他说：“‘克里普托斯’密码应该要用纸和笔解开才对。”[20]

然而，吉洛格利并不同意斯坦的上述说法。他反驳道：

> 这次密码破译挑战并没有任何成文的规则。根据我自己的理解，只要解开密码就可以了。我选择使用什么样的工具并不重要，如何使用手头的工具才重要。[21]

中央情报局前局长威廉·韦伯斯特也对此事发表了评论：

> 对于此事，中央情报局前局长韦伯斯特昨天表示，他早已忘记了这段密码的答案。他说：“我对这件事情完全没有印象了，记忆模糊。”
>
> 对于吉洛格利用计算机破译密码一事，韦伯斯特表示：“谁规定不可以用的？专业机构破译密码时正是使用计算机的呀。”[22]

看来，吉洛格利受到了斯坦的前上司的支持。雪上加霜的是，斯坦很快也失去了“破译‘克里普托斯’密码第一人”的头衔，就跟此前的吉洛格利一样。

最终，中央情报局的主要竞争对手——国家安全局宣布，他们的雇员早在斯坦之前就解开了“克里普托斯”密码的前三部分。然而，“克里普托斯”密码的第四部分至今仍是一个未解之谜。如果有人已经破译了的话，这一信息也并未向公众公布。

创造“克里普托斯”密码的艺术家桑伯恩（他设计了一系列包含谜语元素的雕塑作品）表示：他相信“克里普托斯”密码中未被解开的秘密永远不会被破译出来。这段密码的创造者是中央情报局密码中心前主任爱德华·M. 沙伊特（Edward M. Scheidt）。[23]

在过去的10年中，公众对“克里普托斯”密码的兴趣变得越来越高涨，这在很大程度上是因为丹·布朗在两本小说中提到了“克里普托斯”密码。在精装版的《达·芬奇密码》的防尘套上写有一串数字，给出了“克里普托斯”雕塑所处位置的精确经纬度。在《达·芬奇密码》之后，丹·布朗又出版了另一本小说——《失落的秘符》。在这本书中，丹·布朗更加直白地提到了“克里普托斯”密码，并对其中已经破译出来的部分给出了一种非常奇怪的解读，这并不是艺术家桑伯恩的本意，事实上，这件事情甚至激怒了桑伯恩。《失落的秘符》一书提到了“克里普托斯”密码中的以下这句话：“Who knows the exact location? Only WW.”（谁知道确切的地点？只有WW。）在丹·布朗的这本小说中，故事的情节是这样的：上述这句话中的WW应该被反过来理解，也就是这两个字符实际上应该是MM，代表的是“抹大拉的玛丽亚”（Mary Magdalene）。然而，桑伯恩的原意是，WW代表威廉·韦伯斯特，也就是“克里普托斯”雕塑建成揭幕时的中央情报局局长。我在前文中曾提过，桑伯恩只将“克里普托斯”雕塑的秘密告诉过他一个人，因此才会有“只有WW知道”的说法。当然，现在知道这个秘密的已经不止韦伯斯特一个了，WW将这个密码的解答任务交给了中央情报局的下几任局长。

除了丹·布朗之外，伊隆卡·杜宁（Elonka Dunin）对“克里普托斯”密码的宣传也极大地提高了该密码在公众中的知名度。杜宁在她自己的网站上收集整理了大量关于“克里普托斯”密码的资料，这其中有些资料来自第一手来源，而有些资料则来自第二手来源。此外，她还经常十分热心地举办关于“克里普托斯”雕塑、“克里普托斯”密码，以及创造“克里普托斯”密码的艺术家桑伯恩的讲座。

虽然“克里普托斯”密码获得了公众的大量关注，并激起了研究者的极高热情，但是至今仍然没有人能够破译“克里普托斯”密码的第4部分。这一部分似乎使用了一种与前3部分完全不同的加密系统。当然，第4部分的破译工作也获得了一些进展，但是这些进展主要来自密码的创作者桑伯恩。我将在下文中详细介绍这些进展。

2010年11月20日，桑伯恩为尝试破译“克里普托斯”密码第4部分（K4）的研究者们提供了一条线索：第4部分中的密文“NYPVTT”对应的是明文“BERLIN”（柏林）。

我怀疑K4密码使用的是矩阵加密的技术，我将在下文中详细解释我做出上述判断的理由。在桑伯恩给出上述提示以后，有一些研究者对矩阵加密的理论进行了测试。但是，在向读者介绍这些测试的结果之前，我首先必须向读者解释一下矩阵加密技术的工作原理。与前文中的许多情况一样，要说清楚矩阵加密系统的工作原理，最好的方法似乎是举一个具体的例子。

为了介绍矩阵加密系统，首先必须定义什么是矩阵。对于我们要解释的问题而言，我们可以简单地认为矩阵就是一个由数字组成的矩形。当然，这是一种经过简化后的定义，但是在我们的说明过程中，这个定义已经足够了。我们可以用正方形的矩阵来完成加密工作。在这个例子中，我将使用如下矩阵：$\mathbf{M}=\begin{pmatrix}8 & 11\\ 15 & 3\end{pmatrix}$。在上述公式中，字母**M**代表整个矩阵。

因为矩阵**M**有两行和两列，所以我们说**M**是一个2乘2（2×2）的矩阵。接下来，我们要用矩阵**M**来加密以下这段明文信息：

A REPO MAN SPENDS HIS LIFE GETTING INTO TENSE SITUATIONS.

（追债人的生活就是要不断进入各种紧张的情况中。）

加密的第一步是用数字取代明文中的所有字母，字母和数字的对应关系是：A=0，B=1，C=2，…Z=25。在完成这一步以后，我们就会得到以下结果：

A	R	E	P	O	M	A	N	S	P	E	N	D	S	H	I
0,	17,	4,	15,	14,	12,	0,	13,	18,	15,	4,	13,	3,	18,	7,	8,

S	L	I	F	E	G	E	T	T	I	N	G	I	N	T	O	T
18,	11,	8,	5,	4,	6,	4,	19,	19,	8,	13,	6,	8,	13,	19,	14,	19,

E	N	S	E	S	I	T	U	A	T	I	O	N	S	X
4,	13,	18,	4,	18,	8,	19,	20,	0,	19,	8,	14,	13,	18,	23

在上述步骤中，我们在明文信息的末尾加上了一个字母X，我将在下文中解释这么做的理由。接下来，我们开始加密信息中的第一对数字，也就是0和17。加密的方法如下：

$$\begin{pmatrix}8 & 11\\15 & 3\end{pmatrix}\begin{pmatrix}0\\17\end{pmatrix}=\begin{pmatrix}8\times 0+11\times 17\\15\times 0+3\times 17\end{pmatrix}=\begin{pmatrix}187\\51\end{pmatrix}.$$

上述操作究竟是如何完成的呢？首先，我们取矩阵第一行的 8 和 11，用这两个数字分别乘以右边的两个数字。也就是说，用矩阵第一行的第一个数字 8，乘以右边的第一个数字 0。然后，用矩阵第一行的第二个数字 11，乘以右边的第二个数字 17。接着，我们将以上两次乘法的乘积相加，这样就得到 $8\times 0+11\times 17$。接下来，我们对矩阵的第二行做同样的处理，于是就得到 $15\times 0+3\times 17$。

完成上述运算以后，我们最终得到这样一个矩阵：$\begin{pmatrix}187\\51\end{pmatrix}$。

由于最终我们必须把数字重新转化为字母，而字母所对应的数字不能超过25。因此，如果我们得到的数字大于或者等于26，我们就要从这个数字中减去26（我们可以多次减去26，直到最终的得数小于26为止）。在以上例子中，完成这项操作以后，我们得到的是：$\begin{pmatrix}187\\51\end{pmatrix}=\begin{pmatrix}5\\25\end{pmatrix}$。然后，我们再次根据原始的替换规则把数字重新转化为字母，得到 F 和 Z。

在上述步骤中，我们进行了一种算术操作：从一个较大的数字中多次减去26，直到最终的得数在0~25之间为止。这种算术操作被称为“模除26”，简称“模26”，也写作“mod 26”。要想得到模26的结果，一共有两种方法，第一种方法是像上面描述的那样，不断从大数中减去26，而另一种方法是用这个大数除以26，然后写下除得的余数（除得的商则不是我们关心的结果）。

接下来，我们继续加密信息中的第二对字母，加密的方式和刚才对第一对字母进行的操作完全一样：

$$\begin{pmatrix}8 & 11\\15 & 3\end{pmatrix}\begin{pmatrix}4\\15\end{pmatrix}=\begin{pmatrix}8\times 4+11\times 15\\15\times 4+3\times 15\end{pmatrix}=\begin{pmatrix}197\\105\end{pmatrix}=\begin{pmatrix}15\\1\end{pmatrix}$$

因此，明文信息中的第二对字母被加密为P和B。我们继续用以上方法加密明文中的所有字母，最终就会得到以下这段完整的密文信息：

FZPBKMNNXDTVOVOZFRPFUAHNGXOFXDUPOLQPWKNVEOGXVPHB.

把明文信息和密文信息对应起来，我们就会看到，明文中重复出现的字母在密文中被加密成了不同的字母：

A REPO MAN SPENDS HIS LIFE GETTING INTO TENSE
F ZPBK MNN XDTVOV OZF RPFU AHNGXOF XDUP OLQPW

SITUATIONSX.
KNVEOGXVPHB.

这种矩阵加密系统具有和普莱费尔加密系统相同的特征：它们都把明文信息中的字母每两个组成一对，一起加密，只是加密的方式有所不同。然而，矩阵加密系统不仅可以将字母两两组对进行加密，我们还可以用 3×3、4×4 甚至更大的矩阵，来每次加密3个、4个甚至更多个字母。接下来，我举一个非常简单的例子来展示如何用 3×3 的矩阵加密信息。假设我们要加密的明文信息的前3个字母分别是A、R、E，也就是0，17，4，对前3个字母进行加密的公式如下：

$$\begin{pmatrix}13 & 2 & 3\\ 4 & 0 & 5\\ 7 & 21 & 2\end{pmatrix}\begin{pmatrix}0\\ 17\\ 4\end{pmatrix}=\begin{pmatrix}13\times0+2\times17+3\times4\\ 4\times0+0\times17+5\times4\\ 7\times0+21\times17+2\times4\end{pmatrix}=\begin{pmatrix}46\\ 20\\ 365\end{pmatrix}=\begin{pmatrix}20\\ 20\\ 1\end{pmatrix}\pmod{26}$$

将最后得到的3个数字重新转化成字母，我们就得到UUB。

我认为，在K4密码中，加密者使用3×3矩阵进行加密的可能性远高于使用2×2矩阵的可能性。

我不是第一个提出K4密码可能使用了矩阵加密的人。在互联网上，这个理论已经被各路研究者反复讨论了许多年，但是我并不清楚究竟是谁首次提出了这个理论。有3点理由使我怀疑K4密码是用矩阵加密系统加密的，下面我将逐一解释。记住，最重要的东西永远会被留到最后！

理由1：前三段密码中都出现了故意拼写的错误。

桑伯恩曾表示，在破译“克里普托斯”密码的第四部分时，会用到密码的前三部分（K1、K2、K3）。根据这一线索，我怀疑前三部分中故意出现的拼写错误实际上可以通过某种方式被转化成数字，而这几个数字就是第四部分用来加密矩阵的数字。如果我们把密文看成一个长方形的阵列，拼写错误出现的位置就分别对应于阵列中的一个坐标。雕塑第2行的拼写错误K对应的坐标是（25，2），第6行的拼写错误R对应的坐标是（24，6），第14行的拼写错误L和E对应的坐标分别是（11，15）和（15，15）。因此，“克里普托斯”密码前三部分中的4个拼写错误一共为我们提供了8个数字（这里我们没有考虑这4个错误字母本身对应的数字，如果把这4个数字也考虑进去的话，我们就一共有12个数字了）。2×2的矩阵中需要4个数字，因此这8个数字已经是所需数目的两倍了；而3×3的矩阵中需要9个数字，因此这8个数字看起来数量又不够。即使我们考虑上述括号中讨论的情况，即我们一共有12个数字，这12个数字也没有办法组成一个正方形矩阵。也许，密码前三部分的拼写错误只提供了第四部分使用的加密矩阵中的部分数字。对于这个问题，我还没有形成完整的解答，但是可以肯定的是，密码前三部分中故意出现的拼写错误一定具有某种重要的意义。

理由 2：桑伯恩给出的提示。

如果K4密码使用的加密矩阵是一个3×3的矩阵，那么破译者先要把字母每3个组成一组，再把各组字母转化成数字，然后用破译矩阵逐一乘以每组数字。如果我们把K4的密文每3个字母组成一组，就可以看出桑伯恩破译好的6个字母正好属于其中的两组，即NYP和VTT（对应于明文“BERLIN”）。桑伯恩给出的这6个字母并没有分别位于几个不完整的分组中，因此这一点可以在一定程度上支持3×3矩阵的理论。

OBK RUO XOG HUL BSO LIF BBW FLR VQQ PRN GKS SOT WTQ SJQ SSE KZZ WAT JKL UDI AWI NFB **NYP VTT** MZF P KW GDK ZXT JCD IGK UHU AUE KCA R

当然，上述情况也可能只是一个巧合。如果不是巧合的话，那么桑伯恩可能是想要为我们提供一个或几个完整的密码块的解答，这样的线索比普通密文的解答更有价值。

理由 3：“HILL”。

支持矩阵加密理论的最佳线索是：在“克里普托斯”雕塑右侧面板最右侧竖列的中部出现了“HILL”一词，而中央情报局网站上登出的“克里普托斯”密码中根本看不到这个词。在前文中我曾说过，中央情报局网站上登出的“克里普托斯”密码有错误，这就是那个非常关键的错误。读者可以在图9–8中看到正确的密码。

当然，这也可能是桑伯恩在雕塑中无意犯下的一个错误，毕竟在“克里普托斯”密码前面的部分中也有无意中留下的错误。但是上述

	密码	密钥
K1	EMUFPHZLRFAXYUSDJKZLDKRNSHGNFIVJ	ABCDEFGHIJKLMNOPQRSTUVWXYZABCD
	YQTQUXQBQVYUVLLTREVJYQTMKYRDMFD	AKRYPTOSABCDEFGHIJLMNQUVWXZKRYP
K2	VFPJUDEEHZWETZYVGWHKKQETGFQJNCE	BRYPTOSABCDEFGHIJLMNQUVWXZKRYPT
	GGWHKK?DQMCPFQZDQMMIAGPFXHQRLG	CYPTOSABCDEFGHIJLMNQUVWXZKRYPTO
	TIMVMZJANQLVKQEDAGDVFRPJUNGEUNA	DPTOSABCDEFGHIJLMNQUVWXZKRYPTOS
	QZGZLECGYUXUEENJTBJLBQCRTBJDFHRR	ETOSABCDEFGHIJLMNQUVWXZKRYPTOSA
	YIZETKZEMVDUFKSJHKFWHKUWQLSZFTI	FOSABCDEFGHIJLMNQUVWXZKRYPTOSAB
	HHDDDUVH?DWKBFUFPWNTDFIYCUQZERE	GSABCDEFGHIJLMNQUVWXZKRYPTOSABC
	EVLDKFEZMOQQJLTTUGSYQPFEUNLAVIDX	HABCDEFGHIJLMNQUVWXZKRYPTOSABCD
	FLGGTEZ?FKZBSFDQVGOGIPUFXHHDRKF	IBCDEFGHIJLMNQUVWXZKRYPTOSABCDE
	FHQNTGPUAECNUVPDJMQCLQUMUNEDFQ	JCDEFGHIJLMNQUVWXZKRYPTOSABCDEF
	ELZZVRRGKFFVOEEXBDMVPNFQXEZLGRE	KDEFGHIJLMNQUVWXZKRYPTOSABCDEFG
	DNQFMPNZGLFLPMRJQYALMGNUVPDXVKP	LEFGHIJLMNQUVWXZKRYPTOSABCDEFGH
	DQUMEBEDMHDAFMJGZNUPLGEWJLLAETG	MFGHIJLMNQUVWXZKRYPTOSABCDEFGHI
K3	ENDYAHROHNLSRHEOCPTEOIBIDYSHNAIA	NGHIJLMNQUVWXZKRYPTOSABCDEFGHIJL
	CHTNREYULDSLLSLLNOHSNOSMRWXMNE	OHIJLMNQUVWXZKRYPTOSABCDEFGHIJL
	TPRNGATIHNRARPESLNNELEBLPIIACAE	PIJLMNQUVWXZKRYPTOSABCDEFGHIJLM
	WMTWNDITEENRAHCTENEUDRETNHAEOE	QJLMNQUVWXZKRYPTOSABCDEFGHIJLMN
	TFOLSEDTIWENHAEIOYTEYQHEENCTAYCR	RLMNQUVWXZKRYPTOSABCDEFGHIJLMNQ
	EIFTBRSPAMHHEWENATAMATEGYEERLB	SMNQUVWXZKRYPTOSABCDEFGHIJLMNQU
	TEEFOASFIOTUETUAEOTOARMAEERTNRTI	TNQUVWXZKRYPTOSABCDEFGHIJLMNQUV
	BSEDDNIAAHTTMSTEWPIEROAGRIEWFEB	UQUVWXZKRYPTOSABCDEFGHIJLMNQUVW
	AECTDDHILCEIHSITEGOEAOSDDRYDLORIT	VUVWXZKRYPTOSABCDEFGHIJLMNQUVWX
	RKLMLEHAGTDHARDPNEOHMGFMFEUHE	WVWXZKRYPTOSABCDEFGHIJLMNQUVWXZ
	ECDMRIPFEIMEHNLSSTTRTVDOHW?OBKR	XWXZKRYPTOSABCDEFGHIJLMNQUVWXZK
K4	UOXOGHULBSOLIFBBWFLRVQQPRNGKSSO	YXZKRYPTOSABCDEFGHIJLMNQUVWXZKR
	TWTQSJQSSEKZZWATJKLUDIAWINFBNYP	ZZKRYPTOSABCDEFGHIJLMNQUVWXZKRY
	VTTMZFPKWGDKZXTJCDIGKUHUAUEKCAR	ABCDEFGHIJKLMNOPQRSTUVWXYZABCD

图 9–8 “克里普托斯”密码的一种表示方法，在这种表示方法下能看到“HILL”一词

解释成立的可能性不大，因为这类错误一般会产生无意义的文字，或者产生一个和密码学无关的单词，而我们看到的却是一个和密码学有关的单词。

那么“Hill”一词究竟与密码学有什么关系呢？矩阵加密方法是1929年由莱斯特·希尔（Lester Hill）发明的，因此，矩阵加密系统也常常被称为“希尔密码”。如果这样理解的话，桑伯恩给出的线索实际上是非常直白和明显的了。

以上我已经举出了支持K4密码是用矩阵加密系统加密的各种证据。接下来，我要向读者介绍在上述假设的基础上我在K4密码破译方面取得的结果。这项工作是由我与我的前同事格雷格·林克（Greg Link），以及我的学生丹特·莫勒（Dante Molle）合作完成的。

假设我们根据以下规则让每个字母对应于一个数字：A=0，B=1，C=2，…Z=25。在这一前提下，可以用计算机程序来测试所有可逆的 2×2 矩阵，看有没有哪个矩阵能够破译出有意义的文字——要写出这样的程序代码并不困难。“可逆”矩阵是指这样的矩阵：它可以完全抵消原始矩阵给每对数字带来的变化。在 2×2 的矩阵中，可逆的矩阵一共只有157 248个。

然而，我们的测试结果是：所有 2×2 的可逆矩阵都不能破译K4密码。由于有了桑伯恩提供的线索，我们没有必要逐一检查全部157 248个矩阵产生的文字是否有意义，我们只要让计算机在破译结果中搜索“BERLIN”一词就行了。然而，在我们的测试中，这个词一次也没有出现过。

但是，除了A=0，B=1，C=2，…Z=25以外，我们还可以用其他方式把字母和数字一一匹配起来。有些密码学家喜欢让A=1，B=2，C=3，…X=24，Y=25，Z=0。读者可能还记得，在做模除运算时，26=0。然而，即使改用上述配对关系，所有 2×2 的矩阵仍然无法产生“BERLIN”一词。

在前文中我曾经提过，“克里普托斯”雕塑右侧面板上的内容可能是字母表。接下来，我们测试了如下打乱后的字母表：

K R Y P T O S A B C D E F G H I J L M N Q U V W X Z

也就是说K=0，R=1，Y=2，…Z=25，或者K=1，R=2，Y=3，…W=24，X=25，Z=0。

然而，在上述两种设定下，仍然没有任何 2×2 的矩阵能破译出“BERLIN”一词。然后，我们又测试了用关键词“KRYPTOS”打乱字母表的所有移位方式。换句话说，我们对“克里普托斯”雕塑右侧面板上的所有字母表都进行了测试。在每一种情况下，我们都分别测试以0和1开头的两种不同设定。再接下来，我们又试了完全按顺序排列的字母表（A B C D E F G H I J K L M N O P Q R S T U V W X Y Z）的所有移位方式，同样分别测试了以0、1开头的两种不同设定。然而，仍然没有任何一种情况能给出正确的解答。在上述所有测试中，“BERLIN”一词也没有出现过。

我们考虑到，“克里普托斯”雕塑右侧面板上的“HILL”一词可能有不止一种用途。因此我们找到了莱斯特·希尔1929年首次提出矩阵加密方法时发表的论文，并研究了希尔在该论文中使用的字母—数字配对方法。希尔在这篇论文中采用了一种随机打乱字母表的做法，其配对关系如下[24]：

K	P	C	O	H	A	R	N	G	Z	E	Y	S	M	W	F	L	V	I	Q
0	1	2	3	4	5	6	7	8	9	10	11	12	13	14	15	16	17	18	19

D	U	X	B	T	J
20	21	22	23	24	25

我对这种配对方式也进行了测试，但并没有得到更好的结果。

如果要考虑3×3的矩阵，那么需要测试的可逆矩阵的数目就会高达1 634 038 189 056个。但是，这个巨大的数目对于计算机程序而言并不算一个很大的挑战，我的合作者林克和莫勒写了一个程序，可以在大约10小时内完成对任何给定字母表的测试。我们尝试了在前文中提到的所有字母表。

我们测试过的所有2×2的矩阵都不能在正确的位置上产生“BERLIN”一词（事实上，在任何位置都不行）。而在我们测试的3×3的矩阵中，有许多矩阵都能在正确的位置上出现了“BERLIN”一词。在通常情况下，对于每种我们测试的字母表，我们大约能发现17 000~18 000个符合上述条件的矩阵。[25]虽然“BERLIN”一词确实出现了，但是在所有我们检查过的情况下，这个词周围的文字都没有意义。然而，由于能产生“BERLIN”一词的矩阵数量巨大，我们只能检查其中的少部分结果。显然，我们需要一种更好的方法进一步处理这个问题。

幸运的是，在2014年11月20日，正当我们对上述问题冥思苦想的时候，“克里普托斯”雕塑的创作者桑伯恩又公布了另一条线索。这条新线索是对已有线索的一种扩展，他称：密文中的“NYPVTTMZFPK”对应的明文信息是“BERLIN CLOCK”（柏林钟）。也就是说，桑伯恩在上一条线索的基础上又扩大了提示的范围。这是一条非常有用的线索！我们希望，在这条新线索的帮助下，我们可以把此前取得大量可能解法缩减为一个解法。然而，事实却是，这条新线索把我们此前取得的大量可能解法缩减为了0个解法。在我们测试过的所有字母表的3×3矩阵中，没有一个矩阵能够产生“BERLIN CLOCK”一词。

那么，为什么桑伯恩要不断地向“克里普托斯”密码的破译者们提供线索呢？虽然“克里普托斯”雕塑为桑伯恩带来了大量的公众关注，但是从许多迹象来看，桑伯恩似乎希望“克里普托斯”密码能够被破译出来。比如，桑伯恩的工作室已经发生了两起撬门入室事件，这两起事件中的犯罪分子并不想偷窃珠宝和现金，而是想在工作室中找出K4密码的解法。尽管发生了这些骇人听闻的入室盗窃事件，但桑伯恩似乎把K4密码的解法隐藏得很好——到目前为止，没有任何人成功找到了K4密码的解法，包括我自己，以及和我一起进行此项研究的合作者。

尽管上文中提到的这些破译工作都以失败告终了，但是我并不准备就此放弃矩阵加密法的理论。因为，我们毕竟只测试了所有可能的字母—数字配对方式中的极少的一部分。此外，还有另外一种可能性，那就是：K4密码确实使用了矩阵加密法，但是桑伯恩使用矩阵加密法的方式比我们研究过的方式要更复杂。假设我们用字母**M**来代表矩阵，然后用字母P和字母C来分别代表3个明文字母和3个密文字母，那么我们之前考虑过的方法可以被描述为C=**M**P。也就是说，用矩阵**M**去乘以明文字母的数值P，就得到明文所对应的密文字母数值C。然而，桑伯恩所使用的公式也许并不是C=**M**P，而是C=**M**P+B。又或者，桑伯恩可能使用了某种连锁加密方式，也就是说，每一个密码块对应的密文不仅取决于他所用的矩阵，还取决于之前的某个密码块的明文或者密文的字母数值。许多现代加密系统都会使用这类连锁加密的技巧。

还有一种可能性是，桑伯恩是用一个4×4的矩阵来加密K4密码的。能够用于加密的可逆4×4矩阵的数量超过了1.2×10^8个，假如考虑更大的矩阵的话，需要测试的矩阵数量还会变得更加巨大！桑伯恩真的会把他的这条密码设计得如此难吗？此外，还有另外一种可能性会让问题变得更加复杂，那就是：桑伯恩可以在K4密码中使用不止一种数字—字母匹配关系。比如，在用关键词“KRYPTOS”打乱移位字母表以后，桑伯恩可以根据这些字母表为每一个字母分配一个数值。

还有一种可能性是，桑伯恩可以用两个2×2的矩阵来加密K4密码。虽然我们已经排除了串联使用两个2×2矩阵的可能性，但是要使用两个2×2矩阵还有其他方式。在K1、K2和K3中，一共有4个故意拼写的错误，如果考虑这4处错误的横坐标和纵坐标，就会产生8个数字。而两个2×2的矩阵正好包含8个数字。

然而，即使某位研究者能够在未来成功地破译K4密码，“克里普托斯”雕塑的谜题仍然不能算是完全解开了。

桑伯恩先生本周表示，“克里普托斯”雕塑中还包含一个谜中谜——只有先解开全部4段密码，才能够开始破解这个谜中谜。[26]

在此前的一次访谈中，桑伯恩也谈到了这个问题：

他们也许能够读懂我写下的话，但我写下的话本身也是一个谜题。即便解开了全部4段密码，破译者仍需要去发现更多的东西。在“克里普托斯”雕塑中有一些东西，破译者永远无法找到它们的真正意义。人们永远会说：“这到底是什么意思？我所写下的这些文字只是一个更大的谜题的线索而已。”[27]

虽然从“克里普托斯”密码已经解开的部分来看，这座雕塑深深植根于只用纸笔解开密码的世界中，但是我们所面临的下一个挑战却需要既使用古老的解密方式又使用更加现代的破译技术。在解开这个挑战的过程中，我们需要拥抱密码学的整个历史——一段超过2 000年的丰富历史。

蝉3301密码

蝉3301（Cicada 3301）密码被称为“互联网时代最精巧、最神秘的谜题”，并被《华盛顿邮报》列为“互联网时代五大最诡异的未解之谜”之一。关于蝉3301密码的设计目的，存在各种各样的猜测。很多人认为，蝉3301密码是美国国家安全局、美国中央情报局、英国军情六处，或者某个网络黑客雇佣兵组织的招聘工具。而另外一些人则认为，蝉3301密码是一种“另类现实游戏”（alternate reality game）。然而，既没有任何组织或个人宣称自己是蝉3301密码的创造者或者用它来赚钱，也没有任何人公开表示自己已经解开了这组谜题。上述两项事实使得大多数人既不相信蝉3301密码是某种招聘工具，也不相信“另类现实游戏”的猜测。还有一些人认为，蝉3301密码是某个正在研究加密货币的银行组织的杰作。[28]

蝉3301密码的谜题始于2012年1月4日发表在www.4chan.org网站上的一个帖子。该网站允许用户匿名发表少量文字，还可以给文字配上一张图片，用户也可以以上述方式回应其他人发表的帖子。在www.4chan.org网站上，有多个针对不同兴趣组的留言板，其中一个比较活跃的留言板被称作/b/。/b/留言板中的内容是“随机的”。

> 这个留言板的目的就是让用户贴出一些内容，但这个过程是不可逆的。在/b/留言板上，有一半的内容是为了让用户再也不想回到/b/留言板而贴出的。
>
> ——格雷格·胡什（Greg Housh），
>
> 一位与“匿名”黑客组织有关的网络活动家[29]

但是，接下来我们要讨论的这幅出现在/b/留言板上的图片并不是为了上述目的而产生的。我将这幅图片复制在图9–9中。

一般来说，用文字编辑器软件打开一幅图片并不会得到有趣的可读文字。然而，如果用文字编辑器软件打开上面的这幅图片，最后就会得到这样一条附属信息：

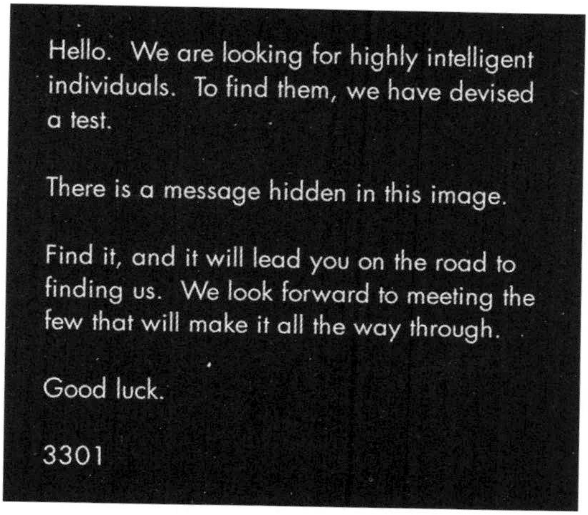
Hello. We are looking for highly intelligent individuals. To find them, we have devised a test.
There is a message hidden in this image.
Find it, and it will lead you on the road to finding us. We look forward to meeting the few that will make it all the way through.
Good luck.
3301

图 9-9　这张 2012 年 1 月出现的图片是蝉 3301 谜题的开端。图中文字翻译为：你们好！我们在寻找高智商的人。为了找到这些人，我们设计了一个测试。在这幅图中有一条隐藏的信息。如果你能找到这条信息，它就会带领你找到我们。只有极少数的人能够沿着这条路一直走到终点，我们热切地盼望和这些人会面。祝大家好运。3301

T I B E R I V S　C L A V D I V S　C A E S A R　s a y s "lxxt>33m2mqkyv2gsq3q= w]O2ntk"

提比略 · 克劳狄 · 恺撒说："lxxt>33m2mqkyv2gsq3q=w]O2ntk"

这条信息中提到的"提比略 · 克劳狄 · 恺撒"是罗马帝国的第四任皇帝，而接下来的内容似乎是一段密码。这种设定让人们很自然地联想到，这段密码可能是恺撒移位密码。对于引号中文字的前4个字母，如果把每个字母向后移动4位，lxxt 就变成了 http。因此，这段密码看起来似乎是一个网址。

对字母进行移位的方法是很显然的，但是如何处理引号中的数字和如">""=""）"之类的符号呢？要解答这个问题，就要用到一种计算机科学家们非常熟悉的东西——ASCII码。ASCII码是"美国信息交换标准代码"

（American Standard Code for Information Interchange）的缩写，这套代码为一系列更多的字符一一分配了对应的数值。下一页上的表9–1就是一种ASCII码对照表。

有了ASCII码对照表，我们就能用数值表示大写字母、小写字母、数字、标点符号，以及其他符号了。比如，第32号字符表示一个空格，其他不会在屏幕上显示任何内容的键盘操作也都有其对应的数值。这些操作包括“水平制表符”（第9号）、“转义”（第27号）等。除了ASCII表以外，还有一套扩展ASCII表，用从128~256的数字代表另外128个（较不常用的）符号。

于是，根据ASCII码对照表，“>”向前移动4位变成“:”，“]”向前移动4位变成字母Y。数字也可以简单地向前移动4位。完成这些操作以后，引号中的密文最终变成了以下明文：

http://i.imgur.com/m9sYK.jpg

只要在浏览器中输入上述网址，就可以看到另一幅图片，我将这幅图片复制在图9–10中。

然而，这幅图并不是一个死胡同，而是另一条线索。

在文件的结尾附上一段信息只是在数字图片中隐藏信息的一种方式。事实上，还可以通过许多更加微妙的方式在图片中附加信息。在一幅图片中，每一个像素的颜色都是用一连串的数字0和1来表示的。如果对像素的颜色做轻微的改变（比如将某种颜色的比重从2 743盎司提高到2 744盎司），就会产生一幅新图片，而人类的肉眼完全无法区分新图和原图。通过这种方式，我们就可以按照自己的意愿改变多个像素的数值，把这些经过修改的二进制数值放在一起就可以组成一段信息。

表 9-1　常规ASCII码对照表

十进制数值	名称	字符	十进制数值	字符	十进制数值	字符	十进制数值	字符
0	空字符	NUL	32	空格	64	@	96	`
1	标题开始	SOH	33	!	65	A	97	a
2	正文开始	STX	34	“	66	B	98	b
3	正文结束	ETX	35	#	67	C	99	c
4	传输结束	EOT	36	$	68	D	100	d
5	请求	ENQ	37	%	69	E	101	e
6	确认	ACK	38	&	70	F	102	f
7	响铃	BEL	39	‘	71	G	103	g
8	退格	BS	40	(	72	H	104	h
9	水平制表符	HT	41	)	73	I	105	i
10	换行键	LF	42	*	74	J	106	j
11	垂直制表符	VT	43	+	75	K	107	k
12	换页键	FF	44	,	76	L	108	l
13	回车键	CR	45	-	77	M	109	m
14	不用切换	SO	46	.	78	N	110	n
15	启用切换	SI	47	/	79	O	111	o
16	数据链路转义	DLE	48	0	80	P	112	p
17	设备控制 1	DC1	49	1	81	Q	113	q
18	设备控制 2	DC2	50	2	82	R	114	r
19	设备控制 3	DC3	51	3	93	S	115	s
20	设备控制 4	DC4	52	4	84	T	116	t
21	否认	NAK	53	5	85	U	117	u
22	同步空闲	SYN	54	6	86	V	118	v
23	传输块结束	ETB	55	7	87	W	119	w
24	取消	CAN	56	8	88	X	120	x
25	媒介结束	EM	57	9	89	Y	121	y
26	替换	SUB	58	:	90	Z	122	z
27	转义	ESC	59	;	91	[	123	{
28	文件分隔符	FS	60	<	92	\	124	|
29	分组符	GS	61	=	93	]	125	}
30	记录分隔符	RS	62	>	94	^	126	~
31	单元分隔符	US	63	?	95	_	127	DEL[1]

① “DEL”即“delete”，表示删除。——编者注

图 9–10 进了死胡同？图中的文字翻译为：哎呀，这只是一个诱饵。看起来你没能够猜出真正的信息

以上这种方式只是在一幅数字图片中隐藏信息的方式之一，此外还有许多其他方式。一般来说，这个领域被称为“数字信息隐藏技术”。据说，发动“9·11”恐怖袭击事件的恐怖分子就是通过这种方式互相联系的——他们把与恐怖袭击相关的信息隐藏在一些色情图片中，然后在网上分享这些色情图片。当然，上述说法只是一种传闻，关于“9·11”恐怖袭击事件的细节目前仍然属于保密信息。

我们可以通过一种名为“OutGuess”的工具向数字图片中嵌入或提取信息（关于这种工具的详细情况参见以下网站：http://www.outguess.org/）。图9–10所展示的图片中包含两个英文单词：“out”和“guess”。因此，如果某位读者不仅熟悉数字信息隐藏技术，还知道“OutGuess”这种软件的话，他就不难猜出接下来的一步。而任何不具备上述知识的人恐怕都很难找到接下来的线索！

只要使用“OutGuess”软件，就可以从上述图片中获得另一个网站的网址：http://pastebin.com/aXYZzzcv。这个网址中包含以下内容：

这里有一段书本密码。要找到这本书以及更多相关的内容，请访问以下网址：

http://www.reddit.com/r/a2e7j6ic78h0j/

1:20	14:4	27:5
2:3	15:8	28:1
3:5	16:4	29:2
4:20	17:5	30:18
5:5	18:14	31:32
6:53	19:7	32:10
7:1	20:31	33:3
8:8	21:12	34:25
9:2	22:36	35:10
10:4	23:2	36:7
11:8	24:3	37:20
12:4	25:5	38:10
13:13	26:65	39:32
40:4	53:18	66:18
41:40	54:4	67:45
42:11	56:6	68:10
43:9	56:4	69:2
44:13	57:24	70:17
45:6	58:64	71:9
46:3	59:5	72:20
47:5	60:37	73:2
48:43	61:60	74:34

49:17

50:13

51:4

52:2

62:12

63:6

64:8

65:5

75:13

76:21

祝你好运。

3301

如果我们访问上述页面中提供的这个Reddit网站的链接，就会看到若干个帖子，这些信息看起来像是一些加密文字。然而，这几条密码可比恺撒移位密码难破译多了。在页面的顶端有几条提示，我将其中一条复制于图9–11中。

图 9–11　关于如何破译 2012 年蝉 3301 密码挑战的一条提示

图9–11中的符号是玛雅数字。这些数字中的点表示1，而横线表示5。倒数第二个位置的看起来像橄榄球的符号表示0。因此，我们可以将图9–11中的内容翻译为以下这串数字：

```
10, 2, 14, 7, 19, 6, 18, 12, 7, 8, 17, 0, 19
```

在这个页面中，还包含这样一串字符：a2e7j6ic78h0j7eiejd0120。如果我们把这串字符和玛雅数字对齐，就会发现两者形成了一种较佳的匹配关系：

```
10,2,14,7,19,6,18,12,7,8,17,0,19
 a,2, e,7, j,6, i, c,7,8, h,0, j,7,e,i,e,j,d,0,1,2,0
```

第一行和第二行中的一位数是完全一样的，而较大的数字则被转换成了字母。也就是说：a=10，c=12，e=14，以此类推。虽然包含字母和数字的这串字符比用玛雅数字表示的数字更长，但是上述对应关系是非常清楚的。我们可以把第一行的数字补全，就会得到以下的对应关系：

```
10, 2, 14, 7, 19, 6, 18, 12, 7, 8, 17, 0, 19, 7, 14,
 a, 2,  e, 7,  j, 6,  i,  c, 7, 8,  h, 0,  j, 7,  e,

18, 14, 19, 13, 0, 1, 2, 0
 i,  e,  j,  d, 0, 1, 2, 0
```

以上这串数字可以用来破译Reddit网页上的密码。比如，该网页上的第一条消息是：

Ukbn Txltbz nal hh Uoxelmgox wdvg Akw; hvu ogl rsm ar sbv ix jwz[30]

把上述密钥与这段信息对齐，然后把信息中的每个字母向后移动相应的位数，我们就得到：

```
 U   k   b   n   T   x   l   t   b  z ...
10, 2, 14, 7, 19, 6, 18, 12, 7, 8 ...
 K   i   n   g   A   r   t   h   u  r ...（亚瑟王……）
```

于是我们可以看出，这段信息的开头与“亚瑟王”有关。

用这种方法破译Reddit网页上的所有信息，就可以得到一本“书”。这本“书”就是上一个网页上的那段“书本密码”所用的书本。有了这本“书”，我们就可以破译那段书本密码。这本书的开头是这样的[①]：

1. King Arthur was at Caerlleon upon Usk; and one day he sat in his
2. chamber; and with him were Owain the son of Urien, and Kynon the son
3. of Clydno, and Kai the son of Kyner; and Gwenhwyvar and her
4. handmaidens at needlework by the window. And if it should be said
5. that there was a porter at Arthur's palace, there was none. Glewlwyd
6. Gavaelvawr was there, acting as porter, to welcome guests and
7. strangers, and to receive them with honour, and to inform them of the
8. manners and customs of the Court; and to direct those who came to the
9. Hall or to the presence-chamber, and those who came to take up their lodging.
10. In the centre of the chamber King Arthur sat upon a seat of green
11. rushes, over which was spread a covering of flame-coloured satin, and
12. a cushion of red satin was under his elbow.
13. Then Arthur spoke, "If I thought you would not disparage me," said
14. he, "I would sleep while I wait for my repast; and you can entertain
15. one another with relating tales, and can obtain a flagon of mead and
16. some meat from Kai." And the King went to sleep. And Kynon the son
17. of Clydno asked Kai for that which Arthur had promised them. "I,
18. too, will have the good tale which he promised to me," said Kai.

① 这段文字的具体意思与密码无关，因此没有翻译。——译者注

19. "Nay," answered Kynon, "fairer will it be for thee to fulfill
20. Arthur's behest, in the first place, and then we will tell thee the
21. best tale that we know." So Kai went to the kitchen and to the mead-
22. cellar, and returned bearing a flagon of mead and a golden goblet,
23. and a handful of skewers, upon which were broiled collops of meat.
24. Then they ate the collops and began to drink the mead. "Now," said
25. Kai, "it is time for you to give me my story." "Kynon," said Owain,
26. "do thou pay to Kai the tale that is his due." "Truly," said Kynon,
27. "thou are older, and art a better teller of tales, and hast seen more
28. marvellous things than I; do thou therefore pay Kai his tale."
29. "Begin thyself," quoth Owain, "with the best that thou knowest." "I
30. will do so," answered Kynon.
31. "I was the only son of my mother and father, and I was exceedingly
32. aspiring, and my daring was very great. I thought there was no
33. enterprise in the world too mighty for me, and after I had achieved
34. all the adventures that were in my own country, I equipped myself,
35. and set forth to journey through deserts and distant regions. And at
36. length it chanced that I came to the fairest valley in the world,
37. wherein were trees of equal growth; and a river ran through the
38. valley, and a path was by the side of the river. And I followed the
39. path until mid-day, and continued my journey along the remainder of
40. the valley until the evening; and at the extremity of a plain I came
41. to a large and lustrous Castle, at the foot of which was a torrent.
42. And I approached the Castle, and there I beheld two youths with
43. yellow curling hair, each with a frontlet of gold upon his head, and

clad in a garment of yellow satin, and they had gold clasps upon their insteps. In the hand of each of them was an ivory bow, strung with the sinews of the stag; and their arrows had shafts of the bone of the whale, and were winged with peacock's feathers; the shafts also had golden heads. And they had daggers with blades of gold, and with hilts of the bone of the whale. And they were shooting their daggers.

"And a little way from them I saw a man in the prime of life, with his beard newly shorn, clad in a robe and a mantle of yellow satin; and round the top of his mantle was a band of gold lace. On his feet were shoes of variegated leather, fastened by two bosses of gold. When I saw him, I went towards him and saluted him, and such was his courtesy that he no sooner received my greeting than he returned it. And he went with me towards the Castle. Now there were no dwellers in the Castle except those who were in one hall. And there I saw four-and-twenty damsels, embroidering satin at a window. And this I tell thee, Kai, that the least fair of them was fairer than the fairest maid thou hast ever beheld in the Island of Britain, and the least lovely of them was more lovely than Gwenhwyvar, the wife of Arthur, when she has appeared loveliest at the Offering, on the day of the Nativity, or at the feast of Easter. They rose up at my coming, and six of them took my horse, and divested me of my armour; and six others took my arms, and washed them in a vessel until they were perfectly bright. And the third six spread cloths upon the tables and prepared meat. And the fourth six took off my soiled garments, and placed others upon me; namely, an under-vest and a

69. doublet of fine linen, and a robe, and a surcoat, and a mantle of

在前一个网页上的那段书本密码中，每一对数字表示这本“书”中某一行中的某个字符。比如，第一对数字是1:20，也就是第1行的第20个字符（包括空格在内）。我们找到第1行的第20个字符是c。密码中的第二对数字是2:3，也就是第2行的第3个字符，这个字符是a。以此类推，整段密文就可以被破译为：

```
    Call us at us tele phone numBer two one four
three nine oh nine six oh eight
```

给我们打电话，电话号码是2143909608

但是，请读者先不要着急给这个号码打电话，因为Reddit网页上还有一些秘密在等待我们发现。这个网页上有两幅图，见图9–12和图9–13。

图 9–12　能沿着之前的线索找到这个网页的人会看到一个写着“欢迎”字样的脚垫

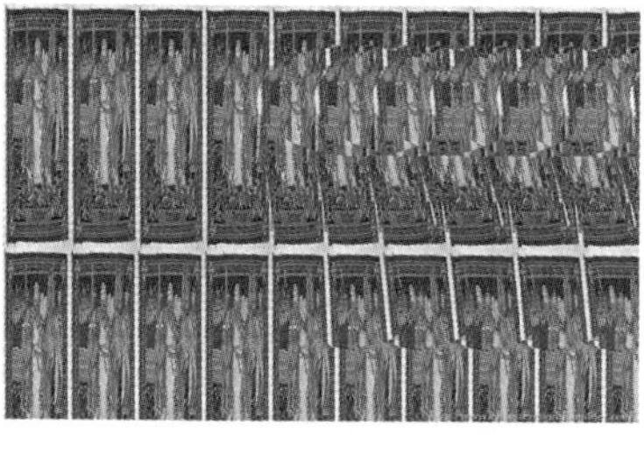

图 9–13　Reddit网页上还有一张墙纸，上面画了一幅三维立体图

图9–12中这幅写着“欢迎”字样的图片里，通过“数字信息隐藏技术”还藏着另一条信息。我们可以用OutGuess软件来提取这条信息，这条信息是：

-----BEGIN PGP SIGNED MESSAGE-----

Hash: SHA1

从现在开始，我会用这个密钥对所有信息进行数字信息隐藏技术的签名。这个密钥可以在mit公共密钥服务器（mit keyservers）上找到。Key ID 7A35090F，就像我在“a2e7j6ic78h0j”中贴出的一样。

耐心是一种美德。

祝你好运。

3301

-----BEGIN PGP SIGNATURE-----

Version: GnuPG v1.4.11 (GNU/Linux)

iQIcBAEBAgAGBQJPBRz7AAoJEBgfAeV6NQkP1UIQALFcO8
DyZkecTK5pAIcGez7kewjGBoCfjfO2NlRROuQm5CteXiH3
Te5G+5ebsdRmGWVcah8QzN4UjxpKcTQRPB9e/ehVI5BiBJ
q8GlOnaSRZpzsYobwKH6Jy6haAr3kPFK1lOXXyHSiNnQby
dGw9BFRIfSr//DY86BUILE8sGJR6FA8Vzjiifcv6mmXkk3
ICrT8z0qY7m/wFOYjgiSohvYpgx5biG6TBwxfmXQOaITdO
5rO8+4mtLnP//qN7E9zjTYj4Z4gBhdf6hPSuOqjh1s+6/
C6IehRChpx8gwpdhIlNf1coz/ZiggPiqdj75Tyqg88lEr66fV
VB2d7PGObSyYSpHJl8llrt8Gnk1UaZUS6/eCjnBniV/BLfZPV
D2VFKH2Vvvty8sL+S8hCxsuLCjydhskpshcjMVV9xPIEYzwSE
aqBq0ZMdNFEPxJzC0XISlWSfxROm85r3NYvbrx9lwVbPmUpL
KFn8ZcMbf7UX18frgOtujmqqUvDQ2dQhmCUywPdtsKHFLc1x
IqdrnRWUS3CDeejUzGYDB5lSflujTjLPgGvtlCBW5ap00cfIH
UZPOzmJWoEzgFgdNc9iIkcUUlkee2WbYwCCuwSlLsdQRMA//

PJN+a1h2ZMSzzMbZsr/YXQDUWvEaYI8MckmXEkZmDoARL0xkbH
EFVGBmoMPVzeC
=fRcg
-----END PGP SIGNATURE-----

后面这段看起来毫无意义的字串并不是需要破译的密码。要想理解这段奇怪的文字，我们必须先详细了解一些现代密码学的知识。

20世纪末，密码学的任务已经不仅是对信息进行加密了。人们不仅要求一套加密系统能对信息进行加密，还要求加密系统能保证真实性（authenticity）和完整性（integrity）。真实性是指加密系统能保证密码来自发送人，而完整性是指加密系统能保证密码在传输过程中没有被任何人修改过。能同时保证真实性和完整性的加密系统才是一个好的加密系统。我们将在本书的第11章中进一步讨论“真实性”的问题，但是在这里我会先向读者提供一些必要的背景知识。

为了满足上述两条要求，有些现代加密系统采用两套不同的密钥。其中一条密钥是对公众公开的信息，称为公钥，而另一条密钥是私有的信息，称为私钥。如果你想用这类系统向某人发送加密信息，你要做的就是找到收信人的公钥，并用这条密钥来加密你的信息。收到信息以后，收信人会用他的私钥来破译你发出的信息。因为要从公钥判断出私钥是极其困难的，因此即使有人截获了你发的信息，他也不太可能在没有私钥的情况下破译出这条信息。

有了这种加密系统，即使两个人不能事先见面商定一套密钥，也可以轻松地实现加密通信。但是，如果我们用另一种方式来应用这类加密系统，就能做更多的事情。不管“蝉3301”这个名字背后的人究竟是什么身份，此人在发送上述信息时在信息正文后面加了一段“签名”。简单来说，这段签名是发信人用他的私钥来加密原始信息时产生的。在这条信息中，发信人提示收信人去麻省理工学院密钥服务器上找他的公钥。找到公钥以后，收信人就可以用这个公钥去破译这段看似无意义的文字（签名），而破译所得的结果就是签名上的这段明文信息。任何人都可以破译这段信息，但是只有知道私钥的人才可能加密出这段信息。因此，这段“签名”确实起到了签名的作用——收信人可以通过这段签名验证发信人确实拥有私钥。

在上一段中我过于简化了，现在我来详细解释一下。如果一段信息很长，用私钥加密整段信息从而产生一个“签名”将非常费时。此外，这还会导致需要发送的信息长度翻倍。因此，在一般情况下，信息的发送者会用另一段信息来制造签名，这段信息相当于是原始信息的“指纹”。也就是说，发信人不会用私钥加密整段原始信息，而是加密一段比原始信息短很多的文件来作为签名，这个文件就是我上面说的“指纹”。但是，怎样把一段较长的信息缩短呢？“散列函数”（hash function）这种工具可以帮助我们做到这一点。散列函数的具体设计非常复杂，在此我们没有必要了解得太深，读者只要简单地理解它的基本功能就行了。它可以把一段长文字转化成更短的文字，并保证另一段长文字被转化成同样的短文字的概率极小。如果收到信息的人篡改了部分内容，然后声称修改过的内容就是原发信人签过名的内容，那么收信人只要查看信息的签名部分就能揭穿这个谎言。对信息的内容做任何修改都会使得该信息经散列函数处理后的值发生变化，而用私钥对这个修改过的值进行加密则会产生一个和原始签名不同的签名。用这种方式作假很容易被收信人识破。在上述过程中，收信人必须知道散列函数的具体形式，这一点很重要。发信人不需要对他使用的函数的形式保密。

在蝉3301密码中，发信人首先使用SHA-1（Secure Hash Algorithm 1，即安全散列算法-1）缩短信息，然后再用一个只有蝉3301本人知道的私钥对缩短后的信息进行加密。自这条消息出现以来，蝉3301发布的所有真实信息都附有这种签名。通过这种方法，可以很好地防止冒名顶替者以蝉3301的名义发布虚假的信息。由于蝉3301一直使用同一种私钥来给信息签名，所以可以确认所有这些信息都是来自同一个人（或者同一个组织，该组织的多个成员可能共享蝉3301的私钥）。

也许有的读者会提出这样一个问题：既然原始信息已经用散列函数缩短了，那为什么签名还这么长呢？这是因为蝉3301组织使用一种非常强大的密钥来对经散列函数处理后的信息进行加密。在这一系统中，这种安全度较高的密钥就会给出较长的最终结果。

接下来，我们讨论Reddit页面上的第二幅图。通过分析，破译者可以发现这幅图中藏着两个秘密。首先，不使用任何密码学的技巧，破译者也可以看出这张墙纸是一幅三维立体图。《魔法眼睛》系列书籍中有一些更加著名的三维立体图（当然该书中的图属于另一种类型——随机点三维立体图。只要一边盯着这幅图，一边试着放松眼睛，不让眼睛的焦点落在一处，我们就会看到图中似乎有一些物体飘浮在背景图案的前方。而对于这幅图来说，我们看到的物体是一个杯子。

此外，对这幅图使用OutGuess软件还会得到一条信息。这条信息的内容如下：

```
-----BEGIN PGP SIGNED MESSAGE-----
Hash: SHA1
```

密钥一直就摆在你们眼前。

这又不是寻找圣杯。不要把它搞得更难了。

祝你好运。

3301

```
-----BEGIN PGP SIGNATURE-----
Version: GnuPG v1.4.11 (GNU/Linux)
iQIcBAEBAgAGBQJPCBl3AAoJEBgfAeV6NQkPo6EQAKghp7ZKY
xmsYM96iNQu5GZVfbjUHsEL164ZLctGkgZx2H1HyYFEc6FGv
cfzqs43vV/IzN4mK0SMy2qFPfjuG2JJtv3x2QfHMM3M2+dwX
30bUD12UorMZNrLo8HjTpanYD9hL8WglbSIBJhnLE5CPlUSBZ
RSx0yh1U+wbnlTQBxQI0xLkPIz+xCMBwSKl5BaCb006z43/HJ
t7NwynqWXJmVVKScmkpFC3ISEBcYKhHHWv1IPQnFqMdW4dExX
dRqWuwCshXpGXwDoOXfKVp5NW7Ix9kCyfC7XC4iWXymGgd+/
h4ccFFVm+WWOczOq/zeME+0vJhJqvj+fN2MZtvckpZbcCMfLjn
1z4w4d7mkbEpVjgVIU8/+KClNFPSf4asqjBKdrcCEMAl80vZor
ElG6OVIHaLV4XwqiSu0LEF1ESCqbxkEmqp7U7CHl2VW6qv0h0G
xy+/UT0W1NoLJTzLBFiOzyQIqqpgVg0dAFs74SlIf3oUTxt6IU
pQX5+uo8kszMHTJQRP7K22/A3cc/VS/2Ydg4o6OfN54Wcq+8IM
ZxEx+vxtmRJCUROVpHTTQ5unmyG9zQATxn8byD9Us070FAg6/
vjGjo1VVUxn6HX9HKxdx4wYGMP5grmD8k4jQdF1Z7GtbcqzDsx
P65XCaOYmray1JyFG5OlgFyOflmjBXHsNad
=SqLP
-----END PGP SIGNATURE-----
```


许多试图破译蝉3301密码的人常常急于向前推进，而忽略了以上有趣的细节和线索。在有些情况下，为了继续前进，走一些回头路是有必要的。现在，让我们来拨打之前的书本密码提示的电话号码。在拨通(214)390–9608以后，我们会听到一段由电脑产生的语音，这段语音的内容是：

很好。你做得不错。原始的图片与三个质数有联系。3 301是其中一个质数，你必须找到另外两个质数。将这三个质数相乘，再加上.com，你就会找到下一步了。祝你好运。再见。[31]

质数是指有且仅有两个正整数因数的数字。比如，数字17就是一个质数，因为数字17只能被1和17整除。而数字18则不是一个质数，因为18可以被1、2、3、6、9、18整除。最小的几个质数分别是：2、3、5、7、11、13、17、19和23。数字1不是质数，因为1只能被1整除。记住，质数是指有且仅有两个正整数因数的数字。

在蝉3301密码挑战中，质数的概念以许多不同的形式出现多次。“蝉3301”这个名字中就包含几个质数。不仅3 301是一个质数，而且把这个数字反过来得到的1 033也是一个质数。具有这种性质的数被称为“可逆质数”或者“反质数”（emirp①）。而上述名称中提到的昆虫——蝉又与更多的质数有关。蝉这种昆虫会在地下蛰伏13年或者17年（具体的年数取决于蝉的种类），然后才回到地面进行繁殖，然后死去。其中的13和17都是质数。

破译者需要在上述这幅图片中再找出两个质数。这幅图片的宽度是509个像素，高度是503个像素，509和503都是质数。把3 301、509和503这三个质数相乘，得到的乘积是845 145 127。

如果你在麻省理工学院密钥服务器上寻找蝉3301的公共密钥，也会同样发现“845 145 127”这个数字。目前仍可以找到这条信息，我将这条信息复制如下[32]：

① “质数”的英文单词是“prime”，“反质数”的英文单词“emirp”刚好是把“prime”的字母颠倒而成。——编者注

'0x7a35090f' 的搜索结果

类型	二进制数/密钥名	日期	用户名
公共	4096R/7A35090F	2012-01-05	Cicada 3301 (845145127)

网址为http://845145127.com/ 的网页上有一幅图片，图上画着一只蝉和一个倒计时。用OutGuess软件可以从这幅图中提取一条信息，这条信息的内容如下：

-----BEGIN PGP SIGNED MESSAGE-----

Hash: SHA1

能走到这么远说明你表现得很不错。

耐心是一种美德。

请于协调世界时 2012 年 1 月 9 日（星期一）17：00 再来查看这里。

3301

-----BEGIN PGP SIGNATURE-----

Version: GnuPG v1.4.11 (GNU/Linux)

iQIcBAEBAgAGBQJPCKDUAAoJEBgfAeV6NQkPf9kP/19tbTFEy
+ol/vaSJ97A549+E713DyFAuxJMh2AY2y5ksiqDRJdACBdvVN
JqlaKHKTfihiYW75VHb+RuAbMhM2nNC78eh+xd6c4UCwpQ9vSU
4i1Jzn6+T74pMKkhyssaHhQWfPs8K7eKQxOJzSjpDFCSFG7o
Hx6doPEk/xgLaJRCt/IJjNCZ9l2kYinmOm7c0QdRqJ+VbV7Px
41tP1dITQIH/+JnETExUzWbE9fMf/eJl/zACF+gYii7d9ZdU8
RHGi14jA2pRjc7SQArwqJOIyKQIFrW7zuicCYYT/GDmVSyILM
03VXkNyAMBhG90edm17sxliyS0pA06MeOCjhDGUIwQzBwsSZ
QJUsMJcXEUOpHPWrduP/zN5qHp/uUNNGj3vxLrnB+wcjhF8Z

```
OiDF6zk7+ZVkdjk8dAYQr62EsEpfxMT2dv5bJ0YBaQGZHyjT
EYnkiukZiDfExQZM2/uqhYOj3yK0J+kJNt7QvZQM2enMV7jb
aLTfU3VZGqJ6TSPqsfeiuGyxtlGLgJvd6kmiZkBB8Jj0Rgx/
h9Tc4m9xnVQanaPqbGQN4vZF3kOp/jAN5YjsRfCDb7iGvuEcFh
4oRgpaB3D2/+Qo9i3+CdAq1LMeM4WgCcYj2K5mtL0QhpNoeJ/
s0KzwnXA+mxBKoZ0S8dUX/ZXCkbOLoMWCUfqBn8QkQ
=zn1y
-----END PGP SIGNATURE-----
```

在协调世界时2012年1月9日（星期一）17：00，这个网址显示了如下信息：

从以下地点中挑选一个离你最近的地点，在那里可以找到我们的符号：

坐标为：

52.216802, 21.018334

48.85057059876962, 2.406892329454422

48.85030144151387, 2.407538741827011

47.664196, −122.313301

47.637520, −122.346277

47.622993, −122.312576

37.577070, 126.813122

37.5196666666667, 126.995

36.0665472222222, −94.1726416666667

33.966808, −117.650488

29.909098706850486 −89.99312818050384

25.684702, –80.441289

21.584069, –158.104211

––33.90281, 151.18421

在上述时刻，网页上原有的蝉的图片也发生了变化。新的图片提供了以上14个坐标点中的12个：

```
-----BEGIN PGP SIGNED MESSAGE-----
Hash: SHA1
52.216802, 21.018334
48.85057059876962, 2.406892329454422
48.85030144151387, 2.407538741827011
47.664196, -122.313301
47.637520, -122.346277
47.622993, -122.312576
37.5196666666667, 126.995
33.966808, -117.650488
29.909098706850486 -89.99312818050384
25.684702, -80.441289
21.584069, -158.104211
- -33.90281, 151.18421
3301

-----BEGIN PGP SIGNATURE-----
Version: GnuPG v1.4.11 (GNU/Linux)
```

iQIcBAEBAgAGBQJPChn7AAoJEBgfAeV6NQkPZxMP/05D9TkS
pwRaBXPqYthuyqxxuo+ZDyr/yVIlAdurTBiWb3aGxKJjtWg/
vlcHcatK0TGL2qaHwB/FFZQAaqOyU7ZfDXdpWr8PWoWhpWNYUK
8IrOaYu1SmWlJnkTdUSzGrX0lbwjwMmJJoPNS7CJuO6MaA2GIw
pv2G7lYqnH3xeX3kzGlPMsVb/wucKRjobsbdbreh1SNuQuRnhf
e4s+oHTTqsXjtGL/VhBI0DUAdfLqW7z4C+Gvbx6okC8x5Sj2N2
UTJOiyMYXz5+QyHoA6fo9g5V6zodNpx/RvxuZP2Ssc9TqERgTo
5FjRBpON1vjDalHgg0H2Fus2LK3gh+NZfj1i5bOqa4Cqd9epI2
pe+glXn86j9crS+2BEAr1cguqAFepvI9sdFEornDja4VXwDtUd
M89hMVkU5NiTUYfvxZbL6W7rHIF7wxjGUwpe1ViuixG+cKNfv0
enrt60PrtDByBOWI9LLIUE0cB5HDT1xrczZ/55CtuM3Zf07/l0
nLFdmgR0oa8KUA9gWcPs6S1EpBa185VcyOTqbpIPiT8neiJEkX
arbJeFk15m1P73Fr8XZxdj7EHK0aOwGYcc8e4PmW/dShgcrSNX
iePCbcRVRD2n9L47C0LkNyRpoBkmjvtpcRyp5ISe+0xcx/QI+g
c1lkSijC89qV+ymCHae1RiSDxVbd
=ZJ37
-----END PGP SIGNATURE-----

这14个坐标给出的地点遍布全球。将上述14个坐标分别翻译为具体地址，我们得到如下结果：

波兰，华沙市，奥莱安德鲁夫路 6 号，邮编：01-001
法国，巴黎，平原路 89—91，邮编：75020
法国，巴黎，马莱谢路 36 号，邮编：75020
美国，华盛顿州，西雅图市，大学道东北 4739 号，邮编：

98105

美国，华盛顿州，西雅图市，克罗克特街 514 号，邮编：98109

美国，华盛顿州，西雅图市，15 街东 428 号，邮编：98112

韩国，首尔，江西区傍花洞 830-8

韩国，首尔，龙山区西冰库洞 287-1

美国，阿肯色州，费耶特维尔市，阿肯色大学，迪克森街西 853—899 号，邮编：72701

美国，加利福尼亚州，奇诺市，欧几里得路 15717—15735 号，邮编：91708

美国，路易斯安那州，新奥尔良市，407 号州内高速公路，邮编：70131

美国，佛罗里达州，迈阿密市，152 街西南 8718—8798 号，邮编：33193

美国，夏威夷州，哈莱伊瓦镇，卡美哈美哈高速路 66—420 号，邮编：96712

澳大利亚，新南威尔士州，厄斯金维尔市，乔治街 143 号，邮编：2043

在上述14个地点中，每个地点都出现了一条线索。这14条线索被贴在电线杆或者路灯柱上。具体的情况如图9–14所示。

至此，挑战蝉3301密码的人才发现，在这个谜题背后藏着一个资源广大的组织。

虽然上述14个地点都有以二维码形式出现的线索，但是这些线索中有许多是雷同的。这14个地点一共只有两种不同的信息。其中一条信息如下：

图 9-14　位于波兰华沙市的蝉 3301 线索

-----BEGIN PGP SIGNED MESSAGE-----

Hash: SHA1

在 29 卷中曾经含有知识。

在马比诺吉昂暂停的时候，还有多少行密码留存?

从最开始的地方出发，走那么远，去找我的名字。

1:29

6:46

前两个质数的乘积

2:37

14:41

17:3

27:40

第一个质数

2:33

1:1

7:45

17:29

21:31

12:17

前两个质数的乘积

22:42

15:18

24:33

27:46

12:29

25:66

7:47

到此为止你已经分享了太多。我们只想要最好的，不想要跟风的人。因此，最初的几个人会获得奖品。

祝你好运。

3301

-----BEGIN PGP SIGNATURE-----

Version: GnuPG v1.4.11 (GNU/Linux)

iQIcBAEBAgAGBQJPB1luAAoJEBgfAeV6NQkP9oAP+gLu+FsRDf

3aRcJtBk-COU2MXr/dagOTvCKWtuV+fedy0enWUZ+CbUjXOr98
m9eq2z4iEGqKd3/MBXa+DM9f6YGUEjPum4wHtQDSJlZMazuYqJ
OVZGw5XmF25+9mRM6fe3H9RCiNDZpuXl3MzwdivYhcGB5hW14
PcdHHteQf3eAUz+p+s06RDs+q1sNGa/rMQIx9QRe71EJwLMMk
Mfs81kfJCtCt21+8ud0Xup4tjUBwul7QCcH9bqKG7cnR1XWsD
gdFP6a4x9Jl2/IUvp1cfeT7BYLS9W3lCM8thMemJr+ztQPZrp
DlaLIitAT2L0B3f/k4co89v5X2I/toY8Z3Cdvoihk0AdWzMy/
XLDgkPnpEef/aFmnls53mqqe9xKAUQPMrI73hiJ+5UZWuJdzCp
vt+FBjfQk15EJoUUW16K2+mBA1cSd+HJlnkslUTsjkq0E36XKC
hP+Cvbu/p6DLUMM2Xl+n3iospCkkHR9QDcHzE4Rxg9A435yHq
qJ/sL2MXG/CY8X4ec6U0/+UCIF9spuv8Y7w66D05pI2u9M/081
L7Br0i0Mpdf9fDblO/6GksskccaPkMQ3MRtsL+p9o6Dnbir6Z2
wH2Kw1Bf0Gfx4VcpHBikoWJ5blCc6tfvT+qXjVOZjWAL7DvRea
vSEmW1/fubNC3RWcjeI4QET2oKmV2NK
=LWeJ
-----END PGP SIGNATURE-----

如果有人曾经完整地解开以上这条线索，那么至少他并没有将答案公之于众。瑞典的一位安全专家约埃尔·埃里克松（Joel Eriksson）在网上发布了一种可能成立的部分解法。他在这种解法的开头写道：

> “在29卷中曾经含有知识”，这句话可能是指第11版《大英百科全书》。这套书正好有29卷。现在该书的内容已经向公众开放，任何人都可以随意下载这本书。此书最初发行的年份是1910—1911年。[33]

而另一条信息也和上述信息类似：

-----BEGIN PGP SIGNED MESSAGE-----

Hash: SHA1

一首关于逐渐消逝的死亡的诗歌，以一位国王的名字命名。

这首诗只能读一遍，然后就会消失不见。

唉，它不可能一直保持未被看到的状态。

1:5

152:24

前两个质数的乘积

14:13

7:36

12:10

7:16

24:3

271:22

10:7

13:28

12:7

86:17

93:14

前两个质数的乘积

16:7

96:4

19:13

47:2

71:22

75:9

77:4

到此为止你已经分享了太多。我们只想要最好的，

不想要跟风的人。因此，只有最初的几个人会获得

奖品。

祝你好运。

3301

-----BEGIN PGP SIGNATURE-----

Version: GnuPG v1.4.11 (GNU/Linux)

iQIcBAEBAgAGBQJPB/nmAAoJEBgfAeV6NQkPEnEQAKl5qtb3Z

E5vs+c08KuzAi4atQEE71fvb65KQcX+PP5nHKGoLd0sQrZJw1

c4VpMEgg9V27LSFQQ+3jSSyan7aIIgSDqhmuAcliKwf5ELvHM

3TQdyNb/OnL3R6UvavhfqdQwBXCDC9F0lwrPBu52MJqkAns93

Q3zxec7kTrwKE6Gs3TDzjlu39YklwqzYcUSEusVzD07OVzhIE

imsOVY+mW/CX87vgXSlkQ69uN1XAZYp2ps8zl4LxoaBl5aVtI

OA+T8ap439tTBToov19nOerusB6VHS192m5NotfQLnuVT4EITf

loTWYD6X7RfqspGt1ftb1q6Ub8Wt6qCIo6eqb9xmq2uVzbRWu0

5b0izAXkHuqkHWV3vwuSfK7cZQryYA7pUnakhlpCHo3sjIkh1

FPfDcxRjWfnou7TevkmDqkfSxwHwP5IKo3r5KB87c7i0/tOPu

QTqWRwCwcWOWMNOS7ivYKQkoEYNmqD2Yz3Esymjt46M3rAuaz

xk/gGYUmgHImgcu1zzK7Aq/IozXI7EFdNdu3EoRJ/UL9Y0l0/

PJOG5urdeeTyE0b8bwgfC2Nk/c8ebaTkFbOnzXdAvKHB03KEeU

PtM6d6DngL/LnUPFhmSW7K0REMKv62h9KyP/sw5QHTNh7Pz+C6

3OO3BsFw+ZBdXLhGqP6XptyZBsKvz2TLoX
=aXFt
-----END PGP SIGNATURE-----

以上两段信息中都出现了成对的数字，这些数字的形式与此前蝉3301给出的那段书本密码中的数字形式类似。事实上，这些数字确实也是书本密码。两段信息开头的文字部分为我们提供了一些线索，这些线索告诉我们应该去哪些书本中寻找这些密码的解法。

第2条信息中包含以下文字：

一首关于逐渐消逝的死亡的诗歌，以一位国王的名字命名
这首诗只能读一遍，然后就会消失不见。
唉，它不可能一直保持未被看到的状态。

这段文字指向一首威廉·吉布森（William Gibson）的诗。吉布森不仅发明了“网络空间 ”（cyberspace）一词，还创造了科幻小说门类下的“赛博朋克”（cyberpunk）子门类。

这首诗的题目是《阿格里帕（死者之书）》[*Agrippa (A Book of the Dead)*]，作者并不打算让这首诗长时间存在。吉布森将这首诗存储在一张3.5英寸软盘上，在使用一次之后，这张软盘就会自动加密。而这张3.5英寸软盘又嵌在一本书里，这本书事先用光敏化学物质处理过，因此书中的内容见光后会自动消退。

尽管如此，这首诗歌在2012年还可以在网络上找到。[34]用这本书的内容来翻译上述信息中的密码，就可以得到以下信息：

```
sq6wmgv2zcsrix6t.onion
```

我们生活中常见的网站一般以“.com”（商业网站）、“.gov”（政府网站）、“.edu”（教育网站）等后缀结尾，而大部分人并不了解网址还可以以“.onion”结尾。这类网站的运作方式和常见网站完全不同。普通的网络浏览者是没有办法访问这种网站的，为了访问这类网站，用户必须先安装一种特定软件，这种软件可以在以下网址找到：https://www.torproject.org/。有了这种软件，就可以做以下三件事：

1. 访问以“.onion”结尾的网站。
2. 匿名浏览互联网（同时隐藏地点和身份双重信息）。
3. 匿名建立网站（同时隐藏地点和身份双重信息）。

上述网址中的“Tor”是“The Onion Router”（洋葱路由器）的缩写。因为实现上述三个目标的程序包含许多“层”，就像一个洋葱一样。

当用户通过洋葱路由器连接到互联网时，他们并不是像普通用户一样通过一个代理服务器直接连到想要访问的网页，而是经过一系列随机选择的代理服务器最终连到他想要访问的网页。除了最后一个出口节点的连接以外，其他所有连接都经过加密处理。这种层层加密的方式就像一个有许多层的洋葱，它保证了任何一台给定的计算机都只知道与其直接连接的上游计算机和下游计算机的地址，这样就保证了用户的匿名性。

洋葱路由器系统中有数千台代理服务器，即使是同一个用户的数据，每次会话在该网络中经过的路径也是不同的。事实上，即使是在同一次会话中，系统也会每10分钟自动改变数据的路径。这些服务器是由一些志愿者运营的，他们认为，隐私和匿名是网络用户人权的一部分。

通过洋葱路由器，人们可以匿名在服务器上建立网站，提供“隐藏服务”。这类提供隐藏服务的网站从2004年开始就存在了，它们都属于暗网（Dark Web）的一部分，而暗网又属于深网（Deep Web，指搜索引擎搜不到的网站）的一部分。在洋葱路由器上，最著名的隐藏服务网站是“丝绸之路”（Silk Road），犯罪分子通过这个网站出售大量毒品。而另外一些用户则通过洋葱路由器系统中的网站发表反对本国政府的言论。这些洋葱路由器系统中的网站可以成为秘密发声的途径，持不同政见者可以通过这些网站秘密地联合起来。

但是，蝉3301所指示的洋葱路由器网站既不提供毒品，也不讨论被政府禁止的政治意见。这个网站只是向破译者提供了一些进一步的说明。奇怪的是，在这段信息中，一些单词出现了拼写错误，这是此前没有发生过的。

-----BEGIN PGP SIGNED MESSAGE-----

Hash: SHA1

祝贺你！

请通过某种公开、免费、基于网页的服务器创立一个新的电子邮箱地址，一个你从来没用过的邮箱地址，并在下面输入这个地址。为了保持匿名，我们建议你在完成上述任务的过程中继续使用Tor。

在接下来的几天中，我们会通过电子邮件向你发送一个数字（根据你们来到这个页面的先后顺序依次发送）。一旦你收到这个数字，请你重新回到这个网站，并在这个网址后面加上一条斜线和你收到的数字。（比如，如果你收到的数字是“3294894230934209”，那么你就应该访问以下网址：“http://sq6wmgv2zcsrix6t.onion/3894894230934209”。）

3301

```
-----BEGIN PGP SIGNATURE-----
Version: GnuPG v1.4.11(GNU/Linux)
iQIcBAEBAgAGBQJPCikYAAoJEBgfAeV6NQkPtCkQAIEa2HC20S
duG8zWtH/0LSJ9dnQLfjTI/MAPsnp0KXpzcQJ4p2UbNdRb5qtq
p3HZQ5qwKK3b/MT+2eB5X+h5q5v/bVOsPKXN7W+krTi8a2v5KA
1fRj8mKcFTXkx9hdq6z6qeePVpeOGXeitU0jgO4uXu1N9hor4A
QFQoXvuELnlqS4YX+/nNgK+Y1Vu1ekvyn9LusidMpSxS/ciiZL
D5d3TlTJeUOQzdw9TlUmmdeigfBRoqZDnm5UESnS8qS1vd7zg9
Ulpw2GfCV8QQl0zWp0lSKCrArG0Qg4K4shZRQ2BopGM95oFVw2
pOnzYr9pp8kpipBWQ6URRF7IEcuG1k5fBtsfiz7Eb7WWr4Cwk/1
Ri625FMyo0Z2KDc1Wo8nkFXZ0/Xle6tyyhhJUZGRbZuHLauLR5
+oQEKKYTRrJFNhqswcXYsCQBBYM2XS+v3nbmTDVKyGNmMpp3/
yC/RR91hQ7frFqMd5qHKSD0pc5tU5G29J+Ez3CDVN7E60Xy3I1
I5ugGE8GP5DmqHn4k/x272lroXBtiIRTh4f+ghcC6DLdPBclPR
AfLtRbpVR3GVTSCJ/EwBZrH4QJTEGcuXfPQryHbF0BO795yyXx
M6lQ9QnU4dkAAQepKLHgcvDlrEJYJzMWo+BOCu7MChoOyo1Zq4
eeuIcPx04KJd4sGtMVWolA3/
=zGQW
-----END PGP SIGNATURE-----
```

从这里开始，蝉3301密码挑战赛中的领先选手和比赛组织者之间的交流方式开始变得更加私密了，我们也就没有办法继续清楚地了解接下来的谜题和线索了。关于接下来的谜题和线索，存在各种谣言和传说。最后，Reddit网站上贴出了如图9–15所示的信息。

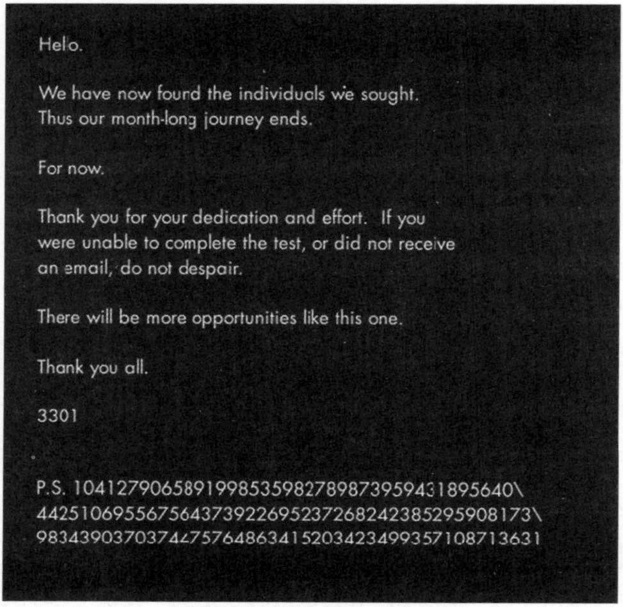
Hello.
We have now found the individuals we sought.
Thus our month-long journey ends.
For now.
Thank you for your dedication and effort. If you
were unable to complete the test, or did not receive
an email, do not despair.
There will be more opportunities like this one.
Thank you all.
3301
P.S. 1041279065891998535982789873959431895640\
442510695567564373922695237268242385295908173\
9834390370374275764863415203423499357108713631

图 9-15　2012 年的蝉 3301 密码挑战宣告结束。图中文字意为：大家好，现在我们已经找到了我们想找的人，因此我们这次长达一个月的旅途也就宣告结束了，至少目前暂时结束了。感谢你的贡献和努力。如果你没能完成这项测试，或者没有收到我们的电子邮件，请不要失望。以后还会有更多这样的机会。感谢你们所有的人。3301（又及：1041279065891998535982789873959431895640\442510695675643739269523726824238529590 8173\983439037037427576486341520342349935710871 3631）

要想完整地讲述蝉3301密码挑战的全部故事，可能必须单独再写一本书才行。为了节省篇幅，在这里我就不向读者介绍2013年和2014年的蝉3301密码挑战的细节了。不过别忘了：谜题总是一道比一道更难！

2015年没有出现蝉3301的挑战密码，但该组织在2015年发布了一条消息，否认自己与黑客攻击“计划生育”（Planned Parenthood）网站的活动有关[1]。攻击上述网站的黑客组织也自称为“3301”。2016年1月，蝉3301又发布了一条新信息，但并未再次组织新的密码挑战赛。也许，2017年我们又会迎来新一届的蝉3301密码挑战赛，也有可能不会[2]。不过，除了蝉3301密码挑战以外，在互联网上还有另一项十分神秘的密码挑战。截至笔者撰写本书时，这种密码挑战已经举办了两次，而且活动背后的组织者称，他们还会再次举办新的密码挑战赛。

圣殿骑士团密码

这次密码挑战始于2010年的某个时候，当时互联网上出现了一个奇怪的网站，其地址是www.pccts.com。现在这个网站已经不存在了，上述域名正在公开出售，也无法找到任何与密码谜题相关的痕迹。但是，我们仍然可以看到这个网站2010年10月20日的存档版本，这个存档版本保存在以下网址中：https://web.archive.org/web/20101020055419/http://pccts.com/。如果你在2010年10月20日访问上述网站，就会看到以下图像和文字，我将这些图像和文字复制在图9–16中。

图中文字是拉丁文，意思为“基督和所罗门圣殿的贫苦骑士团”（简称PCCTS），这是圣殿骑士团的全称。而这段文字下方标出的罗马数字是

① “计划生育”组织是美国一家为妇女提供各种生育健康服务的非营利性组织。因为该组织的诊所提供堕胎等服务，所以经常受到反堕胎人士以及其他保守派人士的抗议甚至攻击。——译者注

② 2017年，蝉3301密码挑战赛并未举办。——编者注

图 9-16　来自圣殿骑士团的官方网站?

该组织成立的年份——1119年。如果访问者点击这张图片上的盾牌图案（包括中间的图案中骑士手中拿着的那个小盾牌），就会被链接到新的网页。新网页上会有一些加密信息、隐藏文字，以及隐藏的链接。比如，其中一个网页上出现了“itanimullI”的字样，将这个单词中的字母从右向左反过来读，就会得到“Illuminati”（意为“光明”）一词。

在上述网页可以链接到的网页之一上（这个网页现在被保存在以下网址中：https://web.archive.org/web/20100809120304/http://www.illuminatiorder. info/illuminati/babilu-d.html）画着4个狮身人面像，它们都看着屏幕中央的一块区域，这块区域和网页背景一样，都是纯黑的。然而，屏幕中央的这个纯黑区域是可以被点亮的。点亮这个区域以后，就可以看到以下德语文字：

Am Osten der Welt
steht der Mitternachtsberg.
Ewiglich wirkt sein Licht.
Des Menschen Auge kann ihn nicht sehen,
und doch ist er da.

Ueber dem Mitternachtsberg strahlt
Die schwarze Sonne.

Des Menschen Auge kann sie nicht sehen,
Und doch ist sie da.

Im Inneren leuchtet ihr Licht.
Einsam sind die Tapfren und die Gerechten—
Doch mit ihnen ist die Gottheit

—Inschrift aus Babilu

中文翻译为[①]：
在世界的东方伫立着
黑暗的午夜之山。
他永恒地发出光芒。
人类的眼睛看不到他，
但这位巨人仍在那里。

在午夜之山上面
是黑色的太阳在照耀。
人类的眼睛看不到她，
但她仍在那里。

她的光芒照亮了内部。
勇敢和公正的人与世隔绝，
但神仍与他们同在。

——巴比鲁的铭文

① 根据英文翻译转译。——编者注

在这里我不打算向读者进一步详细介绍这次密码挑战中的各种网页链接和谜题，读者可以自行对这些存档网页进行探索。这些网页和谜题究竟是否真的与某种秘密社团有关？我认为答案并不重要，探索和解谜本身已经给我们带来了许多乐趣。

然而，对于运营这个网站的组织而言，事情后来变得非常严重。2011年7月25日，这个组织贴出了以下信息：

免责声明（2011年7月25日）：我们与那个挪威白痴绝无任何联系！
这个网站建立于2010年，网站的初衷是为了向世界展示制造一个阴谋有多么容易！ CARIPS.COM也是这个恶作剧的一部分!!
这个项目是QFF——QUO FATA FERUNT①的一部分，HTTP://WWW.QUOFATAFERUNT.COM是一个阴谋/另类媒体网站或论坛。
那个傻瓜给受害者带来了巨大的痛苦和折磨。
我们的心与所有的受害者，
以及受害者的家庭成员、亲朋好友同在。

上述声明中提到的“挪威白痴”是指安德斯·贝林·布雷维克（Anders Behring Breivik）。布雷维克是一名“右翼极端基督教分子”，他“对穆斯林怀有仇恨”。[35] 2011年7月22日，布雷维克在挪威首相延斯·斯托尔滕贝格（Jens Stoltenberg）的办公室所在的大楼外引爆化肥炸弹。首相并未受伤，但有8人被炸死。引爆炸弹以后，布雷维克又搭乘轮渡登上了于特于亚岛。在岛上，布雷维克身着警察制服，对挪威工党青年营活动的参加者开枪扫射，他在岛上共射杀了69人，后来才被挪威精英警察行动组逮捕。[36]

① “QUO FATA FERUNT”是拉丁文，意为“无论命运将我们载向何方”。——译者注

布雷维克自称是圣殿骑士团的成员。

就在这次恐怖事件的同一天，网上出现了一本书，包含超过1 500页充满仇恨的内容。布雷维克的律师称，布雷维克花了数年的时间撰写这本书。[37]以下是摘录的内容：

> 欧洲军令和刑事法庭（圣殿骑士团）判定，所有欧洲A类、B类和C类叛徒犯下了一系列针对欧洲人的罪行（罪名1–8）。所有欧洲A类和B类叛徒被判处死刑。刑罚将由首席政法官骑士执行。

也许读者已经猜到，他在这本书中除了宣扬仇恨以外，还写了一些奇怪的文字：

> 虽然圣殿骑士团是一个宣传泛欧本土民族权利的组织，但是我们向所有欧洲人（不管他们是什么肤色）提供成为首席政法官骑士的机会。只要你是基督教徒、基督教不可知论者，或者基督教无神论者，你就符合我们的条件。
>
> 有一点我们必须说清楚：如果你不愿意为了组织的目标牺牲自己，那么圣殿骑士团并不适合你。
>
> 在这种情况下，未来圣殿骑士团将考虑与欧盟/美国霸权集团的敌人合作，比如伊朗（我们不太可能和朝鲜合作）、基地组织、索马里青年党，以及其他虔诚的伊斯兰人民。合作目标是在西欧各国首都以及其他重要地点使用小型核武器、放射性武器、生物武器，或者化学武器。首席政法官骑士和其他欧洲基督徒烈士可以躲过通常针对阿拉伯裔进行的筛查，因此我们可以保证在我们选择的地点成功使用并引爆上述武器。

> 但是，我们在任何情况下都不接受使用地表爆炸当量超过200吨的核武器，因为这类武器会造成过多伤亡。然而，当量小于等于200吨的核武器是歼灭A类和B类叛徒的理想武器（适用于叛徒总部）。上述做法可以摧毁叛徒密集度较高的一至两个城市街区/总部，因此非常适合用来达到我们的目的。
>
> 作为欧洲圣殿骑士团的首席政法官骑士，我们反对纳粹。在我们的政策中，有大约40%的部分与纳粹党的政策不同。

在这本书中，布雷维克还描述了他要设置的一个网站，但是他对这个网站的细节描述与我们之前列出的密码网站并不相符。

2011年8月5日，上文中提到的这个网站正式被关停。

升级版密码

2013年年末，一个十分类似的网页出现在了一个新地址中，这个新地址是：http://jahbulonian.byethost7.com/ 。

这个组织宣布，他们“回来”了。读者可以在图9–17中看到这个网站的截屏。和之前的网站一样，该网站列出的组织建立年份也是1119年。在这个年份之后还出现了“KT PCCTS”字样，因此我们可以判断，这个网站也与圣殿骑士团有关。

网站上的“No-Lub-Haj”看起来像是某种外国语言文字，但是，其实这个词只是把“Jahbulon”一词反过来写，并且加上连字符而已

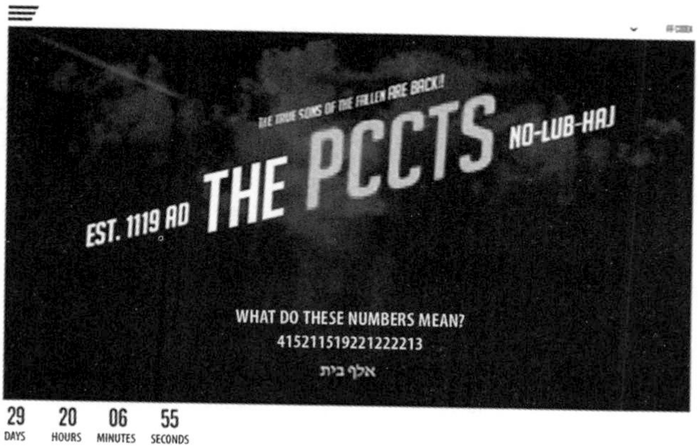

图 9–17　宣布 PCCTS 回归的网站截屏

（“Jahbulon”一词出现在该网站的网址中）。一般来说，希伯来语是要从右向左读的，因此该网页底部的文字可能是一条帮助读者解读“No-Lub-Haj”的线索。希伯来语字符串“אלףבית”连在一起表示“字母表”的意思。至于希伯来字母上面的这串数字，我们下面很快就会讨论到。

图 9–18　该网站更新后的截屏

上述网站经历了几次更新和修改。该网站的下一个版本如图9–18所示。图中间的小字为："你看到的内容是正确的，PCCTS又回来了。大部分人最后一次听到我们的消息应该是在2011年左右。网上有各种各样试图揭穿我们身份的视频。然而，从来没有人真正发现我们的秘密，因此我希望你们所有人都能再尝试一次。"

根据这段文字我们似乎可以推断，设立这个网站的组织与2010—2011年期间恶作剧的组织是同一个。如果这种判断正确的话，似乎这个组织并没有吸取先前的教训，这实在是一件令人惊讶的事。更糟糕的是，这个新网站上还有一些反对伊斯兰的言论。新网站中提到了"2011年"，如果这个新网站的建立者是另一个组织或者个人的话，这至少说明他对旧网站比较熟悉。此外，新网站上使用的图标与旧网站完全一样，因此新网站的建立者肯定看过旧网站。到底为什么会有人在挪威发生恐怖袭击事件以后再次设立这样一个网站呢？为了回答这个问题，接下来我们首先要对这个网站上的谜题和密码做一些研究，之后，我们再来更深入地研究2013—2014年期间的网站背后的组织和谜团。

在图9–18底部的小字部分中，网站的建立者邀请浏览网站的人接受"破译密码"的挑战。待破译密码应该就是415211519221222213，这串数字在图9–17和图9–18中都出现过。事实上，这是一条十分简单的密码。我们可以把它分割为以下形式：4 15 21 15 19 22 1 22 22 13。接着，我们将每个数转换成字母表中相应位置的字母，就得到以下结果：

```
D O U O S V A V V M
```

然而这串字母的意义并不明确。如果你能够破译出这几个字母的意思，你就解开了一个更大的谜题！

这个更大的谜题是由托马斯·安森（Thomas Anson，1695—1773）创造的，他生前曾是全英格兰最富有的人之一。他翻修了自己家族的庄园——位于斯塔福德郡的沙格伯勒庄园。在翻修的过程中，安森在庄园中修建或者置入了许多纪念碑和艺术品。其中，有一座被称为“牧羊人纪念碑”，其外观如图 9–19 所示。

图 9-19 牧羊人纪念碑

在牧羊人纪念碑上，有一段难解的文字。人们认为，这是一段加密过的题词（参见图 9–20）。但是，这段题词究竟是为了谁而题的呢？

图 9-20 沙格伯勒铭文

戴夫·拉姆斯登（Dave Ramsden）提出了这段加密题词的一种解法，虽然没有办法证明这就是正确的解法，但它看上去还是比较合理的。任何对这段加密题词有兴趣的读者都应该去读一读拉姆斯登的书。[38]

当然，也有另外一些研究者提出了和拉姆斯登不同的解法。那么，创建新网站的人究竟认为沙格伯勒铭文代表着什么意思呢?

后来，新网站开始放映一段电视剧《宋飞正传》(*Seinfeld*）的片段。在这个片段中，剧中的角色乔治·科斯坦萨（George Costanza）做了一切与他内心倾向截然相反的事，这令他非常激动。视频的音轨上加入了吓人的音乐。视频播完后，画面变成了黑色背景，上面有两个白色的单词“Satenetas Rotenetor”。这两个单词是所谓的“回文字”，也就是正着读（从左向右）或倒着读（从右向左）都一样的单词。

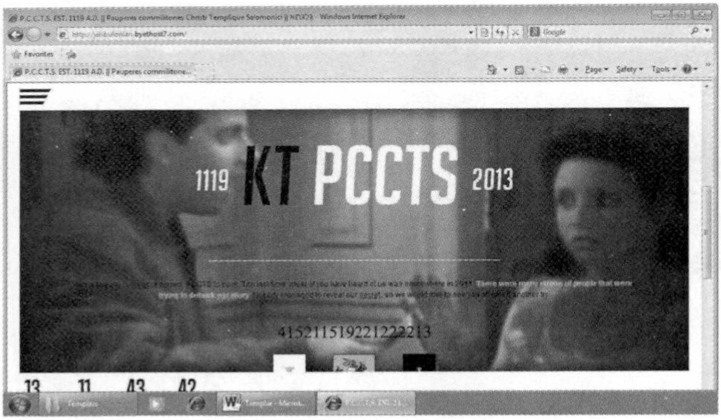

图 9-21　网站上播放的《宋飞正传》的片段

这段视频上交替出现了4条不同的文字信息。它们分别是：

1. 17 是来自东方和西方的人……

   数字 7 和数字 10 的魔力

2. D17

   石头可以滚动……

   他们在罗尔德藏了什么?

3. 约整数 2

只有 11 个人知道我们的事

在线条和钩子之间的 G

4. 危机?

什么危机 ??!!

VAT TOG SUNK

读者可以尽情享受研究上述线索的乐趣。而我只想提示一点：在第 3 条信息中提到“在线条和钩子之间的 G”，这句话可能是指共济会的符号。我在图 9–22 中为读者画出了共济会的符号。

图 9–22　共济会的方尺和圆规标志

总的来说，这个网站上贴出的密码挑战更像是某种与共济会有关的老式谜题，而不像是一位计算机安全专家的作品。比如，只要在页面上移动光标、点击特定的点（没有特殊提示），就会转向各种“隐藏链接”。虽然在这个页面上找全所有的隐藏链接可能会很有趣，但是事实上通过一种简单的方式就可以一次性地找到所有链接。访问该网站的用户只需要在网页上的任何一处右击鼠标，然后选择“查看源代码”，就可以看到所有隐藏链接的地址（以及其他一些信息）。于是，一个问题产生了：这个网站的设计者难道不知道这个捷径吗？显然，蝉3301密码背后的设计者呈现给挑战者的任务比这难多了！

上述网页连向另一个网站，网址是：http://jahbulonian.byethost7.com/666partsofhell.html。新网站的页面是纯黑的，但纯黑色的背景上其实有一张黑色图片是可以被复制粘贴的。这个黑色图片的名称是："lookhere evenifitlooksdark"（看这里，就算这里是黑的），见图9–23。

图9–23　一个黑方框

只要我们复制这个黑色图案，并将它粘贴到微软图片管理器中，然后再点击“自动修正”，我们就会得到如图9–24所示的图案。

图9–24　隐藏的图案显现出来

完成上述任务根本不需要用到OutGuess这样的软件，从这方面看，这个任务的难度也比不上蝉3301密码挑战。

图片上这行符文可以被破译为“YOU ARE CLOSER”（你离答案更近了）。

后来，这个网站又进行过更新，《宋飞正传》的视频片段最终被换成了另一段视频。这段视频中称：“我们拥有长矛。”这句话中的“长矛”似乎是指罗马士兵卡斯卡·隆吉努斯（Casca Longinus）用来刺戳耶稣侧腹的那支传说中的“命运之矛”。[①]

如果我们在这个新页面上点击“查看源代码”，就会看到以下内容：

```
<meta name="keywords" content="knights templar,
pccts, templars, templar, templer, templars,
templar knights, crusades, p.c.c.t.s., 1119,
carips, saint martin, anti islam, anti=islamic,
pro christ, christus, christ, war, last crusade,
322 ">
```

以上内容中，我尤其注意到了“anti islam”（反对伊斯兰）的字样。为什么“反对伊斯兰”会出现在这里？难道这个网站的建立者还没从挪威的恐怖袭击事件中学到教训吗？这一点令我更想要找出这个网站背后的组织或者个人的身份。

① 根据基督教的传说：耶稣在十字架上受刑以后，罗马士兵卡斯卡·隆吉努斯为了确认耶稣已经死亡，用“命运之矛”刺戳了耶稣的侧腹。这支长矛因为沾上了耶稣的鲜血而被一些信徒视为圣物。——译者注

此外，由于这个网站声称它是2011年的PCCTS网站的回归版本，所以我不禁好奇这个新网址在2011年的时候究竟有些什么内容。在时光机网站https://archive.org/web/上，我找到了上述问题的答案。

在2011年7月19日、2011年8月20日，以及2011年9月19日（现在只能查到这个网站在以上3个日期的存档），http://jahbulon.com网页中包含一篇讨论股票市场投资问题的文章。此外，这个网页当时还包含以下这段文字[39]：

那么“加布隆”（Jahbulon）一词到底是什么意思？

据说，这个词是共济会秘密语言中对上帝的称呼。最近，这个词受到了许多非共济会成员的批评，因为据说这个词来自三个神的名字。“加布隆”一词本来应该被分成以下几个音节：Jah、hy、Baal、leb、On、Na，据说这些音节代表耶和华（Yahweh）、巴力（Baal）、俄塞里斯（Osiris）这三个神。因为巴力和俄塞里斯在犹太教–基督教的共济会语境下属于异教之神，所以某些评论者才会有异议。然而，把“On”这个音节和俄塞里斯联系起来也太牵强了。俄塞里斯的埃及名字是阿萨尔（Asar），要把“阿萨尔”这个词音译成“On”看起来也没有什么道理。就算是对翻译问题一向很随便的希腊人，也将“阿萨尔”翻译成了发音还算接近的俄塞里斯。

最合理的解释是：音节“On”是指圣经中的“On”，即赫利奥波利斯城（Heliopolis），这个词来自埃及语中的An或者On。“Baal”不一定是指巴力神，也可以指“主”或者“王”。根据这种新的解读方式，我们就会得到一个这样的短语：“耶和华是赫利奥波利斯之主（Yahweh is Lord of Heliopolis）”。当然，对于敏感的犹太教–基督教极端主义者而言，这种描述恐怕也同样不可接受。因此共济会总会所对这个话题保持了沉默，所以这种情况是完全可以预见的。

接着，我又查了该网站2012年的存档。2012年时，这个页面上已经查不到上述这段文字了，只剩下与股票有关的内容。然而，在2012年12月10日的存档中有一个链接，在这个链接指向的网页中，不仅有上述这段解释“加布隆”意义的文字，还有一段解释主站的任务和目标的文字[40]：

> **任务**：本站的任务很简单，就是靠激进地使用基本面和技术分析来将我的收益最大化。我通过以下方式完成上述任务：每天找出目前市场上的趋势和基本面信息，然后以适当的方式做出回应，既要提高我的收益，又要确保资本不流失。这个网站是我的交易日记。
>
> **目标**：在15年内，或者在2025年1月之前赚到500万美元。我建立这个网站并不是为了指导你如何操作自己的账户。我的目标是要赚到500万美元，请不要照抄我的做法！在这个网站上，我不会给出任何投资建议。这是我的个人交易日记，是我的学习资料。

此外，当时这个网站还包含以下内容：

> **隐私政策：**我对我的个人隐私非常在意，我认为你也非常关注你的隐私。我**绝对**不会分享或者出售该网站访问者的名字或者电子邮件信息，**句号**。

许多黑客会把此人“非常在意”的个人隐私当作一项挑战。

在这个网站2013年3月2日的版本中，能够看到一个电子邮箱地址：HolyGrailResearchLab@yahoo.com。这个邮箱地址在互联网的其他地方也出现过几次。如果对这一线索继续深挖，就有可能找出网站背后的个人身份。

在笔者写作此书时，http://jahbulonian.byethost7.com/网页上只有很少的内容了。这个页面上只有一个标示为“No Lub Haj!”的链接，点击这个链接就会转到这个网站：http://www.jahbulon.com/。这个网站上有一幅图片，上面画着一个侧翻的滑板。在这幅图片的下面有一行文字：“我们很快就会上线”（we will be online soon）。我查看了这个网页的源代码，其中没有任何隐藏的链接。源代码显示，网站上这张图片的大小是高375像素，宽666像素。

2011年的旧网站承认，他们只是在搞一个恶作剧。我认为，我们有理由相信这个新的网站也只是一个恶作剧。这个新网站给我们提出的唯一挑战就是找出网站创立人的身份——我们随时可以去寻找这个问题的答案。

我认为，接下来我要介绍的这个挑战密码也很可能是一个恶作剧，但是，我还是把做最终判断的权力交给本书读者。

比尔密码——没有解法的挑战密码

在介绍比尔密码（Beale ciphers）之前，让我们先来考虑一下我们度量各种商品所使用的单位。我们以加仑为单位购买汽油（但有些国家的人以升为单位购买汽油），以箱为单位购买啤酒（但有些国家的人以桶为单位购买啤酒），以袋为单位购买覆盖地面用的园艺材料，以磅为单位购买虾。但是，如果我要购买的商品是黄金，我们就必须使用一种比上述单位小得多的度量单位。黄金的度量单位是“金衡盎司”，一个“金衡盎司”[①]等于1/12磅（约0.04克）。我们肯定不能以吨为单位来度量黄金——除非我们谈论的是诺克斯堡[②]或者比尔密码！据说在比尔密码中，藏有一个关于宝藏埋藏地点的密码，这里埋藏着超过1吨的黄金（准确地说是2 921磅）和超过2吨的白银（约5 100磅）。此外，这个地点还埋藏着大量其他珠宝。我们很快就会讨论到这些具体细节。

比尔密码的故事始于1885年出版的一本23页的小册子，其出版人是詹姆斯·沃德（James Ward）。这本小册子的作者的身份一直未被公开，但是这出好戏中其他几位“演员”的名字却是公开的。其中一位是罗伯特·莫里斯（Robert Morriss，1778—1863），他在弗吉尼亚州的林奇堡经营一家旅馆。根据这本小册子中的说法，莫里斯于1862年首次将宝藏的故事告诉了这位姓名不详的作者。在这本小册子中，莫里斯从自己的角度讲述了这批财宝的故事。作者声称，他让莫里斯写下了这段故事，而小册子中引号内的内容只是将莫里斯的原话复制了一遍而已。我们很快就会看到，这个细节非常重要。

莫里斯讲述的故事大致包括以下内容：1820年1月，一位名叫托马斯·J. 比尔（Thomas J. Beale）的人首次来到了莫里斯经营的旅馆。比尔在同年3月末离开了旅馆，但是，1822年1月，比尔又再次来到了这家旅馆。这次比尔仍是在3月份离开，但是在离开前，比尔交给莫里斯一个盒子，并给莫里斯讲了一段故事。根据莫里斯的描述，比尔声称这个盒子中装有一些价值很高、非常重要的文件，他希望由莫里斯来保存这些文件，直到他来索回。之后，莫里斯收到了一封比尔寄给他的信，这封信来自圣路易斯，信上标记的日期是1822年5月9日。从此以后，莫里斯再也没有

---

① 1金衡盎司约为31.3克。——编者注

② 诺克斯堡是美国国库黄金的存储地点。——译者注

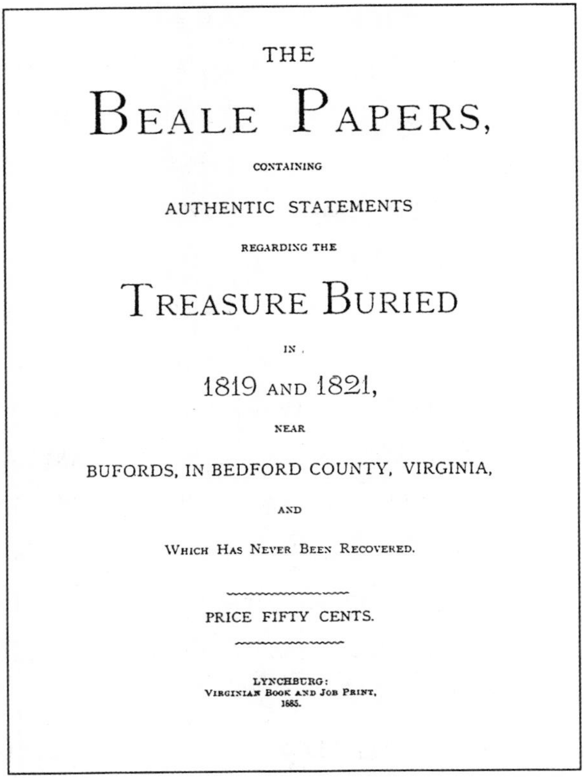
THE

BEALE PAPERS,

CONTAINING

AUTHENTIC STATEMENTS

REGARDING THE

TREASURE BURIED

IN

1819 AND 1821,

NEAR

BUFORDS, IN BEDFORD COUNTY, VIRGINIA,

AND

WHICH HAS NEVER BEEN RECOVERED.

PRICE FIFTY CENTS.

LYNCHBURG:
VIRGINIAN BOOK AND JOB PRINT,
1885.

图 9-25 一个谜题的诞生？小册子的标题为：比尔文件，包含关于 1819—1821 年间埋在地下的宝藏的真实声明，宝藏的位置靠近弗吉尼亚州贝德福德县的比福德酒馆，从未被发现

收到来自比尔的消息——当然，在收到这封信时，莫里斯并不知道比尔会从此杳无音讯。莫里斯把这封信和比尔交给他的盒子放在了一起。几十年之后，这封信的内容被全文收录进了沃德出版的这本小册子中。由于这封信很长，下面我只为读者摘抄其中比较重要的部分：

关于我交给你保管的那个盒子，我有一些话要说。如果你允许的

话，我还想给你一些关于这个盒子的指示。这个盒子中包含的文件不仅会对我个人的命运产生极大的影响，也会对和我一起做生意的人的命运产生极大的影响。如果我死了的话，盒子丢失所产生的损失将是不可挽回的。因此，为了防止这种巨大的灾难，你必须谨慎小心地保管这个盒子。这个盒子中还包含一些写给你的信，可以帮助你了解我们经营的生意。如果没有人回来找你，请你小心地将这个盒子保管10年（从这封信的日期开始算起）。如果10年后，我或者任何经过我许可的人没有向你索回这个盒子，你就可以打开锁，开启这个盒子。到时你会发现，在这个盒子中除了给你的文件以外，还有另一些加密文件。如果没有密钥的话，是无法读懂那些加密文件的。我已经把密钥交给了此处的一位朋友，这个密钥被我放在一个密封的信封里，信封上写着你的姓名和地址。我在信封上注明，到了1832年6月才能寄出这封信。通过这种方式，你就会理解我要求你做的所有事情了。

如果莫里斯完全按照比尔的上述指示行动的话，他应该于1832年打开这个盒子。然而由于某种原因，事实上莫里斯直到1845年才打开这个盒子。为了打开这个盒子，莫里斯不得不撬开盒子上的锁。但是，莫里斯一直没有把这件事告诉过任何人，直到1862年他最终决定将这个秘密告诉这本小册子的匿名作者。

在听到这个故事以后，这位匿名作者又等了几年，直到1885年，他才把这个秘密向全世界公布。在这本小册子中，作者详细介绍了莫里斯保管的盒子中的内容。盒子中的第一份文件是一封比尔写给莫里斯的信，日期在比尔最后一次离开莫里斯旅馆前的几个月。在这本小册子中，作者全文复制了比尔写给莫里斯的信，但是由于这封信很长，我在这里只向读者提供一个概要，只有信中特别重要的部分我才会做摘抄引用。

在这封信中，比尔向莫里斯描述了他如何组织起一个30人的队伍，去西部大平原猎杀水牛、灰熊，以及其他野生动物。1817年4月，这支队伍离开了“老弗吉尼亚”，出发前往密苏里州的圣路易斯。在圣路易斯市买齐他们需要的所有东西以后，这支队伍于5月19日再次动身，出发前往新墨西哥州的圣塔菲。他们当时的计划是于当年秋季到达圣塔菲。由于此行可能会遇上一些潜在的危险，他们的向导建议建立一个军事化组织，推选出一名领袖，所有其他成员都必须服从领袖的命令。比尔被选为这个组织的领袖。最终，这支队伍在当年的12月1日左右到达了圣塔菲。

第二年3月初，这支队伍的部分成员离开了大部队。这本来应该只是一次为期几天的短途任务，然而这几位队员却没有按时回来。直到一个月或者更长时间以后，上述小分队的两名成员才终于回来，并向一直在为他们担心的其他队员解释了他们长时间不归队的原因。根据这两名队员的说法，队伍向北行进了若干天，找到了大量猎物，但是当他们准备归队的时候，却碰到了一大群水牛。于是，小分队决定跟着这群水牛走。在两个星期或者更长的时间中，他们捕猎到多头水牛。某一天，这支小分队扎营在距离圣塔菲北侧大约250~300英里处的一个山涧里，其中一名队员在那里发现了黄金。这两名队员是回来报告发现黄金的消息。

于是，比尔在这两名队员的带领下来到了这个山涧，其他队员还在那里继续挖掘黄金。队员们约定，大家对这批宝藏享有同等的权利，不管他们最后能挖出多少东西，都会平均分配。接下来，挖掘宝藏的工作持续了18个月或者更长时间。在这段时间中，他们挖出了大量黄金和白银。1819年夏天，队员们决定把他们挖出的宝物运到一个更加安全的地点。他们认为，在这样一个危险的区域中持有如此巨额的财富可能会给他们带来生命危险。最终，他们决定由比尔带领，将这批宝物运往弗吉尼亚州，并埋在贝德福德县比福德酒馆附近的一个洞穴里。

然而，比尔和长途跋涉来到弗吉尼亚州的其他10名队员发现，原定用于埋藏宝藏的洞穴不够安全，因为当地农民经常会在这个洞穴里储存土豆和其他蔬菜。因此，这队人马在贝德福德县找了另外一个地方，并将宝物藏在了那里。在贝德福德县，比尔还需要代表队里的其他人完成另一项任务，那就是把埋宝藏的地点交给一个可靠的人。万一全队人马在继续挖掘黄金的过程中不幸遇难，他们就需要依靠这位代理人了。如果所有人都遇难了，这名代理人应该负责把藏在贝德福德县的黄金和白银重新挖掘出来，分给各位队员的继承人们。

比尔在信中说，他正是在这个时候选中莫里斯作为代理人的。当然，当时的莫里斯完全不知道比尔交给他的任务究竟是什么。离开莫里斯的旅馆以后，比尔再次回到了宝藏的发现地点，之后他将另一批黄金运到了弗吉尼亚州，和第一批宝物放在一起。在藏好第二批宝物以后，比尔将上文中提到的那个盒子交给了莫里斯，但完全没有向莫里斯透露盒子里装着什么。

接下来，在这封比尔写给莫里斯的信中出现了一行非常重要的内容。因为这句话对我们的分析至关重要，我将这句话摘抄如下。比尔在这封信中写道：

> 然而，在圣路易斯，我却非常想要写信给你，告诉你这个盒子的重要性。

接下来，比尔又写道：

> 这里附上的文件在没有密钥的情况下是无法读懂的。我会将密钥送到你处。这些文件只是为了告诉你以下信息：我们埋藏的东西的具体内容、这些东西的具体地点、我们队员的名单，以及他们的住址等。

读者可以回忆一下，莫里斯此前还收到另外一封来自比尔的信，在那封信中，比尔写了这么一段话："我已经把密钥交给了此处的一位朋友，这个密钥被我放在一个密封的信封里，信封上写着你的姓名和地址。我在信封上注明，到了1832年6月才能寄出这封信。通过这种方式，你就会理解我要求你做的所有事情了。"

由于比尔的这封信寄自圣路易斯，所以信中所说的"此处"应该就是指圣路易斯。看起来，比尔计划在10年之后让一位朋友把密钥寄给莫里斯。这个计划看起来足够安全，因为要求一个人在某个特定的时间帮助你寄出一封信并不会引起此人的特别怀疑。比尔说不定向这位负责寄信的朋友编造了一个理由，以解释委托他寄信的原因以及信件的内容。这位负责寄信的人有什么理由要打开比尔托付给他的这封信吗？他又不可能预见到这封信中藏着不同寻常的内容。但是，不管怎么样，如果比尔的计划确实是这样，那么这个计划并没能成功地实施，因为莫里斯从来没有收到这封包含密钥的信件。

在写给莫里斯的信中，比尔解释说，如果过了10年（事实上这段等待的时间比10年长了许多，因为莫里斯并没有完全按照比尔的指示行动）他还没有回来的话，他和他的队员一定已经全部死亡了。比尔要求莫里斯，如果发生这种情况就要从埋藏宝藏的地点取出这批金银，并将其平均分成31份。其中的30份应该分发给队员们的继承人，密码中列出了这些人的名单；而最后一份则是为了感谢莫里斯的帮助而留给莫里斯本人的。在这封信的最后一段，比尔又把分配这批宝藏的方式重述了一遍。我将这封信的最后一段复制如下，读者很快就会知道我为什么要摘抄这段内容。

> 在其中一份加密文件中，你可以找到所有队员的名字，这些人有权均分这批宝藏。在队员名字的另一边，是他们的亲人或者继承人的

名字以及住址，每位队员都把他们的那份宝藏留给了他们选定的继承人。根据这份名单，你就能执行所有人的遗嘱，将每个人应该享有的宝藏交给他们选定的继承人。我已向你提供了这些继承人的住址，因此这项任务并不会太困难，这些人很容易找到。

比尔在写给莫里斯的信中附上了几页写满数字的纸——待破译的密码。但是，由于没有密钥，莫里斯完全无法读懂纸上的内容。最终，莫里斯决定将这个秘密告诉这本小册子的匿名作者，而这位作者则通过这本小册子将这个秘密告诉了整个世界。在发行这本小册子的时候，这位匿名作者已经读出了“2号文件”的内容，也就是说他已经解开了第2段密码。第2段密码的内容如下：

115, 73, 24, 807, 37, 52, 49, 17, 31, 62, 647, 22, 7, 15, 140, 47, 29, 107, 79, 84, 56, 239, 10, 26, 811, 5, 196, 308, 85, 52, 160, 136, 59, 211, 36, 9, 46, 316, 554, 122, 106, 95, 53, 58, 2, 42, 7, 35, 122, 53, 31, 82, 77, 250, 196, 56, 96, 118, 71, 140, 287, 28, 353, 37, 1005, 65, 147, 807, 24, 3, 8, 12, 47, 43, 59, 807, 45, 316, 101, 41, 78, 154, 1005, 122, 138, 191, 16, 77, 49, 102, 57, 72, 34, 73, 85, 35, 371, 59, 196, 81, 92, 191, 106, 273, 60, 394, 620, 270, 220, 106, 388, 287, 63, 3, 6, 191, 122, 43, 234, 400, 106, 290, 314, 47, 48, 81, 96, 26, 115, 92, 158, 191, 110, 77, 85, 197, 46, 10, 113, 140, 353, 48, 120, 106, 2, 607, 61,

420, 811, 29, 125, 14, 20, 37, 105, 28, 248, 16, 159, 7, 35, 19, 301, 125, 110, 486, 287, 98, 117, 511, 62, 51, 220, 37, 113, 140, 807, 138, 540, 8, 44, 287, 388, 117, 18, 79, 344, 34, 20, 59, 511, 548, 107, 603, 220, 7, 66, 154, 41, 20, 50, 6, 575, 122, 154, 248, 110, 61, 52, 33, 30, 5, 38, 8, 14, 84, 57, 540, 217, 115, 71, 29, 84, 63, 43, 131, 29, 138, 47, 73, 239, 540, 52, 53, 79, 118, 51, 44, 63, 196, 12, 239, 112, 3, 49, 79, 353, 105, 56, 371, 557, 211, 505, 125, 360, 133, 143, 101, 15, 284, 540, 252, 14, 205, 140, 344, 26, 811, 138, 115, 48, 73, 34, 205, 316, 607, 63, 220, 7, 52, 150, 44, 52, 16, 40, 37, 158, 807, 37, 121, 12, 95, 10, 15, 35, 12, 131, 62, 115, 102, 807, 49, 53, 135, 138, 30, 31, 62, 67, 41, 85, 63, 10, 106, 807, 138, 8, 113, 20, 32, 33, 37, 353, 287, 140, 47, 85, 50, 37, 49, 47, 64, 6, 7, 71, 33, 4, 43, 47, 63, 1, 27, 600, 208, 230, 15, 191, 246, 85, 94, 511, 2, 270, 20, 39, 7, 33, 44, 22, 40, 7, 10, 3, 811, 106, 44, 486, 230, 353, 211, 200, 31, 10, 38, 140, 297, 61, 603, 320, 302, 666, 287, 2, 44, 33, 32, 511, 548, 10, 6, 250, 557, 246, 53, 37, 52, 83, 47, 320, 38, 33, 807, 7, 44, 30, 31, 250, 10, 15, 35, 106, 160, 113, 31, 102, 406, 230, 540, 320, 29, 66, 33, 101, 807, 138, 301, 316,

353, 320, 220, 37, 52, 28, 540, 320, 33, 8, 48, 107, 50, 811, 7, 2, 113, 73, 16, 125, 11, 110, 67, 102, 807, 33, 59, 81, 158, 38, 43, 581, 138, 19, 85, 400, 38, 43, 77, 14, 27, 8, 47, 138, 63, 140, 44, 35, 22, 177, 106, 250, 314, 217, 2, 10, 7, 1005, 4, 20, 25, 44, 48, 7, 26, 46, 110, 230, 807, 191, 34, 112, 147, 44, 110, 121, 125, 96, 41, 51, 50, 140, 56, 47, 152, 540, 63, 807, 28, 42, 250, 138, 582, 98, 643, 32, 107, 140, 112, 26, 85, 138, 540, 53, 20, 125, 371, 38, 36, 10, 52, 118, 136, 102, 420, 150, 112, 71, 14, 20, 7, 24, 18, 12, 807, 37, 67, 110, 62, 33, 21, 95, 220, 511, 102, 811, 30, 83, 84, 305, 620, 15, 2, 10, 8, 220, 106, 353, 105, 106, 60, 275, 72, 8, 50, 205, 185, 112, 125, 540, 65, 106, 807, 138, 96, 110, 16, 73, 33, 807, 150, 409, 400, 50, 154, 285, 96, 106, 316, 270, 205, 101, 811, 400, 8, 44, 37, 52, 40, 241, 34, 205, 38, 16, 46, 47, 85, 24, 44, 15, 64, 73, 138, 807, 85, 78, 110, 33, 420, 505, 53, 37, 38, 22, 31, 10, 110, 106, 101, 140, 15, 38, 3, 5, 44, 7, 98, 287, 135, 150, 96, 33, 84, 125, 807, 191, 96, 511, 118, 40, 370, 643, 466, 106, 41, 107, 603, 220, 275, 30, 150, 105, 49, 53, 287, 250, 208, 134, 7, 53, 12, 47, 85, 63, 138, 110, 21, 112, 140, 485, 486, 505, 14, 73, 84, 575, 1005, 150,

```
200, 16, 42, 5, 4, 25, 42, 8, 16, 811, 125, 160,
32, 205, 603, 807, 81, 96, 405, 41, 600, 136, 14, 20,
28, 26, 353, 302, 246, 8, 131, 160, 140, 84, 440, 42,
16, 811, 40, 67, 101, 102, 194, 138, 205, 51, 63,
241, 540, 122, 8, 10, 63, 140, 47, 48, 140, 288
```

这本小册子的匿名作者用《独立宣言》作为密钥解开了上述密码。作者称，创造这些密码的人名叫“托马斯 · J. 比尔”（Thomas J. Beale），但是作者从来没有解释过字母J究竟代表哪个中间名。后来，又有其他作者转述过这段有趣的传说，并将这些密码的创造者称为“托马斯 · 杰斐逊 · 比尔”（Thomas Jefferson Beale）。这种假设是没有根据的。这个故事中的主角比尔的身份从未被确认过，也没有人知道比尔的以J打头的中间名究竟是什么。之所以会出现“托马斯 · 杰斐逊 · 比尔”的误传，可能是因为《独立宣言》的作者就是托马斯 · 杰斐逊，托马斯 · 杰斐逊既是美国总统，也是一名密码学家。这个误传的开端很可能是某位不太细心的作家一边想着《独立宣言》，一边下意识地把比尔的中间名补成了“杰斐逊”，而之后的其他作家则将这个错误一而再，再而三地重复了下去。

那么，究竟如何以《独立宣言》为密钥破译这段数字密码呢？破译的过程其实非常简单！这位匿名作者只不过是把《独立宣言》中的所有单词按顺序标上数字，然后再把这段密码中的每个数字替换成对应单词的第一个字母而已。比如：

```
instituted (115), hold (73), another (24)…
```

用上述方法可以破译出以下的明文信息：

我在贝德福德县距离比福德酒馆约4英里处的一个洞穴或者叫地窖里埋藏了以下这些东西，这个地窖在地表以下6英尺深的地方。这批宝藏归我的所有队员共同拥有，他们的名字参见3号文件。

第一批宝藏是我于1819年11月埋好的，其中包含1 014磅（约460千克）黄金和3 812磅（约1 729千克）白银。第二批宝藏是我于1821年12月埋好的，其中包含1 907磅（约865千克）黄金和1 288磅（约584千克）白银，以及一些珠宝。这些珠宝是我们为了运输方便在圣路易斯用白银换来的，珠宝的价值为13 000美元。

我们将以上这些宝物安全地放在铁罐中，这些铁罐的盖子也是铁的。地窖里粗略地铺上了一层石头，盛放宝物的罐子都放在坚固的石头上，上面再盖上一层石头。1号文件里详细描述了这个地窖的具体位置，你应该不难找到。

多么激动人心的一段信息啊！根据比尔的承诺，只要破译出1号文件中的密码，就可以找到藏宝地点。然而，撰写这本小册子的匿名作者发现，用《独立宣言》去破译1号文件无法产生任何有意义的文字——原来并非所有密码都是用《独立宣言》进行加密的！这位匿名作者表示，他花了大量时间试图破译1号文件里的密码，但是他也承认，这些努力完全失败了。

由于我在上述问题的调查研究上花了太多时间，我从一个相对富裕的人变成了一个赤贫的人。一些我本来应该保护的人对我的上述行为提出了许多劝告和抗议，但我一意孤行，给他们带来了许多痛苦。现在，我又不顾他们的劝阻写了这本小册子。

——匿名作者

这位匿名作者并没有像莫里斯一样只把这个秘密分享给一位他信任的朋友，而是通过出版这本小册子把未解开的密码向全世界公布。以下就是这段密码的内容：

地窖的位置：

71, 194, 38, 1701, 89, 76, 11, 83, 1629, 48, 94, 63, 132, 16, 111, 95, 84, 341, 975, 14, 40, 64, 27, 81, 139, 213, 63, 90, 1120, 8, 15, 3, 126, 2018, 40, 74, 758, 485, 604, 230, 436, 664, 582, 150, 251, 284, 308, 231, 124, 211, 486, 225, 401, 370, 11, 101, 305, 139, 189, 17, 33, 88, 208, 193, 145, 1, 94, 73, 416, 918, 263, 28, 500, 538, 356, 117, 136, 219, 27, 176, 130, 10, 460, 25, 485, 18, 436, 65, 84, 200, 283, 118, 320, 138, 36, 416, 280, 15, 71, 224, 961, 44, 16, 401, 39, 88, 61, 304, 12, 21, 24, 283, 134, 92, 63, 246, 486, 682, 7, 219, 184, 360, 780, 18, 64, 463, 474, 131, 160, 79, 73, 440, 95, 18, 64, 581, 34, 69, 128, 367, 460, 17, 81, 12, 103, 820, 62, 116, 97, 103, 862, 70, 60, 1317, 471, 540, 208, 121, 890, 346, 36, 150, 59, 568, 614, 13, 120, 63, 219, 812, 2160, 1780, 99, 35, 18, 21, 136, 872, 15, 28, 170, 88, 4, 30, 44, 112, 18, 147, 436, 195, 320, 37, 122, 113, 6, 140, 8, 120, 305, 42, 58, 461, 44, 106, 301, 13, 408, 680, 93, 86, 116, 530, 82, 568, 9, 102,

38, 416, 89, 71, 216, 728, 965, 818, 2, 38, 121, 195, 14, 326, 148, 234, 18, 55, 131, 234, 361, 824, 5, 81, 623, 48, 961, 19, 26, 33, 10, 1101, 365, 92, 88, 181, 275, 346, 201, 206, 86, 36, 219, 324, 829, 840, 64, 326, 19, 48, 122, 85, 216, 284, 919, 861, 326, 985, 233, 64, 68, 232, 431, 960, 50, 29, 81, 216, 321, 603, 14, 612, 81, 360, 36, 51, 62, 194, 78, 60, 200, 314, 676, 112, 4, 28, 18, 61, 136, 247, 819, 921, 1060, 464, 895, 10, 6, 66, 119, 38, 41, 49, 602, 423, 962, 302, 294, 875, 78, 14, 23, 111, 109, 62, 31, 501, 823, 216, 280, 34, 24, 150, 1000, 162, 286, 19, 21, 17, 340, 19, 242, 31, 86, 234, 140, 607, 115, 33, 191, 67, 104, 86, 52, 88, 16, 80, 121, 67, 95, 122, 216, 548, 96, 11, 201, 77, 364, 218, 65, 667, 890, 236, 154, 211, 10, 98, 34, 119, 56, 216, 119, 71, 218, 1164, 1496, 1817, 51, 39, 210, 36, 3, 19, 540, 232, 22, 141, 617, 84, 290, 80, 46, 207, 411, 150, 29, 38, 46, 172, 85, 194, 39, 261, 543, 897, 624, 18, 212, 416, 127, 931, 19, 4, 63, 96, 12, 101, 418, 16, 140, 230, 460, 538, 19, 27, 88, 612, 1431, 90, 716, 275, 74, 83, 11, 426, 89, 72, 84, 1300, 1706, 814, 221, 132, 40, 102, 34, 868, 975, 1101, 84, 16, 79, 23, 16, 81, 122, 324, 403, 912, 227, 936, 447, 55, 86, 34, 43, 212, 107, 96, 314, 264, 1065,

323, 428, 601, 203, 124, 95, 216, 814, 2906, 654, 820, 2, 301, 112, 176, 213, 71, 87, 96, 202, 35, 10, 2, 41, 17, 84, 221, 736, 820, 214, 11, 60, 760

接下来的一段密码则应该包含30位有权获得这批宝藏的人的名字，以及他们的继承人的名字和地址。

人名和地址：

317, 8, 92, 73, 112, 89, 67, 318, 28, 96, 107, 41, 631, 78, 146, 397, 118, 98, 114, 246, 348, 116, 74, 88, 12, 65, 32, 14, 81, 19, 76, 121, 216, 85, 33, 66, 15, 108, 68, 77, 43, 24, 122, 96, 117, 36, 211, 301, 15, 44, 11, 46, 89, 18, 136, 68, 317, 28, 90, 82, 304, 71, 43, 221, 198, 176, 310, 319, 81, 99, 264, 380, 56, 37, 319, 2, 44, 53, 28, 44, 75, 98, 102, 37, 85, 107, 117, 64, 88, 136, 48, 151, 99, 175, 89, 315, 326, 78, 96, 214, 218, 311, 43, 89, 51, 90, 75, 128, 96, 33, 28, 103, 84, 65, 26, 41, 246, 84, 270, 98, 116, 32, 59, 74, 66, 69, 240, 15, 8, 121, 20, 77, 89, 31, 11, 106, 81, 191, 224, 328, 18, 75, 52, 82, 117, 201, 39, 23, 217, 27, 21, 84, 35, 54, 109, 128, 49, 77, 88, 1, 81, 217, 64, 55, 83, 116, 251, 269, 311, 96, 54, 32, 120, 18, 132, 102, 219, 211, 84, 150, 219, 275, 312, 64, 10, 106, 87, 75, 47, 21, 29, 37, 81,

44, 18, 126, 115, 132, 160, 181, 203, 76, 81, 299,
314, 337, 351, 96, 11, 28, 97, 318, 238, 106, 24,
93, 3, 19, 17, 26, 60, 73, 88, 14, 126, 138, 234,
286, 297, 321, 365, 264, 19, 22, 84, 56, 107, 98,
123, 111, 214, 136, 7, 33, 45, 40, 13, 28, 46,
42, 107, 196, 227, 344, 198, 203, 247, 116, 19,
8, 212, 230, 31, 6, 328, 65, 48, 52, 59, 41, 122,
33, 117, 11, 18, 25, 71, 36, 45, 83, 76, 89, 92,
31, 65, 70, 83, 96, 27, 33, 44, 50, 61, 24, 112,
136, 149, 176, 180, 194, 143, 171, 205, 296, 87,
12, 44, 51, 89, 98, 34, 41, 208, 173, 66, 9, 35,
16, 95, 8, 113, 175, 90, 56, 203, 19, 177, 183,
206, 157, 200, 218, 260, 291, 305, 618, 951, 320,
18, 124, 78, 65, 19, 32, 124, 48, 53, 57, 84, 96,
207, 244, 66, 82, 119, 71, 11, 86, 77, 213, 54,
82, 316, 245, 303, 86, 97, 106, 212, 18, 37, 15,
81, 89, 16, 7, 81, 39, 96, 14, 43, 216, 118, 29,
55, 109, 136, 172, 213, 64, 8, 227, 304, 611, 221,
364, 819, 375, 128, 296, 1, 18, 53, 76, 10, 15,
23, 19, 71, 84, 120, 134, 66, 73, 89, 96, 230, 48,
77, 26, 101, 127, 936, 218, 439, 178, 171, 61,
226, 313, 215, 102, 18, 167, 262, 114, 218, 66,
59, 48, 27, 19, 13, 82, 48, 162, 119, 34, 127,
139, 34, 128, 129, 74, 63, 120, 11, 54, 61, 73,
92, 180, 66, 75, 101, 124, 265, 89, 96, 126, 274,

896, 917, 434, 461, 235, 890, 312, 413, 328, 381, 96, 105, 217, 66, 118, 22, 77, 64, 42, 12, 7, 55, 24, 83, 67, 97, 109, 121, 135, 181, 203, 219, 228, 256, 21, 34, 77, 319, 374, 382, 675, 684, 717, 864, 203, 4, 18, 92, 16, 63, 82, 22, 46, 55, 69, 74, 112, 134, 186, 175, 119, 213, 416, 312, 343, 264, 119, 186, 218, 343, 417, 845, 951, 124, 209, 49, 617, 856, 924, 936, 72, 19, 28, 11, 35, 42, 40, 66, 85, 94, 112, 65, 82, 115, 119, 236, 244, 186, 172, 112, 85, 6, 56, 38, 44, 85, 72, 32, 47, 63, 96, 124, 217, 314, 319, 221, 644, 817, 821, 934, 922, 416, 975, 10, 22, 18, 46, 137, 181, 101, 39, 86, 103, 116, 138, 164, 212, 218, 296, 815, 380, 412, 460, 495, 675, 820, 952

在这本小册子的最后，这位匿名作者称，他已经将他知道的关于宝藏的一切信息都写在其中了。这位匿名作者还表示，小册子的出版人在看到这份手稿之前对宝藏的事情完全不知情。这位作者这么写似乎是不想让小册子的读者提出任何问题。事实上，由于没有人知道作者的身份，读者就算想提问题也没有地方可以提。

由于能破译上述密码的人有可能找到大量的黄金和白银，所以这本小册子出版后获得了大量的关注。然而，我个人认为小册子里说的宝藏根本不存在。有很多证据可以支持我的这一看法，但我只打算向读者列出与密码本身有关的几点证据。在下文中，我还会用数学工具分析小册子里所写的内容。

理由 1

《独立宣言》有许多不同版本。多年来,《独立宣言》被重印了无数次。在某些重印版本中,印刷者不小心引入了一些小错误。还有的时候,印刷者为了“润色”会有意地做出一些改动。有些改动只会对破译出来的明文产生很小的影响,比如把“inalienable”改为“unalienable”(意为“不可剥夺的”)。然而,也有一些改动会使破译完全失败。比如,有的版本删去了一个单词,有的版本则加入了一个以前没有的单词。进行这种改动以后,删去或者新加入的单词之后的所有单词对应的数字都会发生改变,因此,如果使用错误的版本破译密码,失败的概率将会很高。除非这个被删去或者新加的单词出现在《独立宣言》靠近结尾的部分,否则用错误的版本作为密钥只能产生一些无意义的文字。难道这位匿名作者正好找到了一份和比尔所用的版本完全一样的《独立宣言》吗?这种巧合发生的概率有多大呢?

美国国家安全局数学家托德·马蒂尔(Todd Mateer)在互联网上找到了《独立宣言》的原始版本,并且认真地把该版《独立宣言》中的文字一一标上了数字。接着,他用这个版本的《独立宣言》作为密钥去破译2号文献中的密码,他得到了以下结果:

```
ahaie depos otedt nttte ointt oaitd strsa boapt
hrrmi lestr oabaa ottst tafep coiat ionor iaalt
snpti ntbea owtht ssram wnhst hhfbh ntdth ntoof
hcang mttaw fntbt tonat fgphi otatt ttheo attoe
swott tttdt sabea tiiti ntndb ththr tfttb ewnth
thotn tttde posit eotta stedo otinh itdrt tanda
oigte tfoos estth btlds tittt httia ghtta odbed
indtw efief ornds oosaa ieroi posit tdnhi iiaht
```

eeftt nttee nthes econd watmw dnttu tifht eentw
entth ntwfs aonst ottot htine ttfnh aetre dcnos
eienp oands hsaoa owtot welii oafob edato eigtt
teiah ttfti aierm fsoti wtfso btain idtns tehaa
tinep chang etotw intgo ntftt tatto nttdi alail
tattt gttit rhtss andtt aaigs theab oieit secrb
tatpc eohdi nibhf ohthw ittdg ttchi ittth eiitf
tnith aftlt fined wstts tonta ndthi itste atres
tonth ltdst onett oitee tintt twuar htttb hpioe
rtamd eront ditcg tlttt heopi atloc alitt tsttt
imtdt tttha tftti ostea lttwi lafit adstt indtn
gtn

显然，要读懂这段文字几乎是不可能的。如果某位破译者得到了这样的结果，他肯定会认为密钥错了，转而去继续尝试其他密钥。

理由2

即使是正确版本的《独立宣言》也不可能解开2号文件中的密码。因此，作者在小册子中称自己用《独立宣言》解开了这段密码是完全不可信的。

假设小册子的作者选择的《独立宣言》恰好和比尔使用的版本完全一样，如果这位匿名作者非常认真地把该版《独立宣言》中的单词一一标上数字，并用这份文件来破译密码，他也还会得到一堆无意义的文字！这是因为加密这段密码的人（不管他是谁）在给原始文件标数字时犯了好几个错误。以下是这位加密者所犯的错误[41]：

1. 将“self-evident”（意为“不言而喻”）当作两个单词（如果没有连字符，这是两个单词，但是有了连字符后应视为一个单词）。

2. 在第480个单词处跳过了10个单词。

3. 给两个不同的单词都标上了数字480。

4. 在大约第630个单词处少数了一个单词。

5. 在大约第670个单词处又少数了一个单词。

因为加密者在加密过程中犯了以上错误，所以如果密码的破译者比较认真仔细，他一定会得到一个和加密者使用的密钥不同的密钥，而使用这个“正确”的密钥根本无法将密文信息翻译成有意义的文字。然而，这本小册子的作者却翻译出了有意义的明文信息，也就是说他不仅使用了和比尔版本完全相同的《独立宣言》，还在给单词标数字的时候和比尔犯了完全一样的错误。出现这种巧合的概率有多大呢？

理由3

如果我们用《独立宣言》作为密钥去破译1号文件中的密码，得到的结果显然是不合理的。

1号文件密码里的数字最大可以达到2 906，而《独立宣言》中一共只有1 322个单词。因此，如果用《独立宣言》作为密钥去破译1号文件中的密码，许多数字根本没有对应的字母。基于以上理由，任何一个正常人都可以很轻松地判断出《独立宣言》不可能是1号文件中密码的密钥。如果有人因为某种原因尝试用《独立宣言》作为密钥去破译1号文件中的密码，他就会得到一段由以下这些字符开头的文字：“SCS?E TFA?G CDOTT”。其中“?”代表该处的数字超过了《独立宣言》的长度，无法找到与之对应的字母。因为这几个字符显然不是有意义的文字，一个正常人破译到这里就一定会排除《独立宣言》作为密钥的可能性，转而去尝试别的密钥了。

然而，詹姆斯·吉洛格利并不是一个“正常人”，也许对我们来说这是一件幸运的事情。即使在一开始就得到了上述这些毫无意义的结果，吉洛格利仍然决定继续破译，或者更准确地说，是他的计算机仍然按指令继续破译。吉洛格利编写的程序要求计算机完成整个破译过程，最终他得到了一段很有意思的结果。在破译出来的明文的后半部分中，出现了“ABFDE FGHII JKLMM NOHPP”这样的内容。也就是说，吉洛格利破译出了按顺序排好的大半个字母表！甚至连破译结果中的某些错误都能够找到很好的解释。上述字母表中之所以会在字母C的地方出现字母F，可能是因为加密者在写出字母C对应的数字（194）时，不小心写下了与194相邻的数字195，这样字母C就变成了字母F。

因为《独立宣言》并不是1号文件中密码的正确密钥，所以如果这段密码是真实的，用错误的密钥进行破译应该会得到一段无意义的文字。然而我们看到，用这段错误的密钥去破译1号文件中的密码能够得到按顺序排列的大半个字母表，这种情况出现的概率是极低的。根据这种现象我们可以得到以下结论：1号文件中的密码只是一段用来欺骗读者的没有意义的字符。写下这段“密码”的人只是想制造一段看起来像是密码的随机数字，然而在进行上述操作的过程中，此人要么是觉得无聊而写下了按顺序排列的大半个字母表，要么是认识到很难随手写出真正随机的数字，并且觉得字母表对应的数字看起来更加合理。不管是出于什么样的原因，重要的是：吉洛格利破译出的这个按顺序排列的字母表强有力地支持了“1号文件中的密码是骗局”的理论。

理由4

第3段密码的长度太短，不可能包含作者声称的内容。

通俗侦探杂志《弗林周刊》（*Flynn's Weekly*）刊登过一系列“破译密码谜题”的文章，文章作者是肯戴尔·福斯特·克罗森（Kendell Foster Crossen）。1927年，克罗森在这一系列文章的一篇里提出了一种新方法，不需要对第3段密码进行破译，就可以否定这段密码的真实性。克罗森的原理是研究第3段密码的字母数目以及这些字母所能提供的信息量。第3段密码一共包含618个字母，[42]它们真的能够包含这批宝藏的30名队员的继承人的姓名和地址吗？克罗森假设这30个队员一共有60名继承人，但我们就算假设只有30名继承人，30名队员及其继承人每个人的姓名和地址加起来平均只有10.3个字母。你的姓名和地址能用如此少的字母表达出来吗？

理由5

文体统计分析或文体学的证据。在《比尔文件》这本小册子中，应该一共包含3种不同的写作风格。第1种是小册子作者本人写的内容，第2种是作者大量引用的比尔写的书信，第3种是作者引述的莫里斯的话。我们可以用统计学的工具分别分析这3种写作风格的特点。如果这3部分内容的作者确实是3个不同的真实人物，对应的统计数据就应该有明显的差异。有好几名研究者都对这3部分内容的写作风格进行了对比和研究，其中包括美国国家安全局密码破译专家所罗门·库尔贝克（Solomon Kullback）。第二次世界大战期间，库尔贝克在破译敌方密码的工作中取得过巨大的成功，这一贡献使他在美国国家安全局的名人堂中获得了一席之地。下面，让我们来看看文体统计分析的部分数据。

我在这里列出的统计分析结果均引自乔·尼克尔（Joe Nickell）发表的论文。尼克尔是用科学方法调查超自然现象的专家，他认为，这本小册子的匿名作者就是该书的出版人沃德，并将文中所有不属于比尔和莫里斯的文字标记了出来。在下面的表格中，为了简单起见，我将把这部分标记为“作者”。尼克尔不仅比较了“作者”和比尔的文风，为了增强结果的说服力，他还把小册子中的文字与另外3位生活在19世纪的弗吉尼亚州的人［分别是首席大法官约翰·马歇尔

（John Marshall），罗阿诺克市的约翰·兰道夫（John Randolph），以及约翰·兰道夫·塔克（John Randolph Tucker）］所写的文字进行了对比。表9–2至表9–6中的数字本身已经能够充分说明尼克尔的结论，我想我没有必要对这几个表格做过多的解释。

表9–2 每个句子中的平均单词数目显示：比尔就是作者[a]

	每个句子中的平均单词数目	标准差
比尔	29.95	12.76
作者	29.74	13.75
马歇尔	19.13	11.40
兰道夫	20.11	15.09
塔克	34.49	22.03

a Nickell, Joe, "Discovered, The Secret of Beale's Treasure," *The Virginia Magazine of History and Biography* 90, no. 3 (July 1982), 310–24.

接下来，尼克尔比较了几名不同作者文字中常见单词“the”、“of”以及“and”出现的概率。

表9–3 常见单词出现的百分比[a]

	the	of	and
比尔	5.10%	3.51%	3.66%
作者	5.20%	3.26%	3.47%
马歇尔	3.08%	1.92%	2.69%
兰道夫	4.55%	2.93%	2.87%
塔克	7.35%	5.28%	3.85%

a Nickell, Joe, "Discovered, The Secret of Beale's Treasure," *The Virginia Magazine of History and Biography* 90, no. 3 (July 1982), 310–24.

上面的表 9–3 中列出的是这几个常见单词出现的总体概率，我们还可以比较这几个单词出现在句首的概率。这项分析的结果列在下面的表 9–4 中。

表 9–4　常见单词出现在句首的百分比[a]

	the	of	and
比尔	6.80%	0.00%	0.00%
作者	6.12%	0.00%	0.00%
马歇尔	8.93%	2.86%	0.00%
兰道夫	10.34%	0.00%	1.82%
塔克	2.90%	1.73%	1.54%

a Nickell, Joe, "Discovered, The Secret of Beale's Treasure," *The Virginia Magazine of History and Biography* 90, no. 3 (July 1982), 310–24.

除了以上三项分析以外，尼克尔还研究了不同作者对标点符号的使用情况。

表 9–5　平均每个句子中逗号和分号的数目[a]

	逗号	分号
比尔	2.6	0.06
作者	2.4	0.06
马歇尔	0.32	0.02
兰道夫	1.6	0.16
兰克	2.96	0.18

a Nickell, Joe, "Discovered, The Secret of Beale's Treasure," *The Virginia Magazine of History and Biography* 90, no. 3 (July 1982), 310–24.

最后，尼克尔又提供了一些相对不是很明显的对比结果（参见表 9–6）。

表 9-6　更多支持比尔就是作者的数据[a]

	否定句	否定被动语态	不定词	关系从句
比尔	24	6	44	30
作者	36	7	40	39
马歇尔	15	0	21	8
兰道夫	29[b]	0	18	9
塔克	14	0	16	34

a Nickell, Joe, “Discovered, The Secret of Beale’s Treasure,” *The Virginia Magazine of History and Biography* 90, no. 3 (July 1982), 310–24. 以上测试由肯塔基大学的教授让 · G. 皮瓦尔（Jean G. Pival）完成。

b 尼克尔写道：“为了论证自己参与一次决斗的必要性，兰道夫在一封信中使用了10次否定句。”

以上5项比较结果都强烈支持以下结论：小册子中引述的比尔的信件和号称由匿名作者书写的内容其实是由同一个人写出来的。其实我们还可以再把匿名作者的文风和号称由莫里斯所叙述文字的文风进行比较，如果这两者的文风也一致，就可以为骗局论提供更加强有力的证据。

除了以上5点支持骗局论的证据以外，我还可以给出第6条理由，那就是“专家的结论”。一般来说，我并不喜欢把专家的结论当作可靠的证据，因为专家的结论经常是错的。但是，我相信也有一些读者更加愿意相信专家的说法，因此我决定在这里搬出这些专家的名号。威廉 · F. 弗里德曼曾让弗兰克 · 罗利特（Frank Rowlett，在第二次世界大战中，他在日本外交密码的破译中扮演了非常重要的角色）和他在通信情报部门的同事一起对比尔密码进行分析和破译，这是弗里德曼给他们布置的一项培训练习。那么以上这些密码专家得到了怎样的结论呢？他们的结论是：比尔密码就是一个骗局。

然而，在另一方面，又出现了另外一种支持比尔密码真实性的证据。

根据这本小册子中的说法，比尔留给莫里斯的指示是：如果10年后他还不回来，莫里斯就可以打开这个盒子。这本小册子中还说，比尔计划让一位朋友在同一时间点（10年后）把密码的密钥寄给莫里斯。然而，由于某种原因，莫里斯从未收到这封包含密钥的信件。

1984年，一本目标读者群为电脑爱好者的杂志《运行》（*RUN*）登出了一篇文章，报道了一些与这封丢失的密钥信有关的情况。这篇文章的作者道格拉斯·尼克洛（Douglas Nicklow）称，在1832年8月，《圣路易斯灯塔报》（*St. Louis Beacon*）确实刊出过一条启事，称有一封寄给罗伯特·莫里斯的信在邮局无人认领。尼克洛称，布鲁金斯研究所的J. 索拉里奥（J. Solario）发现了上述启事。后来，一位晚于尼克洛的研究者——加拿大的计算机分析专家韦恩·S. 陈（Wayne S. Chan）似乎对尼克洛提供的细节不满意，因此又对这则启事展开了调查和研究。

陈的调查研究显示，《圣路易斯灯塔报》只在每周星期四出版，这份报纸确实定期会登出一些待领信件的收信人名字。在1832年8月2日、9日和16日的《圣路易斯灯塔报》上，都刊有邮局待领信件收件人的名单，而且陈确实在这几份名单中找到了罗伯特·莫里斯的名字。于是，陈提出了以下问题："'莫里斯'（Moriss）一词末尾双写s的拼法是比较罕见的，加上这则启事刊登的时间，这真的只是一个令人惊奇的巧合吗？"[43]

陈调查了当时在圣路易区域中是否还有其他名叫莫里斯的人，但是，不管是在1821年的城市名录中，还是1840年的城市名录以及人口普查资料中，都找不到姓莫里斯的人，更不要说全名为罗伯特·莫里斯的人了。[44]

再接下来，陈又检查了1832年9月的待领邮件列表，这份清单上也没有罗伯特·莫里斯的名字。此外，同年7月的待领邮件列表上也没有罗伯特·莫里斯的名字。因此，这封信的寄送时间完全符合比尔要求那位不知名的朋友将其从圣路易斯寄往弗吉尼亚州的时间，只是由于某种原因被退回了。对于这封信此后的去向，陈并没有能够调查出什么线索。

以上这个事件真的只是巧合吗？它们会不会具有某种重要的意义呢？

我个人认为，上述事件确实相当有趣，但是仅凭这一事件不足以扳倒支持骗局说的各种强有力的证据。虽然我相信比尔的故事是伪造的，但是我知道的另一个关于地下埋有黄金的故事，可信度比较高一些。下面我将向读者介绍一段可能是真实的藏宝故事！

福里斯特·芬恩的宝藏

福里斯特·芬恩（Forrest Finn）的宝藏可谓比尔宝藏的迷你版，成功破译福里斯特·芬恩密码的人可能获得的奖励不是成吨的黄金和白银，而是20多磅（1磅≈0.45千克）黄金（外加一些珠宝以及其他一些物件）。这个故事会更可信一点儿吗？

福里斯特·芬恩宝藏的故事最先出现于2010年出版的《追寻的战栗》（*The Thrill of the Chase*）一书中，这本书的作者福里斯特·芬恩声称在某处埋下了一些宝藏。根据芬恩的说法，埋下宝藏的想法最早可以追溯到1988年。那一年，58岁的芬恩被诊断患有癌症，医生认为他能活过未来3年的概率只有20%。芬恩没有像许多癌症病人一样不相信医生的诊断，当时，他认为自己会像医生认为的那样很快死去。[45]接下来，我们用芬恩自己的话来继续讲述他的故事：

图 9-26　福里斯特·芬恩的宝藏。照片由艾迪生·多蒂（Eddison Doty）拍摄

后来某一天晚上，我终于彻底接受了自己的命运，于是我产生了一个想法。在过去的这些年中，我不断扩大自己的藏品库，这为我提供了许多乐趣（芬恩喜欢收集），为什么不在我还活着的时候让别人来寻找我的藏品呢？也许在我死后，这些人还会继续寻找我的藏品。于是，我决定在一个藏宝箱里装上黄金和珠宝，然后把这个藏宝箱放到某个秘密的地方——当然，我会给所有愿意玩寻宝游戏的人留下一些线索。这个游戏完全符合我的想法，现在的时机也非常完美。啊，我真喜欢这个主意，但是要设计好这个寻宝游戏需要一些时间来计划，而我剩下的时间已经不多了。不管距离那个日子究竟有多远，在我所处的这种不愉快的情况下，我总觉得它很快就会到来。

幸运的是，我很快说服了一位在博物馆工作的朋友，他把一个漂亮的藏宝箱卖给了我。这个藏宝箱是用锻造青铜制成的，箱子的四面以及箱盖上都装饰有立体的美女图案。我知道我付出的价钱比这个箱子的真实价值高太多了，但是每过一段时间总会出现一些非常特别的物品，因为它们太美了，我已不在乎所有关于价值的逻辑规则了。这个藏宝盒也令一位古董学者感到相当兴奋。他说，这个藏宝盒很可能是一种罗马风格的带锁盒，其历史可以追溯到公元1150年左右。这位学者还认为，这个盒子以前可能是用来装家庭用《圣经》或者《岁月之书》的。现在，我打算用它来装我的古董珠宝和实心金块。我很高兴，这是一个完美的藏宝盒。

除了珠宝和黄金以外，我还想在这个藏宝盒里放一些私人物品，因为也许找到宝藏的幸运儿会想要了解我——一个多么愚蠢的人才会放弃这么丰富的宝藏呢？因此，我写了一份20 000字的自传放在这个藏宝盒里。我将这份自传放在一个小玻璃罐里，玻璃罐的盖子用蜡封好，以防止湿气破坏里面的东西。这份自传用很小的字体打印出来，因此需要借助放大镜才能够看清上面的文字。我想把所有事情都安排妥当。

然后，我往这个藏宝盒里装上金币：大部分是老式美国鹰洋金币和双鹰金币，还有大量来自阿拉斯加的砂矿金块。其中两个较大的金块重量都超过一磅，此外还有数百个较小的金块。此外，这个宝箱里还有前哥伦布时代的黄金动物雕像，以及来自古代中国的玉雕人脸。由于这个宝箱里装满了大量各种不同的东西，我没有办法逐一描述。但是，有几件宝物我想特别提一下：其中有一个17世纪的西班牙金戒指，上面镶嵌有一枚很大的祖母绿宝石，这枚戒指是用金属探测器找到的。还有一件宝物是一枚古董女士龙纹金手镯，上面镶嵌有254枚红宝石，6枚祖母绿宝石，2枚锡兰蓝宝石，以及不计其数的小钻石。

我还不情愿地在藏宝盒里放了一个小银手镯，上面有镶成一排的22个绿松石片珠。这个银手镯正好符合我的手臂尺寸，我非常喜欢它，但是我喜爱它最主要的原因还是它的历史。1898年，理查德·韦瑟雷尔（Richard Wetherell）从一处废墟中挖掘出了这些绿松石珠子，同年，一名印第安纳瓦霍族的银匠把这些珠子打造成这枚手镯。1901年，韦瑟雷尔将这个手镯卖给了旅馆业巨头弗雷德·哈维（Fred Harvey）。64年之后，我在和弗雷德·哈维的继承人拜伦·哈维（Byron Harvey）玩撞球时赢得了这枚手镯。

这个宝盒里还有一件宝物是我的收藏品中的珍品，那是一条来自哥伦比亚的泰罗娜和锡努印第安项链。这条项链上有39个动物雕像，以石英水晶、玉髓、翡翠以及其他不寻常的宝石刻成。但是，真正令这条项链与众不同的是项链上的两个铸金雕像——一个是美洲虎爪，一个是青蛙。这只青蛙双眼突出，两腿翘起，似乎正准备起跳。我最后一次拿起这件有2 000年历史的珠宝时，似乎可以感到它身上散发出来的原始的力量，那种至高无上的美。然而，最终我还是将这件珠宝放进了藏宝盒里，并关上了盒盖。我的一部分似乎也随着它被装到了盒子里。我相信，这个世界上总是有一些像我这样喜欢探索的人，随时准备好将行李扔进小卡车，踏上寻宝的旅程。这些人很有可能会最终发现我的宝箱，以及宝箱里装着的20多金衡磅的黄金。对我来说，最重要的永远是追寻的过程所带来的战栗感。你觉得呢？

我已经想好了藏这个宝箱的具体地点，把宝箱藏在那里的话，既不容易找到，又不是完全找不到。它就位于圣塔菲北部的群山中。优柔寡断总是会带来很多不确定性，也正是因为这个原因，我才会等了这么久还没有藏好我的宝箱。在乔治·伯恩斯（George Burns）100岁的时候，有人问他：你的健康状况如何？伯恩斯回答说："我的健康状况很好。正在杀死我的只是我的年龄而已。"就像埃里克·斯隆（Eric Sloane）一样，在将近80岁的时候，我终于决定：现在是时候采取行动了。于是我写下了一首诗，诗中包含9条线索。如果能够精确地根据这9条线索的提示行动，你就能走到我的彩虹尽头并找到我埋藏的宝物。[46]

既然我已经单独去了那里，
既然我的宝藏如此醒目，
我可以藏好那个地点的秘密，
关于新旧宝物的线索。

从暖水止步的地方开始，
走向下面的山谷，
并不远，但在徒步行走距离之外，
走进布朗之家的下面。

从那里开始就不是弱者该来的地方了，
终点越来越近；
你的溪流上没有船桨，
只有沉重的负载和高高的水位。

如果你聪明地找到了火焰，
快点儿向下看，你的追寻就要结束。
但是请停留一小会，用惊奇的目光凝视。
然后就拿着宝箱平静地离开吧。

然而为什么我必须离开，
却把我的宝藏留给所有人去寻找？
我已经知道这个问题的答案，

我疲惫地做完了所有事，现在我只感到衰弱。

所以，所有人都听我说，好好地听我说，
你的努力值得冒着严寒。
如果你很勇敢，如果你在树林中，
我就把金子的所有权交给你。

好吧，读完了诗中留下的线索，我还是无法立即找出藏宝的地点在哪儿。但是芬恩还写道：“故事里还散落着一些微妙的线索。”[47]这里所说的“故事”是指芬恩的《追寻的战栗》一书，这本书几乎全是芬恩生活中的一些逸事。后来，芬恩又写了续集——《徒步距离之外》（*Too Far to Walk*），其中的大部分内容仍然是芬恩生活中的逸事。

我注意到的一个有趣的细节：芬恩称宝物藏在“圣塔菲北部的群山中”。根据前文中提到的小册子中的说法，比尔和他的队员们正是在这一区域内找到了大量黄金和白银，并将这些宝物运回弗吉尼亚州藏好。这只是一个巧合吗？芬恩会不会是在向这个更古老的藏宝游戏致敬呢？

在读完芬恩的两本自传以后，我的印象是：芬恩是一个非常诚实的人。在自传中写的那些逸事中，只有少数几件出格的事：比如青少年时期曾用伪造的驾照开车，曾在某人的油箱里撒过尿，还有曾回到童年住过的老房子里，挖出以前埋下的玩具并且私自拿走（严格意义上讲，这种行为属于偷窃，因为芬恩一家已经不再是这处房产的所有人了）。

在芬恩的书中，唯一值得我们注意的是关于一位为《国家地理》（*National Geographic*）杂志工作的作家兼摄影师的故事。这个人当时正在创作一组与圣塔菲有关的作品。有一天，摄影师来到芬恩的画廊，芬恩给这组圣塔菲的照片开了价，并说：“我宁愿要知名画家的坏作品，也不要不知名画家的好作品。”[48]他知道这种说法会招致批评，但他对自己的员工解释道：“这是一种经营策略，为的是引起话题，让大家都谈论我们的画廊。”[49]

那么，芬恩是否可能为了宣传某种他认为重要的东西而虚构一个宝藏的故事呢？从芬恩写的书中，我们可以看出他对大自然抱有很深的感情。芬恩有没有可能为了鼓励人们热爱大自然而编造了宝藏的故事呢？

在《追寻的战栗》一书的续集《徒步距离之外》的开头，芬恩写下了以下这段话：

> 哦，告诉我吧，聪明的先生，
> 所有的宝藏究竟在何处？
> 聪明的先生回答道：
> “你在哪里找到宝藏，宝藏就在哪里。”[50]

如果芬恩的计划是用一个虚构的寻宝故事来鼓励人们去登山和露营，那么他的这个计划恐怕产生了一些不太好的副作用。某电视新闻的一则报道称：

> 虽然芬恩对自己获得的大量关注感到十分激动，但芬恩也表示，他有点儿担心这场寻宝游戏已经变得“有些失控了”。在接受《7号行动新闻》节目采访时，这位百万富翁说，据估计目前已经有大约30 000人来到西南部寻宝。他预计，今年还会有50 000人前来寻宝。[51]

芬恩在接受采访时还说，若干名寻宝者其实已经走到了离宝箱不到200英尺的地方，但他们并不知道自己当时离宝箱有多近。

虽然芬恩似乎有过几次稍显不诚实的经历，但他曾在军队任职多年，他在那里表现出的高贵品质足以打消我对他人品的所有怀疑。芬恩1950年加入美国空军，并在越南战争期间继续为国家效力。越战期间，芬恩共驾驶过328架次飞机，其中有两次失去飞机，只身涉险归队。后来，芬恩反思道：

> 我曾参与过的战争从很大程度上来看是由某些人出于哲学上的原因发动的，这些人错误地认为那是正确的、必须做的事情，而完全放弃了自己的道义感。[52]

从上面这段话中，我们可以看出，芬恩是个很有勇气的人——他并不怕对有争议，或者容易激起人们强烈感情的问题发表自己的看法。

后来，芬恩成了一名买卖艺术品的商人。虽然他的话可能曾令一些《国家地理》杂志的读者感到震惊，但芬恩做生意时是非常正直和诚实的。以下这个例子可以证明芬恩的诚实品格：芬恩曾经从一位名叫埃尔米·德奥里（Elmyr de Hory）的著名仿造者那里买过100幅油画，德奥里非常善于仿造法国印象派画家的作品。这批油画仿造了包括德加、雷诺阿、高更、凡·高、莫迪利亚尼、马蒂斯、毕加索、莫奈等画家的作品。他一共花了225 000美元，只要把其中一幅假画当作真画卖出去，就能轻松收回全部成本。然而，芬恩却要求每位买家在购买这批油画前都签署一份文件，文件上写明，虽然画上签着其他画家的名字，但是买家知道他们购买的画是由埃尔米·德奥里所画。

如果芬恩宝藏真的是一个骗局的话，芬恩应该有某种促使他说谎的动机。芬恩会不会是为了描述宝藏的那本书的销售利润而说了谎呢？好吧，针对这个问题，我只能说芬恩过去对待自己的财富一贯十分慷慨。比如，芬恩曾将爱德华·霍珀（Edward Hopper）的一幅大型画作《为和平欢呼》

（*Hail to Peace*）捐献给得克萨斯州坦普尔市的斯科特与怀特医院。既然芬恩愿意如此慷慨地捐出这么多钱，他似乎不太可能为了赚钱而采用不诚实的手段。并且，后来的这条新闻几乎完全打消了我对芬恩为了卖书而造假的怀疑，这条新闻的内容是：

> 芬恩将《追寻的战栗》一书的全部销售所得（以及一笔数额不小的额外私人捐款）都捐献给了慈善机构。这笔钱将用于帮助付不起治疗费用的癌症病人。[53]

因此，我们可以排除芬恩因为贪婪而造假的可能性。也许这个藏宝的故事确实是真实的！

芬恩曾经表示，他在《徒步距离之外》中提供了一些关于宝藏位置的微妙暗示。我仔细地读完了这本书，并且可能找到了一些其他地方都没有提到的线索。在这本书的结尾处，芬恩为我们提供了一幅新地图。由于这幅地图包括了美国很大一部分疆土，所以它本身并没有太高价值。但是，在向读者展示这幅地图的时候，芬恩写下了这样一段话：

> 最近，我们对基准地图公司的地图和地图集产品产生了兴趣。我与他们合作制作了这份藏宝图，这个过程十分有趣。他们独特的制图风格非常符合我们对于探索的共同兴趣。他们邀请我在这幅地图上画上一个“X”，我拒绝了，但我必须承认，在精神上，我已经把这个“X”画在了地图上。[54]

以上这段话引起了我的兴趣，这是因为在《徒步距离之外》一书前面的章节中，曾经出现过一幅以某种方式标有“X”的地图，详见图9–27。

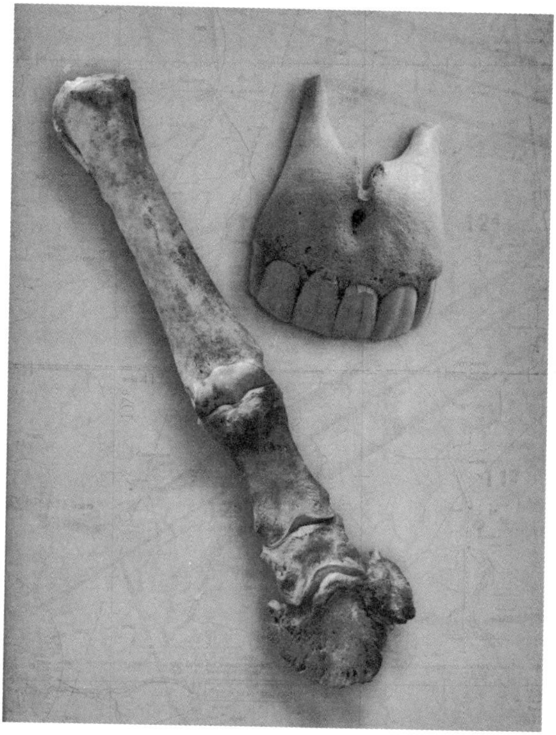

图 9–27　芬恩提供了一个标有“X”的地点?

从图 9–27 中我们可以看出，这张地图上有两条很明显的线，而放在地图上的这根骨头与这两条线都有交叉，交叉处都可以看成组成一个“X”。

虽然不能排除我对这张图进行了过度解读的可能（毕竟读者已经从芬恩所写的书中挖掘出了数千条并不符合作者本意的线索），但是我们也无法排除这确实是一个 X 的可能性。如果我的猜测正确的话，那么宝藏要么埋在科罗拉多州，要么埋在怀俄明州，取决于你认为哪一个 X 符号才是芬恩留下的线索。在某次采访中，芬恩向我们提示了藏宝地点的范围，而上述的两个州都在芬恩提示的范围之内。

芬恩还确认了宝箱埋在以下4个州中的一个，这4个州分别是：科罗拉多州、蒙大拿州、新墨西哥州和怀俄明州。[55]

不管怎么说，芬恩宝藏的故事最后迎来了一个圆满结局。首先我要告诉读者的是，虽然芬恩的医生曾对他的病情做过非常悲观的诊断，但是最终芬恩战胜了癌症。在我写这本书的时候，芬恩仍然活着，已经85岁了。因此，芬恩完全有可能活着看到宝箱被找到的那一天，和成功寻获宝藏的人一起分享胜利的喜悦。

我认为有必要向读者强调一下以下事实：不管比尔和芬恩的宝藏是不是真的存在，在寻宝的历史上，确实有过按照线索提示成功找到宝藏的例子。下面我就为读者列举几个这样的例子：

1.霍克斯尼宝藏

1992年11月，在英格兰的萨福克郡，一个农民请埃里克·劳斯（Eric Lawes）用他的金属探测器帮忙寻找一把丢失的榔头。结果，劳斯并没有找到农民丢失的榔头，而是找到了一些钱币——准确地说是14 780枚钱币（其中包括565枚金币、14 191枚银币和24枚铜币）。除了这些钱币以外，劳斯还找到了数百件金勺子和银勺子、珠宝和雕塑，所有这些宝物都来自罗马帝国时期。按照英国法律，这批宝物属于英国政府，但是政府有义务向宝物的发现者支付宝物的市场价值。于是，根据这条法律，这个农民和埃里克·劳斯分享了政府支付给他们的1 750 000英镑。

而另一个更加激动人心的寻宝故事则发生在美国。

2. 马鞍岭宝藏

2013年2月，在美国北卡罗来纳州，一对不愿透露姓名的夫妇在自家土地上遛狗时发现地面上有一处闪光的地方。这对夫妇挖开了闪光处的地面，发现了8个金属罐，里面装有超过1 400枚金币。这些金币的日期在1847—1894年之间，总面值大约在27 000美元左右。但是，由于这些金币中有很多具有历史价值并且保存状态完好，所以根据专家的估计，如果对这些金币进行拍卖的话，这8罐金币的总价值可能超过1 000万美元。比如，其中有一枚1866–S无题字双鹰金币，仅这一枚金币的价值就将近100万美元。这些金币究竟为何被埋在地下目前仍是一个谜。

在寻宝的历史中，也许曾有过通过密码发现宝藏的故事，但是在下面这个故事中，却出现了相反的情况，人们不是通过密码找到宝藏，而是通过宝藏发现了一些密码。

中国金条密码

中国金条密码的故事始于7根总重超过1.8千克的金条。据说，这7根金条是1933年由某家美国银行发给中国上海的一位王将军作为存款凭证的。在这些金条上，写有一些汉字，这些汉字显示，交易的总价值超过了300 000 000美元。但是，上述说法的真实性存在争议。这7根中国金条上有一些目前尚未被解开的密码，这些密码很可能就是解决上述争议的关键。

一位美国博物馆馆长将这些金条上的密码提交给了国际密码研究协会（IACR）。虽然上述协会的会员中没人能解开这段密码，但是他们将其发布在了该组织的网站上，任何有兴趣的人都可以尝试破译这段密码。关于中国金条密码的信息很少，这里我向读者提供的所有相关信息都来自国际密码研究协会的网站。这些信息也出现在互联网上的其他地方，但是除了国际密码研究协会网站上的原始信息以外，其他信息大部分都是没有根据的猜测。国际密码研究协会的网站上贴出了所有金条上的图案，以及抄写后的密码。为了方便读者了解这批密码的内容，我将其中一些图片以及抄写后的密码复制在了图9–28和图9–30中。网站称，密码中有些字符很难看清楚，因此抄写后的密码中有可能存在错误的字符。

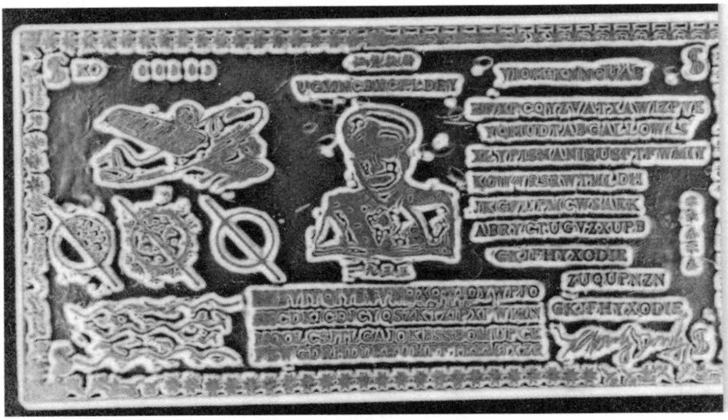

图9-28　一根画有飞机、将军，并写有密码的金条？

在上图中，共包含以下这些密码：

人物图案上方的密码：　　UGMNCBXCFLDBEY.

人物图案下方的密码：

RHZVIYQIYSXVNQXQWIOVWPJO

SKCDKJCDJCYQSZKTZJPXPWIRN

MQOLCSJTLGAJOKBSSBOMUPCE

FEWGDRHDDEEUMFFTEEMJXZR

人物图案右侧的密码：

VIOHIKNNGUAB

HFXPCQYZVATXAWIZPVE

YQHUDTABGALLOWLS

XLYPISNANIRUSFTFWMIY

KOWVRSRKWTMLDH

JKGFIJPMCWSAEK

ABRYCTUGVZXUPB

GKJFHYXODIE

ZUQUPNZN

接下来，在图9–29中，我们可以看到另一根金条上的图案。这根金

图 9-29　另一根神秘的金条

条上除了将军的图案以外，显然还画着另一个人。

图9–29中的金条共包含以下这些密码：

左侧人物图案上方的密码：	ABRYCTUGVZXUPB XLYPISNANIRUSFTFWMIY
左侧人物图案下方的密码：	MVERZRLQDBHQ HLMTAHGBGFNIV
金条中央上方的密码：	MQOLCSJTLGAJOKBSSBOMUPCE FEWGDRHDDEEUMFFTEEMJXZR
右侧人物图案上方的密码：	ZUQUPNZN VIOHIKNNGUAB
右侧人物图案下方的密码：	HFXPCQYZVATXAWIZPVE GKJFHYXODIE HLMTAHGBGFNIV

其中一根金条的背面既有密码，又有汉字，两者交替出现。此外，又出现了另外几行密码，但是相对于上方那段较长的文字，这段密码似乎是上下颠倒的。

图 9-30　其中一根金条上的汉字和密码字符

上图中和汉字交替出现的密码是：

JKGFIJPMCWSAEK

SKCDKJCDJCYQSZKTZJPXPWIRN

MQOLCSJTLGAJOKBSSBOMUPCE

FEWGDRHDDEEUMFFTEEMJXZR

RHZVIYQIYSXVNQXQWIOVWPJO

MQOLCSJTLGAJOKBSSBOMUPCE

FEWGDRHDDEEUMFFTEEMJXZR

SKCDKJCDJCYQSZKTZJPXPWIRN

RHZVIYQIYSXVNQXQWIOVWPJO

MQOLCSJTLGAJOKBSSBOMUPCE

SKCDKJCDJCYQSZKTZJPXPWIRN

请注意，在上面这些密码字符中，除了第一行以外，其他行都重复出现了多次。

接下来的这段上下颠倒的密码是：

```
HLMTAHGBGFNIV
ZUQUPNZN ABRYCTUGVZXUPB
MVERZRLQDBHQ
GKJFHYXODIE UGMNCBXCFLDBEY
VIOHIKNNGUAB
HFXPCQYZVATXAWIZPVE
```

国际密码研究协会的网站上还登出了两个名字和地址，该网站称，如果破译者对中国金条密码的相关信息有疑问，可以联系这两个人。但是我发现这些联系信息都已经过期了。

接下来，在本章的最后，我要向读者介绍另一个与藏宝有关的故事。

海盗的战利品

接下来的这段密码来自一位18世纪的法国人奥利维耶·勒瓦瑟（Olivier le Vasseur），又名奥利维耶·拉布什（Olivier la Bouche），绰号“秃鹫”[①]。这

① 法语的“秃鹫”（la buse）和拉布什（la Bouche）发音相近。——编者注

是一段和宝藏、黄金，以及海盗冒险有关的故事。

1721年4月，拉布什和另一个英国海盗约翰·泰勒（John Taylor）合作“干了一票大的”——他们抢劫了葡萄牙船只“处女角号”。“处女角号”上本来配有70门火炮，但由于该船在风暴中被严重损坏，为了防止整艘船沉没，船员将船上的大部分枪炮都丢进了海里。

由于“处女角号”当时处于这种脆弱和缺乏防御的状态，所以拉布什和泰勒成功地抢到了一大批财宝——这批宝物号称是“海盗历史上抢到的最大一笔财宝之一”，其中包括金条、银条、成箱的金几尼①、珍珠、成桶的钻石、丝绸、艺术品，还有属于果阿大主教（果阿大主教本人当时也在“处女角号”上）的许多珍稀物件。成功抢到这批财宝以后，拉布什和泰勒给他们手下的每一位船员都分发了5 000枚金几尼和42枚钻石作为奖赏。[56]

然而，在1730年年初，拉布什的好运气走到了尽头，他碰上了法国船只“美杜莎号”的船长莱尔米特（L’Ermitte）。在一场激烈的厮杀之后，拉布什被莱尔米特船长用铁链锁住送到了波旁。1730年7月11日，拉布什因其犯下的罪行被处以绞刑，他生前抢到的宝藏此后再也没被找到过。关于拉布什的宝藏，阿索尔·托马斯（Athol Thomas）讲述过这样一个故事，正是这个故事使得拉布什出现在了我这本以密码为主题的书中：

> 浪漫主义作家相信一个这样的故事：在拉布什颈上的绳索收紧之时，他写下了一张小纸条，并将纸条抛到了人群中。他一边抛出这张纸条，一边大喊道：“谁能找到我的宝藏就去找吧！”这一举动虽然很符合拉布什虚张声势的个性，但到目前为止没有人知道他究竟是在和大家开玩笑，还是真的在给出关于宝藏的线索。[57]

① 金几尼是英国当时发行的一种金币。——编者注

据说，拉布什临死前抛出的这张纸条上写着一段密码，见图9–31。

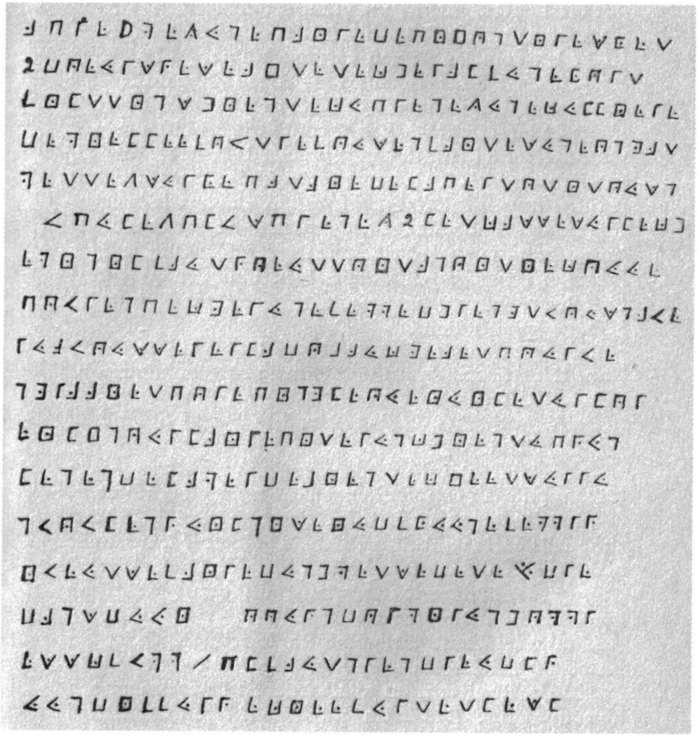

图 9-31 一段海盗的密码?

任何有经验的密码破译者看到这样一段密码后都会喜不自禁，因为上面这段密码使用的似乎是最简单的一种加密方式。在图9–32中，我用一段样本密钥来解释这种加密方式的工作原理。

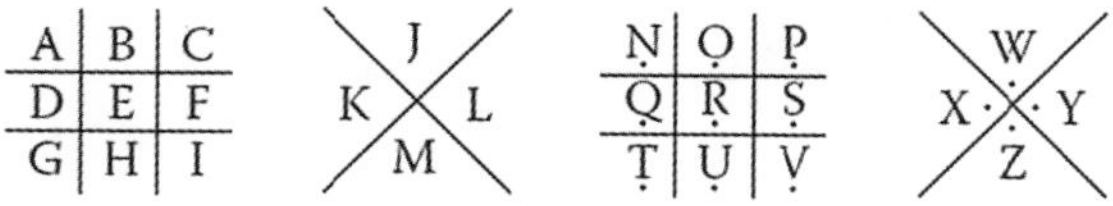

图 9-32 一段可用于加密简单密码的样本密钥

从图9–32中我们可以看到，这种密钥把字母放进了一个井字网格和一个X形中。由于只能放下一半字母，所以必须再次重复上面的步骤——但是在第二个井字网格和第二个X形中，每个格子里不仅有一个字母，还有一个点。接下来，我们用图9–32中的密钥来加密一条消息。加密的方式很简单，只要用每个字母所在网格的形状（包括其中的点）来代替这个字母就可以了。假设我们想要加密的消息是：“THE CODE IS MORE LIKE GUIDELINES, REALLY”（“说真的，这段密码更像是一段指导方针”），那么加密以后就会得到图9–33中这段密文。

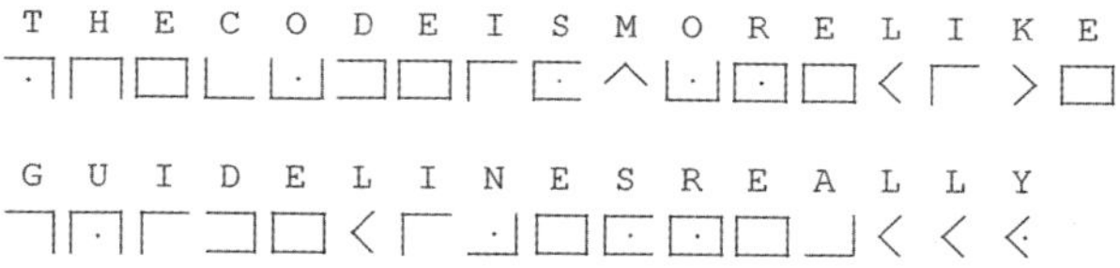

图9–33　用“猪圈加密法”加密后的密文

由于在这种加密系统中会用到一些像猪圈般的围栏，所以这种加密方式有时被称为“猪圈密码”（the pigpen cipher）。由于这种加密系统在共济会中被大量使用，故人们有时也会将这种密码称为共济会密码。

这种猪圈密码的密钥并不一定要采取图9–32中的顺序。在图9–32中，密钥的顺序是一个井字格、一个X形格，接着再是一个有点的井字格和一个有点的X形格。但是其他加密者也完全可以使用其他顺序的密钥，比如先是一个井字格，接着是一个有点的井字格，然后才是一个X形格，最后是一个有点的X形格。还有一些版本的密钥会用到每个小格中有两个点的网格。除了网格的变化以外，加密者还可以用不同的顺序将字母表中的字母放入网格中。

但是，不管网格和字母顺序如何变化，猪圈密码都只是一种MASC密码。在《未解之谜（上）》的第1章中，我们已经解释过，破译MASC密码是一件非常容易的事。但是，由于某些原因，至今也没有人能把拉布什的密码翻译成比较流畅的明文信息。哪怕是目前最好的解法也充满了令人摸不着头脑的错误，根本不能指望靠这样一段话去寻找什么宝物。难道拉布什在这段密码中使用了其他的加密手段？或者，拉布什会不会在加密时犯了太多错误，致使密文根本无法译出？又或者，拉布什密码也许根本就是一个骗局？

虽然猪圈密码应该是一种特别简单、容易破译的密码，但除了拉布什密码以外，我们手头还有另一段至今无人能破译的猪圈密码。这段密码出现在美国俄亥俄州的一块墓碑上。

在《未解之谜（上）》第5章讨论德博斯尼斯密码时，我们曾经提到一个叫布伦特·莫里斯的人。莫里斯不仅是共济会的第33级成员，还是美国国家安全局的一名密码分析专家。莫里斯对俄亥俄州的这块墓碑上的密码做了如下评论：

> 我怀疑墓碑上的共济会密码是一些共济会职位名称的首字母缩写，比如PM（Past Master，前大师）、PHP（Past High Priest，前大祭司）、PIM（Past Illustrious Master，前光荣大师）等。但是，这段密码太短，没法系统地对其破译。[58]

图9-34　俄亥俄州梅塔莫拉南边不远处的一块神秘墓碑

在编辑对本书进行审稿时，我到许多地方进行了演讲，在这些演讲中，我提到了本书中讨论的各种未解之谜。在其中一次讲座中，我提到了俄亥俄州的墓碑上的密码，当时，德国密码专家克劳斯·施梅恰好也在现场。后来，施梅把这段墓碑上的密码发布在他的博客上，在读者的共同努力下，这段密码终于被解开了。读者可以根据尾注中列出的来源找到他们提出的解法，[59]也可以尝试自己破译这段密码。在这里，我可以为打算自行破译密码的读者提供一条提示：最终的解答和莫里斯的判断并不一致。

其实，这座墓碑上的信息的具体内容不那么重要，重要的是，这段墓碑密码的解法也许能给《未解之谜（上）》第5章中提到的蝎子密码的破译工作提供一些启发。如果读者把书翻到《未解之谜（上）》第5章，就会发现蝎子密码中的有些符号看起来很像猪圈密码中出现的符号。如果这些符号确实与猪圈密码有关的话，也许解出这些符号就能找到破译蝎子密码的重要线索，并最终解开蝎子密码的全部明文。毕竟，在破译黄道十二宫杀手的3段密码时，破译者仅通过假设第一个单词是“I”（我）并假设文中出现过“KILL”（杀）这个单词，就成功地打开了缺口，并最终攻破了整段密码。

我必须得承认，当我向读者介绍与大量黄金有关的密码时，我觉得自己有点儿像麦克斯韦·斯马特（Maxwell Smart）①。

我：就在现在这个时刻，在弗吉尼亚州的贝德福德县埋着超过1吨的黄金。你相信吗？超过1吨的黄金！

你：我很难相信有这种事。

① 麦克斯韦·斯马特是美国电视剧及同名电影《糊涂侦探》（*Get Smart*）的主角，是一名笨拙、幼稚，经常注意力不集中的新手特工。——译者注

我：那么你信不信圣塔菲以北的某个地方埋着二十几磅黄金?

你：不信。

我：那你信不信我家后院有两枚25美分的硬币和一枚1美分的硬币?

第10章

未解的长密码

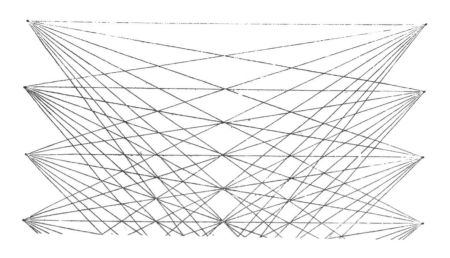

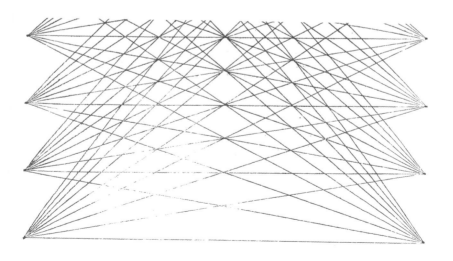

本书的第9章中，我们讨论了蝉3301密码。当时，我们提到了“书本密码”，即以一本书为密钥的密码。而在《未解之谜（上）》的第1章中，我们知道一本书本身也可以是一段密码，比如著名的伏尼契手稿。事实上，伏尼契手稿在这方面并不是唯一的，世界上存在许多长度足以与一本书比肩的密码，这其中当然有一些目前仍未被破解出来。在本章中，我会向读者介绍一些目前尚未被破解的书本长度的密码，以及其他一些长密码（这些密码虽长，但可能还没有达到一本书那么长）。

这些书本长度的密码至今未被破解是一件颇令人意外的事。因为一般来说，密码越长，破解的难度也就越低。事实上，当情报工作者截获一些长度较短的加密信息时，一种常见的破译策略就是积累尽量多的同类加密信息，然后放在一起进行破译。如果这些加密信息都使用同样的密钥，那么这组信息就被称为“深度信息”。从某些方面来看，一组“深度信息”非常类似于一段长密码或者一本用加密文字写成的书。当我们把使用相同密钥的一组信息放在一起时，破译难度会远远低于破译单条信息的难度。

德国密码研究专家克劳斯·施梅整理出了一份清单，上面列出了各种用加密文字写成的书籍。在笔者写作本书时，施梅的上述清单上的书籍还在不断地增加。[1]施梅在这份清单中对所有密码书系统地编了号，这个编号系统很可能会成为这一领域的标准。因此，在本书接下来的部分中，我会用施梅清单上的编号来指代这些加密书籍。

在施梅整理出的这份加密书籍清单上，第一本书是我们已经介绍过的伏尼契手稿，对应的编号是00001。从1前面的4个0中，我们可以看出，施梅做好了这份书单日后会变得非常长的准备。如果当年计算机系统全交给施梅编写的话，“千年虫”就根本不会成为问题了。在施梅的清单上，伏尼契手稿之后的第二本加密书是编号为00002的罗洪齐抄本。

罗洪齐抄本密码

罗洪齐抄本（*Codex Rohonci*）于1838年首次进入公众视野，但它的创作日期不详。罗洪齐抄本可能没有伏尼契手稿那么古老，但它共有450页，比伏尼契手稿长出许多。对于罗洪齐抄本的情况我们了解得并不多，拜奈代克·朗（Benedek Láng）在一篇2014年发表的论文中总结了目前知道的各种关于它的信息：

> 1838年，罗洪齐抄本和已故的古斯塔夫·包贾尼（Gustav Batthyány）伯爵的其他30 000本藏书一起被捐献给了匈牙利科学院。古斯塔夫·包贾尼伯爵的私人图书馆此前位于罗洪齐镇（即今天的奥地利雷希尼茨），因此这份抄本被命名为“罗洪齐抄本”。[2]

截至2005年，研究者已经对罗洪齐抄本提出了3种不同的解法，但这3种解法都没有获得广泛的认可。[3]

在本书的前言部分中我已经说过，除了各种未解之谜以外，本书也会讨论一些已经被破译出来的密码。我个人认为，罗洪齐抄本就属于已经被破译出来的密码。罗洪齐抄本是以基督教为主题的。事实上，最新问世的一种破译结果也的确证实了这一点。破解罗洪齐抄本的研究者已经把他们的破译结果写成了一篇论文，我读过这篇论文的一个早期草稿版，并且非常期待能在未来的某期《密码学》杂志上看到这篇论文的完整版。届时，读者也可以自行阅读这篇论文，做出自己的判断。

在大部分情况下，密码都是“独立”出现的。当然可能有一些关于它们的背景信息，但通常不会伴随大量的“附件”出现。而我们的下一个例子却是一个反例。和罗洪齐抄本一样，下一段密码的内容也与宗教有关。

詹姆斯·汉普顿的域外艺术[1]

1950—1964年间，迈耶·韦特利布（Meyer Wertlieb）将位于美国华盛顿特区的一间车库租给了一位名叫詹姆斯·汉普顿（James Hampton）的房屋管理员。后来，由于没有收到汉普顿的租金，韦特利布前去这间车库检查，却发现了很多相当复杂精巧的艺术作品。

汉普顿创造的这些艺术作品究竟是什么？他又是出于什么原因创造出这些作品的呢？遗憾的是，韦特利布已经没有机会向汉普顿提出以上问题了，因为当他看到这些作品时，汉普顿已经去世了——这也是汉普顿未能及时向韦特利布交纳租金的原因。汉普顿留下的关于他本人的信息并不多，为了理解汉普顿留下的这些作品，我们有必要仔细看一看这些信息。

汉普顿1909年出生于美国南卡罗来纳州的埃洛里镇。1910年的人口普查数据显示，埃洛里镇的总人口只有540人。汉普顿的父亲是一名浸礼

① 域外艺术，英文作“Outsider Art”，是指未经专业训练的人所创作的艺术作品。——译者注

会牧师兼福音音乐歌手，但这个牧师资格是他自己颁发的。后来，汉普顿的父亲抛弃了妻子和4个孩子，开始了四处旅行传教的生活。我们几乎查不到任何关于汉普顿的教育情况的信息，在一份工作申请材料中，汉普顿自称有10年级的教育水平。但是，在这份申请材料中填写的学校里却找不到汉普顿的任何记录。

1928年，汉普顿离开了南卡罗来纳州的乡村，来到了华盛顿特区，这一变化可能给他带来了巨大的文化冲击。在华盛顿特区，汉普顿和他的哥哥李（Lee）住在一间公寓里。1931年，汉普顿开始出现幻视，我们并不清楚这种症状是否与他来到华盛顿特区受到的文化冲击有关。汉普顿相信，他亲眼看见了专程拜访自己的上帝和他的天使，他还相信，上帝和天使们让他为即将再次来临的耶稣基督制造一个宝座。根据汉普顿的说法，在他接下来的日子中，上帝和天使不断来访，要求他制造宝座，他还将上帝和天使访问他的情况做了记录。在其中一份记录中，汉普顿写道："真的，在1931年4月11日，向人们传达十诫的伟大摩西出现在了华盛顿特区。"[4]

目前我们并不清楚汉普顿究竟从何时开始按照上帝的要求制造宝座。如果他在20世纪30年代已经开始的话，那么这一工程显然被第二次世界大战的爆发打断了。从1942年开始，汉普顿一直在第185航空中队服务，直到1945年光荣退伍。在此期间，汉普顿随队到过得克萨斯州、西雅图、夏威夷、塞班岛，以及关岛等地。汉普顿在军队中主要负责木工和机场维护工作，他并没有参加过任何战斗活动。

1946年，退伍后的汉普顿在华盛顿特区找到了工作，他成了政府服务管理局的一名房屋管理员。汉普顿住在一个公寓房间里，但他认为自己需要更大的空间来建造耶稣的宝座——汉普顿把这个宝座称为"国家千禧年联合大会第三天堂宝座"。因此，他才会从1950年开始租用韦特利布的车库。汉普顿似乎将建造宝座当作自己的人生目标。每天午夜下班以后，汉普顿会投入5~6个小时用来建造这些宝座。汉普顿相信，上帝会定期访问他租下的这间没有供暖设备的车库，来监督宝座的建造情况。

当时，汉普顿一个人生活，人们把他形容成一个“贫穷可怜又没有任何朋友的人”。[5]他的哥哥于1948年去世。虽然汉普顿希望有一名“神圣的女人”出现在他的生活中，帮他一起完成宝座的建造工作，但他终其一生也没有找到她。

那么，汉普顿究竟是如何建造耶稣的宝座的呢？因为预算很紧，他使用的材料主要是大量铝箔。在汉普顿死后，研究者搜集整理出了一些关于他的资料，从这些资料中我们可以了解他收集宝座材料的方式。

> 为了寻找金箔和铝箔，他会清扫附近街区的各种垃圾，比如商店的陈列品、香烟盒、厨房纸等。他甚至从附近的流浪汉那里购买他们酒瓶上的那一点儿铝箔。不管走到哪里，他都随身带着一个口袋存放在街上找到的任何他认为有用的材料。此外，他（似乎）还从他工作的政府大楼的垃圾堆里收集东西——电灯泡、桌子上的垫板、塑料布、隔热板、包装纸等。[6]

汉普顿通过上述方法累计创作了180件作品。

据汉普顿的一位女同事说，汉普顿希望在退休后成为一名牧师。汉普顿生前并不是该地区教堂的成员，但有人猜测，汉普顿可能希望退休后能在沿街的教堂里建立一个牧师群体。不管这一猜测是否属实，汉普顿并没能实现自己的梦想——他还没退休就去世了。

汉普顿去世以后，他的一位姐妹为了认领他的遗体而来到华盛顿特区。在这里，她终于首次见到了汉普顿付出毕生精力造出的这些宝座，但她并不想要这些作品。

由于某些原因，美国国家美术收藏馆（现已更名为“史密森尼美国艺术博物馆”）愿意接收汉普顿的作品。国家美术收藏馆支付了汉普顿拖欠的房租，并于1970年接收了全部180件作品。1971年，汉普顿的作品首次在国家美术收藏馆的“隐藏面”（Hidden Aspect）展览中公开展出。

好了，下面我们终于要谈到密码了！汉普顿死后留下了一个奇怪的笔记本，上面写着许多密码。我之所以要如此详细地介绍汉普顿的生平，就是因为这些背景信息有可能帮助我们破译汉普顿留下的密码。汉普顿留下的这个笔记本大约有70页，是博物馆的工作人员在汉普顿租用的车库里发现的。

汉普顿相信，他书写的这些奇怪的文字是上帝告诉他的。在这本笔记中，每一页奇怪的文字下方都写有“REVELATiON（启示）”的字样。此外，汉普顿还在每一页的上方写下了自己的名字——他给自己取名叫“ST JAMES（圣詹姆斯）”。他给自己取头衔的水平似乎超过了他的父亲。

汉普顿制造的宝座由许多部件组成，上面贴着标签，并写着与笔记本相同的密码。但是，在这些标签上一般先写有一个或者多个明文英文单词，然后才出现密码文字。如果说标签上的这些密码对应的明文就是同一个标签上的英文单词，那么我们应该能够通过这种宝贵的配对关系来解释这个密码。然而，到目前为止，尚没有任何人能成功破译汉普顿留下的这些密码！有人怀疑，汉普顿留下的这些奇怪的文字根本没有任何意义，但是，显然这些文字对汉普顿而言是有意义的。汉普顿认为这些文字是什么意思？我们有没有可能找到这个问题的答案呢？

斯蒂芬·杰伊·古尔德（Stephen Jay Gould）称自己是一名“犹太教不可知论者”，[7]因此他显然不会以汉普顿的方式去理解这些宝座。但是古尔德却在自己的论文中对汉普顿的作品滔滔不绝地表示了赞美，并称这些作品是“最优秀的美国民间雕塑作品之一”。[8]古尔德这样解释自己对汉普顿宝座的看法：

> 我并不是一位美术史学家，我不能从审美的角度解读汉普顿的作品，也不能对他的作品给出专业的评价。我只能说，单纯从我个人的角度来看，汉普顿的“宝座”给我带来了震撼，也令我感到愉悦。第一次见到“宝座”时，我正在史密森尼美国艺术博物馆开会，在会间休息时，我恰巧看到了汉普顿的这些作品。此后，我又多次专程去那里观看，每次看到这些作品，我仍会如第一次一样发自内心地感到快乐和敬畏。[9]

我可无法体会古尔德的这种感受。如果我从一位去世的亲戚那里继承了这样一些作品，我会把这些作品放在街边，等垃圾车把它们收走。我唯一感兴趣的是汉普顿留下的密码。

但是，我又有什么资格去评判汉普顿的作品呢？我自己的艺术水平有多高呢？请看一下图10–1：这是占据了我家一整面墙的一件6英尺见方的艺术作品，这件作品是我用瓶盖、纸张、磁板和木头做成的。

图 10–1　这是一件艺术品！[10]

潘尼坦沙手稿[10]

在前文中，我们已经讨论过戈登·鲁格对另一本加密书——伏尼契手稿所做的工作。但是，鲁格还与另外两种至今未解的长密码有关，这次他不是要破译密码的研究者，而是密码的创造者。鲁格将这两种密码分别命名为“潘尼坦沙手稿”（The Penitentia Manuscript）和“里卡德斯手稿”（The Ricardus Manuscript）。2005 年之后，“潘尼坦沙手稿”和“里卡德斯手稿”的全部内容都可以在互联网上找到。

因为鲁格是一名无神论者，所以我们猜测他的两本手稿内容恐怕会和罗洪齐抄本以及汉普顿笔记本密码的内容差别较大。但是，鲁格也曾发表过“圣经内容分析”方面的论文，所以，谁又能说得准呢？不管怎么说，接下来让我们来看看鲁格创造的这些密码。如果能够揣摩出鲁格头脑中的某些想法，我们也许就能猜到他创造这些密码的方式了。

在潘尼坦沙手稿的网页上，鲁格写下了以下这段话：

> 大部分现代密码都是基于一组同样的前提假设。我想知道，如果在创造密码时故意无视这些前提假设会怎么样呢？如果抛弃这些假设的话，会创造出什么样的密码呢？

图 10-2　戈登·鲁格（1955—　）

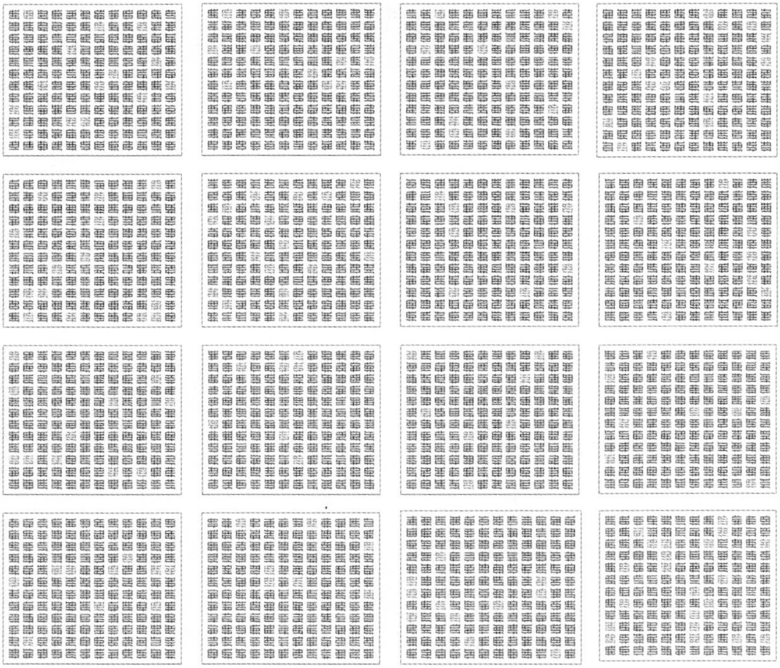

图 10-3　潘尼坦沙手稿

很明显可以看出，鲁格在创造上述密码时肯定抛弃了一个常见的假设，那就是：密码的密文必须由字母、数字，或者字母和数字组成。他创造出的密码形式如图 10–3 所示。

图 10–3 中的内容太多，我们很难看清每个图形的细节。为了方便读者细看这段密码，鲁格对该图进行了特殊设计，只要点击图上的任何部分，就可以看到放大后的局部图。因此，我们点击图 10–3 中的各个部分后，就会得到类似图 10–4 这样的局部图。

图 10–4 中的每个图标都相当复杂，很难想象这么复杂的图标只表示一个字母。我认为图中的每个图标可能表示两个或者更多个字母。此外，

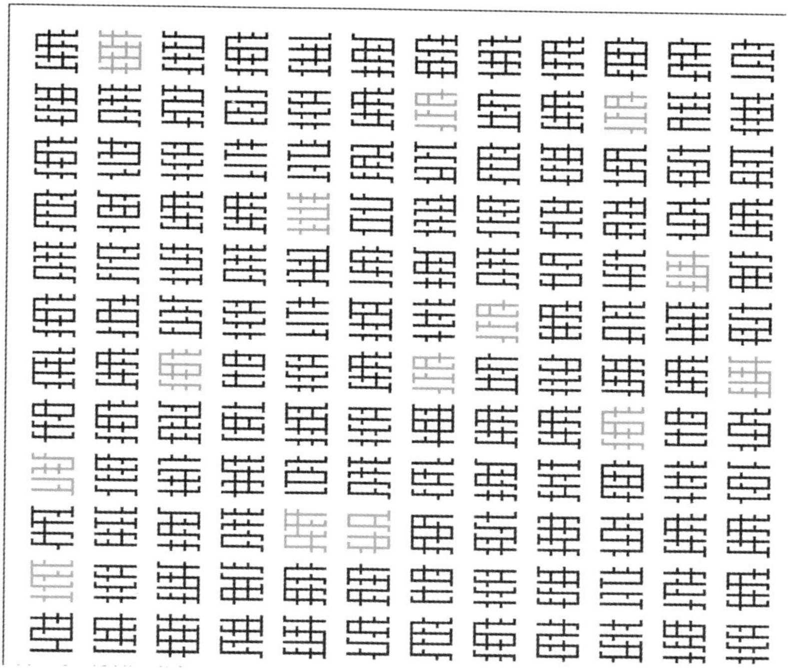

图 10-4　潘尼坦沙手稿位置（1，1）上的内容

我认为鲁格也许还抛弃了文字必须从左向右读或者从上向下读的模式。还记得《未解之谜（上）》第1章中鲁格对伏尼契手稿提出的“格栅”理论吗？也许，鲁格通过把格栅放在不同的位置形成了某种路径，然后再根据这条路径把字符放入格栅中，说不定在上述过程中他还通过旋转格栅来进一步增加破译的难度。

如果我的以上几点猜测都正确的话，我们就需要解决以下两个问题了：

1. 搞清楚每个符号如何与多个字母的组合对应；
2. 搞清楚这些符号应该用怎样的顺序读。

我认为，我们先要研究第1个问题。一旦知道了图标与字母的替代关系，我们就可以把密文的字母概率和英语的字母概率进行比较（假设这段密码真的是用英语写成的话！），从而得到一些有帮助的结论。

当然，我的上述猜想可能是完全错误的！也许，在大致了解鲁格的背景信息以后，读者能够想出一些其他可能性。

在密码学领域发表论文的人通常都有数学或者计算机科学方面的教育背景。当然，只研究历史密码的学者可能并没有上述两个领域的教育背景。然而，鲁格的情况与上述两种情况都不相符。鲁格的第一个学位是雷丁大学的法语和语言学学位。接着，鲁格又在同一所学校取得了实验心理学的博士学位。博士毕业以后，鲁格在诺丁汉大学从事人工智能方面的博士后研究，他还担任《专家系统：知识工程学和神经网络学国际期刊》（*Expert Systems: The International Journal of Knowledge Engineering and Neural Networks*）杂志的主编。

因为有以上这些丰富的教育背景，鲁格参加的工作领域也十分广泛，包括英语授课、实地考古、人工智能、信息检索，以及人因学①。

对自己目前的工作情况，鲁格是这样描述的："我是基尔大学计算机学院的高级讲师，也是开放大学计算机学系的客座高级研究员。"[11]

对某些聪明的人来说，这个世界就像一个充满智力挑战的游乐场。一开始学习法语，最后却成为一名计算机科学教授，这又有什么？永远不要因为自己的教育背景和目前的工作岗位而设置人为的限制。我认识一名非常优秀的计算机科学教授，上过他的许多课。而这位教授却向我坦白说，他自己从来没有上过一节与计算机科学有关的课，他的计

① 人因学是研究人和机器的相互作用的一门科学，该学科的研究目的是使人们设计出的机器系统更适合人的生理及心理特点，从而提高生产效率或者安全性等。——译者注

算机科学知识完全是自学而来的。就在几年前，这位教授还是一位化学系终身教授呢。他之所以选择转行，只是因为厌倦了化学，需要一些智力方面的新挑战。

我让鲁格提供更多关于他的教育背景的细节，在鲁格给我的回复中，有这样一段话[12]：

> 语言学方面的学习经历让我认识到：对于同一个研究课题，人文学科和自然科学学科的研究方式是截然不同的。这一点深刻地改变了我的思维方式。此外，我还发现：即使一群聪明人已经对某个问题研究了若干个世纪，仍有可能存在一种前人都没想到的新的研究方法，而这种新的方法可能比现存的方法强大许多。

鲁格还提到，他曾在尼泊尔担任过英语讲师。

> 在尼泊尔的经历让我看到了什么是真正的贫困，也让我看到了世界上很多人的生活究竟是什么样子。如果我得了重病，我可能会被直升机送到医院，并获得目前世界上最好的治疗。然而如果我在尼泊尔的朋友得了重病，他们却只能等待死亡。我无法相信这样的情况是公平和正确的。对我来说，反对不平等和贫困的信念是我的个人信仰。

除此之外，鲁格还曾经参加过考古学的实地工作。

> 在一次挖掘工作中，我所在的团队发现了英国最早的基督教教堂之一。这座教堂建在一个罗马神殿的旧址上，而这个罗马神殿则位于一个铁器时代的宗教遗址上。

鲁格还谈到了他在教学方面的一些经历，他写道：

我曾经用风筝和一把光剑作为教学道具。有人曾描述我如何用一把弩从学生的头顶上射下一个苹果，这有些夸大其词了。

鲁格还描述了他与学生以及一位同事共同进行的一些研究工作：

我以前教过的几位本科生最近在他们本科研究项目的基础上，开始与我和贾森·多德斯韦尔（Jason Dowdeswell）合作进行研究。多德斯韦尔是好莱坞的一位著名的视觉效果专家。[13]

我和我的同事埃德·德坎塞（Ed de Quincey）博士合作开发了一个搜索视觉化（Search Visualizer）软件，这个软件可以做到谷歌搜索引擎做不到的事情。我们合作研究了如何寻找推理中的谬误，这方面的研究使我们发现了一系列被其他人忽略的可能情况。[14]

以上这些信息是否对你有所启发呢？也许通过研究鲁格的背景信息，我们可以对他加密的内容或者使用的加密方式进行一些猜测。

里卡德斯手稿

下面我们来介绍鲁格创造的第2段未被破译的密码。这段密码也无法直接输入计算机中。鲁格创造的这段密码显然受到伏尼契手稿的启发。这段密码被称为“里卡德斯手稿”。在图 10–5 至图 10–9 中，我复制了一些里卡德斯手稿中的内容。[15]

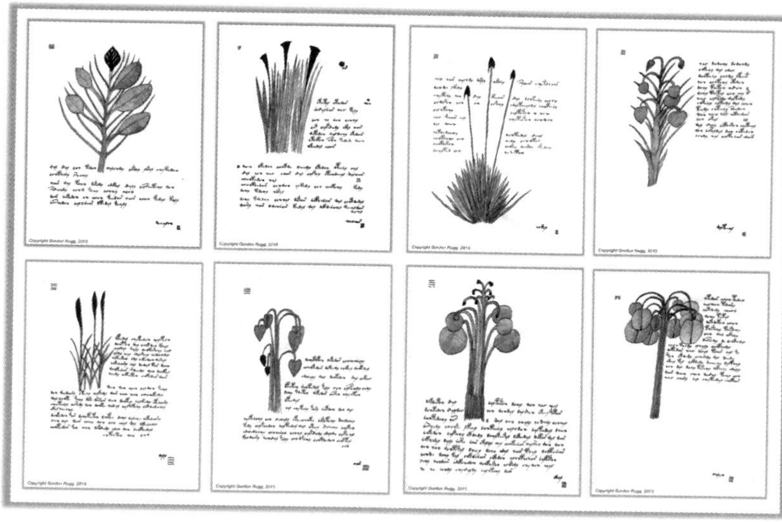

图 10–5　里卡德斯手稿

鲁格为里卡德斯手稿创办了一个网站，在这个网站上可以找到这些图片的放大版本。[16]以下是手稿中的 4 页放大图。

鲁格说他会把潘尼坦沙手稿和里卡德斯手稿分别印在帆布上并亲笔签名，奖给首位破译这两段密码的人。（当然，首位破译上述两段密码的人还会获得向别人炫耀的资本！）可以看出，鲁格的作品具有较强的艺术性。接下来，我要向读者介绍下一位未解长密码的创造者，他的作品也同样将计算机科学和艺术这两个领域结合在了一起。

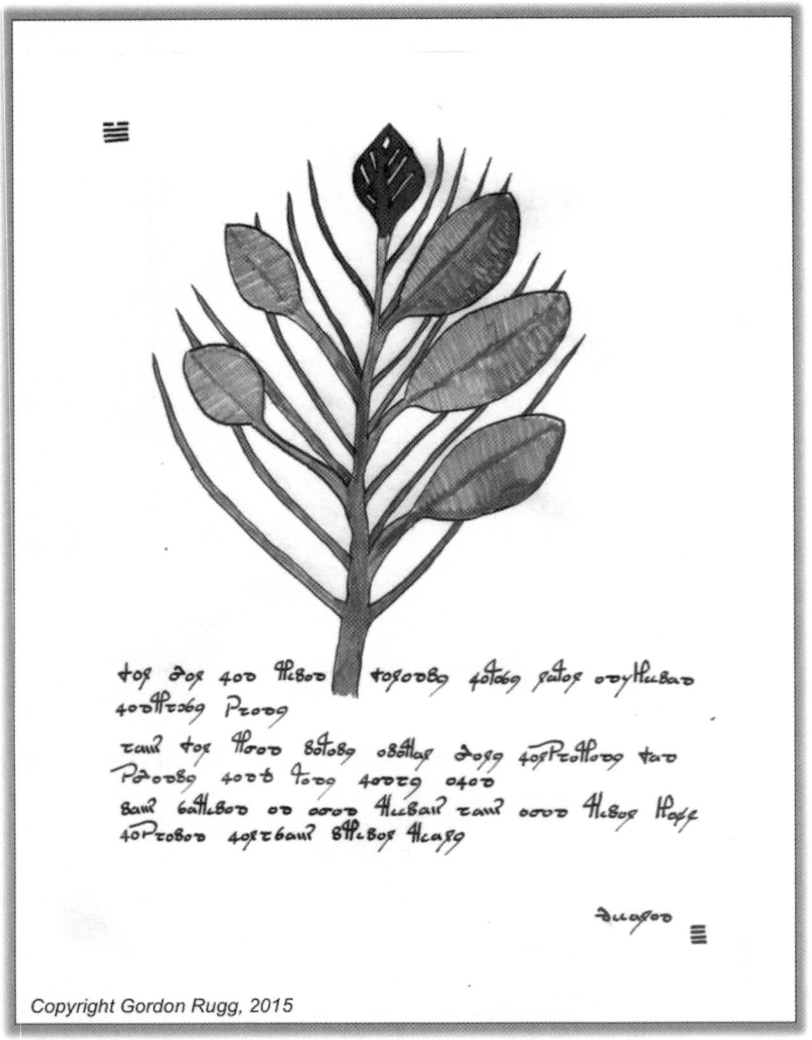

图 10-6 里卡德斯手稿，植物 1

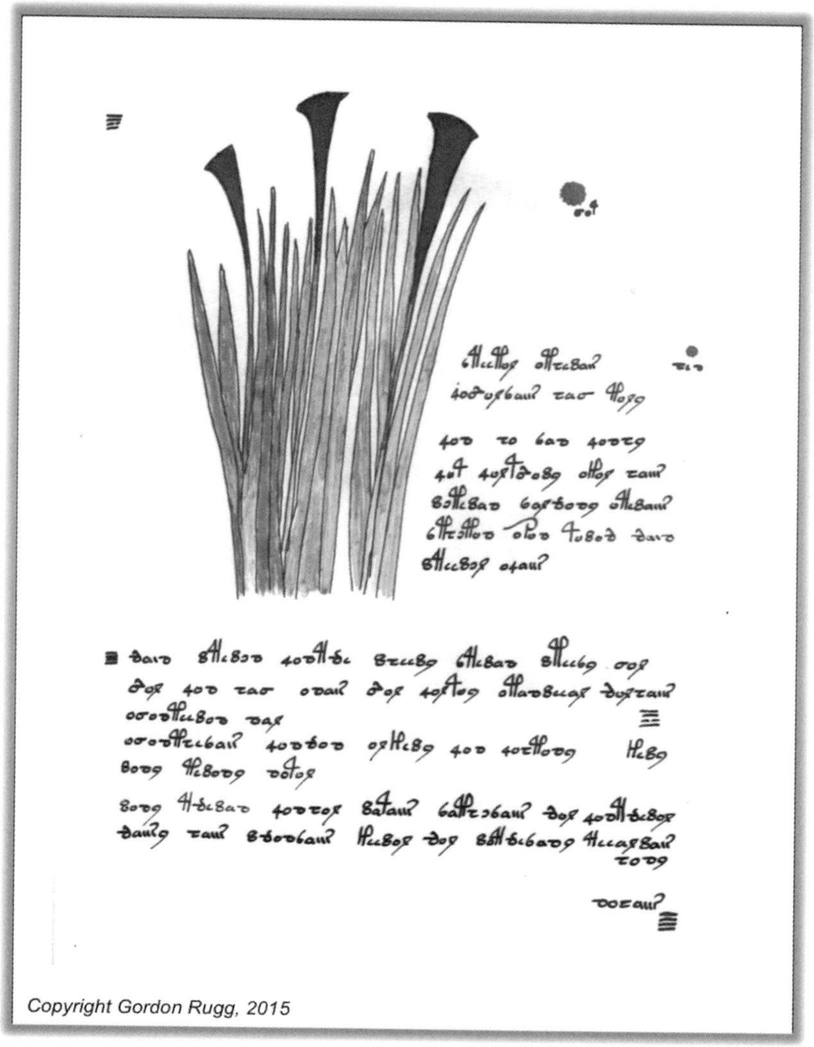

图 10-7　里卡德斯手稿，植物 2

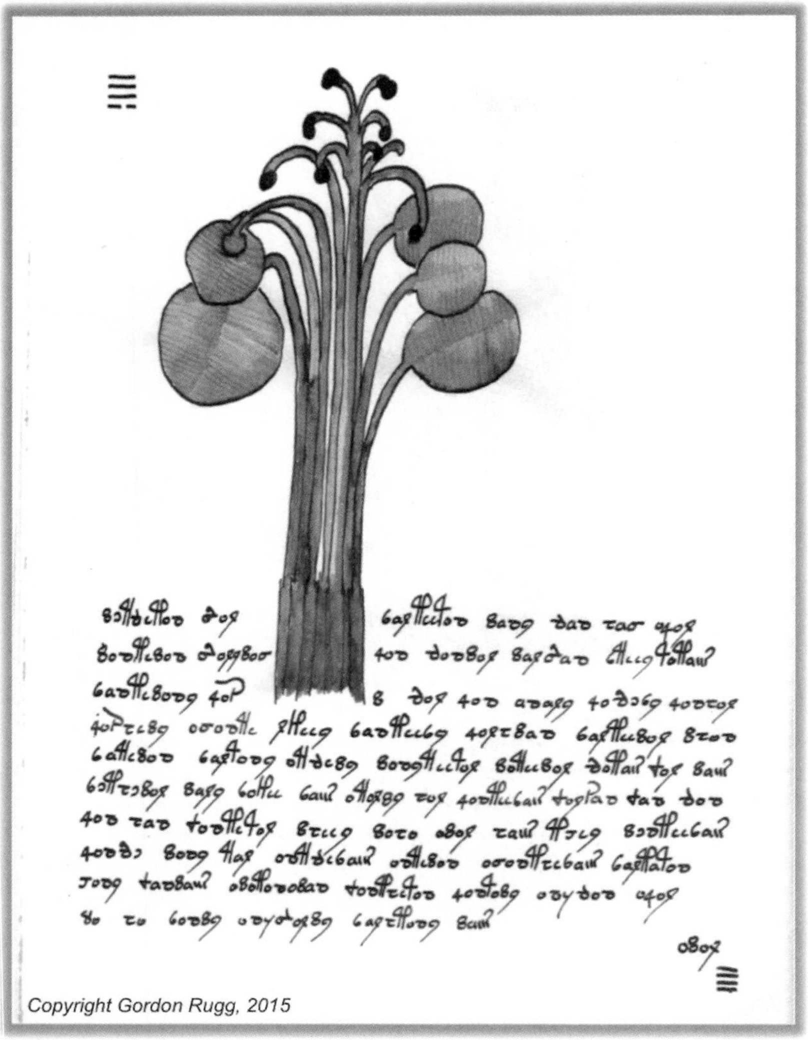

图 10-8　里卡德斯手稿，植物 7

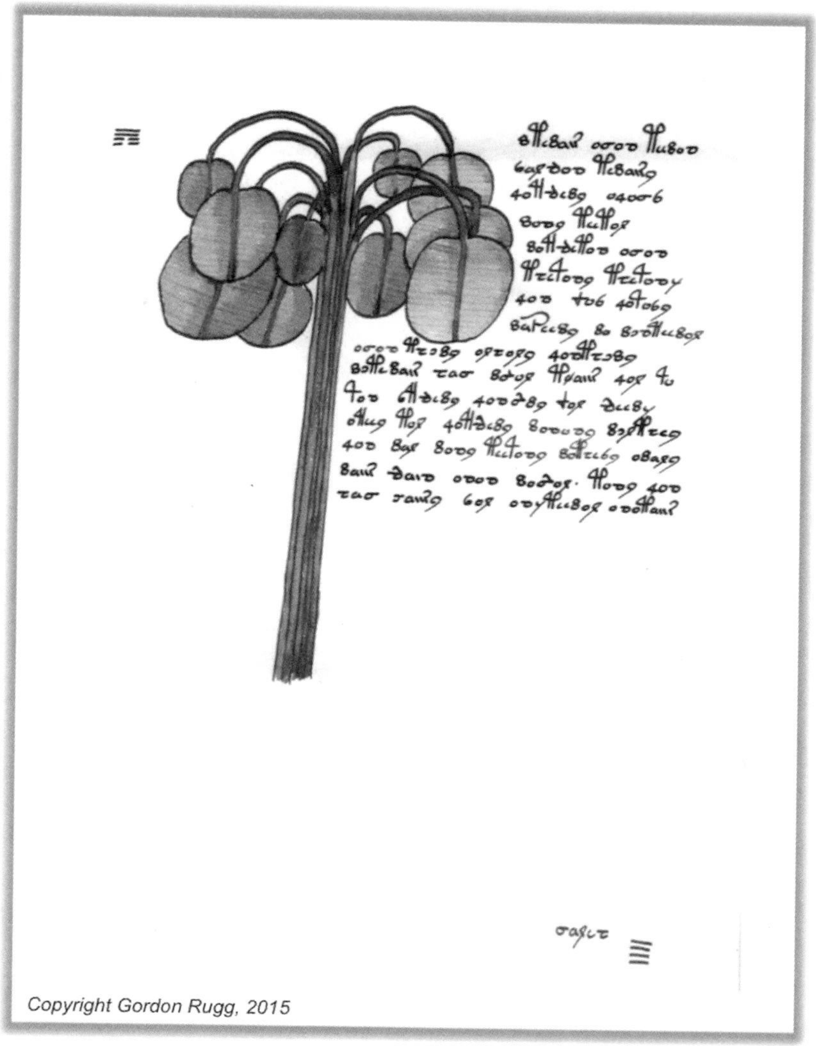

图 10-9　里卡德斯手稿，植物 8

吴之书密码 [17]

《桑德拉和吴》（*Sandra and Woo*）是由奥利弗·克内策尔（Oliver Knörzer）创作、普里·鲍里（Puri Powree）执笔绘制的一组网络漫画。奥利弗·克内策尔是一名拥有计算机科学学位的德国人。在介绍前两种密码时，我们看到了一个拥有法语学位的人在计算机科学领域中的成就。而在下文中，我们将看到一个有计算机科学学位的人如何在艺术世界中展开创作。但是，漫画创作目前并不是克内策尔的全职工作。他解释道："我在法如科技公司（FARO Technologies）担任软件研发工程师，在公司的研发部门工作 ，目前我正在研发一种网络服务，人们可以通过这种网络服务看到由激光扫描仪获取的三维数据。真是令人激动！"[18]

克内策尔通过图10–10中的这幅漫画引入了这段至今未解的密码。图中是该系列漫画的第500期，叫作《吴之书》（*The Book of Woo*）。

图 10–10 《吴之书》密码的简介

我与克内策尔通过电子邮件进行了一些交流。他告诉我，他创造的这幅漫画受到一些其他漫画的影响，其中对他影响最深的漫画之一是《凯文的幻虎世界》(*Calvin and Hobbes*，又译《卡尔文和霍布斯》)。但带领他走进网络漫画世界的作品则是D. C. 辛普森（D. C. Simpson）的《奥兹和米莉》(*Ozy and Millie*)。他还告诉我："我特别喜欢浣熊，也了解很多关于浣熊的知识。在维基百科网站上，关于浣熊物种的几乎所有德语和英语条目都是我写的。"[19]通过维基百科上这些由克内策尔创作的条目，试图破译吴之书密码的人也许能够了解到他在词语选择方面的一些偏好和习惯。

接下来的4页（图10–11至图10–14）是吴之书密码的具体内容。

克内策尔说，他创作的这段密码获得了"压倒性的正面评价"。这段密码发表以后，创下了读者评论数量的新纪录。一位网名叫"Foogod"的读者甚至为吴之书密码创建了一个维基百科页面（维基百科页面允许使用者对其内容和结构进行共同编辑）。这位读者认为，有了这个维基页面，有兴趣破译吴之书密码的人就能够更容易地进行合作了。[20]

吴之书密码的维基页面上提供了许多有用的信息，其中包括一条本人留给破译者的提示。他写道：

> 我决定给你们一个小小的提示。吴之书密码的明文中出现了以下这个单词：
>
> 英语：Potbelly Hill（大肚子山）| 德语：Bauchigen Hügel
>
> 读者应该很容易判断出这个词出现在密码的哪一页中。我希望这条线索能够为本网页应当包含的内容提供一些讨论的素材。[21]

当然，任何一段密码的内容和加密方式都一定会受到密码创造者的背景的影响。克内策尔向我们透露了以下内容：

图 10-11　吴之书密码，第 1 页

图 10-12　吴之书密码，第 2 页

图 10-13 吴之书密码，第 3 页

图 10–14　吴之书密码，第 4 页

> 在大学期间，我通过几门课程了解到了一些密码学的基础知识。我对密码学产生了特殊的兴趣，但却并不想以此作为我的职业。后来，我发现了尼克·佩林的博客“密码谜团”（http://www.ciphermysteries.com/）。这个博客上的内容非常广泛，再次激起了我对密码学的兴趣，尤其是对伏尼契手稿的兴趣。我感到，在这些几十年，甚至几个世纪都未被解开的密码文字中，包含着某些特别迷人的东西，而伏尼契手稿是这些密码中最为神秘的一种。我第二喜欢的密码是黄道十二宫杀手的Z340密码。[22]

克内策尔承诺，如果截至该漫画第1 000期发行时还没有人能够破译吴之书密码，他就会在第1 000期漫画中公布这段密码的解法。他估计，这系列漫画的第1 000期会于2018年春末或者夏初与读者见面①。但是，目前破译吴之书密码的工作已经取得了一些十分重要的进展。克内策尔写道：

> 读者“Satsuoni”已经破译了最后一个加密步骤。此后，大家对部分破译出来的密码文字进行了进一步分析，从而产生了一些十分有趣的统计学结果。然而，虽然有些读者对上述统计结果提出了一些看似很有希望的解读，但到目前为止，我尚未看到任何决定性的突破……如果你有兴趣了解这些分析结果和理论，你应该去看一看吴之书密码的维基网页。[23]
>
> 一定有许多人花费了大量的空闲时间来试图破译我的密码。作为这段密码的作者，这当然让我感到非常高兴。[24]

① 2018年6月28日，克内策尔如约在网站上公布了吴之书密码的解答及加密思路，见http://sandraandwoo.com/2018/06/28/0500-the-book-of-woo-in-english/。——编者注

读者不妨去访问吴之书密码的维基页面，看看你能否为这段密码的破译做出一些贡献。你的贡献或许能让作者感到更加高兴！

读到这里，读者可能已经发现，本章介绍的这些至今未解的长密码都有一个共同特点：这些密码的密文都不是由传统的字母或数字组成的。到目前为止，本章介绍的所有密码都是由一些奇怪的符号组成的。

除了这一点以外，本章介绍的几种密码还有另一个共同特点：破译者都不知道密码的加密方法。如果破译者知道这些密码的加密方法，仅仅不知道密钥的话，很可能本章中介绍的所有密码都已被解开了。在军事密码中，加密方总是希望能对自己使用的加密方法进行保密，但是，他们同时也会假设敌方能够猜出他们所使用的加密方法。这是因为，敌方总是可以通过很多方式判断出密码所使用的加密方法。如果密码采用的加密方式是某种常见的加密系统，那么敌方要猜出加密方式就会变得很容易。

在前文中，我向读者介绍过克劳德·香农在熵、冗余度、唯一点等概念上的贡献。香农有一句著名的格言："敌人知道我们的系统。"但其实早在1883年，奥古斯特·克尔克霍夫（Auguste Kerckhoff，也译作柯克霍夫）就已提出了这句格言所包含的意义。克尔克霍夫提出，好的密码必须满足以下6条原则，称为"柯克霍夫原则"。"柯克霍夫原则"的内容如下[25]：

K1. 加密系统如果不能做到理论上无法破解，也至少要做到实践上无法破解。

K2. 加密系统泄露后不会给我方的通信人员带来不便。（这就是香农那句名言的意思，那就是，即使敌方知道我方的加密系统，我方也不需要做任何改动。而且，在没有密钥的情况下，就算知道我方的加密系统，敌方也应该无法破译我方的密码。）

K3. 选择特定加密系统（密钥）的方式应该易记忆，也易修改。

K4. 密码的密文必须能用电报传输。（今天，这条原则应该改成“密码的密文必须能用计算机传输。”）

K5. 密码设备必须易于携带。

K6. 使用该加密系统不应依赖于太长的规则，也不应该让使用者花费太多心力。

知道了这6条原则以后，我们再回头看看本章中介绍的密码。在本章中介绍的密码中，有几种密码违反了不止一条柯克霍夫原则呢？大部分吗？还是全部？本章中介绍的密码都违反了不止一条柯克霍夫原则！不过这也没有关系，毕竟这些密码不是用来保护重要军事机密的。本章中的大部分密码都是为娱乐目的而设计的（也许前两种密码除外）。

接下来，我要向读者介绍本章的最后一种密码。这段密码本应放在第9章中，因为它是2014年蝉3301挑战密码的一部分。我之所以把这段密码放在本章中，是因为这段密码有58页之长。我将这段密码的第1页复制在了图10–15中。感兴趣的读者可以在网上找到这段密码的其他部分。[26]

蝉3301如尼文秘符书

在前文中，我向读者介绍过不少看上去简单、实际上却极难破译的密码。接下来要介绍的蝉3301如尼文秘符书又是一段这样的密码。它看起来只是一段MASC密码，虽然用了一套比较特殊的字母表。用如尼文字母进行加密的系统几乎随处可见，《魔戒》（*The Lord of the Rings*）的防尘套上就有这种密码（在 J. R. R. 托尔金的写作事业成功之前，他曾受过密码破译方面的训练），奥齐·奥斯本（Ozzy Osbourne）的唱片《说到魔鬼》（*Speak of the Devil*）的封面上也有这种密码。但是，蝉 3301 如尼文秘符书中的密码却和上面这些密码完全不同。

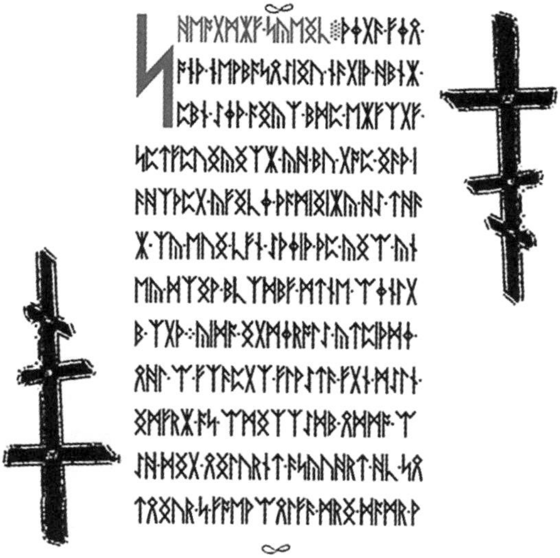

图 10–15　2014 年的蝉 3301 如尼文秘符书的第 1 页

2014 年的蝉 3301 密码比赛步骤烦琐，如果要详细介绍该比赛的所有细节的话，非得单独写一本书不可。在这次密码比赛中，好几种密码的页面上都有看起来像是如尼文字母的字符，和图 10–15 中的密码看起来非常类似。但是，那些密码并不是 MASC 密码，而且那些密码已经被成功破译了。然而，成功破译那些密码的技巧只适用于我们讨论的这本如尼文秘符书的其中一页内容，后来，整本书也出现在了 2014 年的蝉 3301 密码比赛中。

对于蝉3301如尼文秘符书，我将再次把探索谜题的机会留给各位读者。在本章结尾，我想提醒读者注意，未解密码的规模正在快速地增长。在我写作本书时（2017年2月20日），克劳斯·施梅整理的书单中一共包含85本加密书籍。在这85本加密书籍中，有许多已经被成功破译，但我相信，当本书英文版印刷出版时，施梅的这份书单肯定会变得比现在更长，而且，其中一定会包含更多的未解之谜。感兴趣的读者可以去看一看！[27]

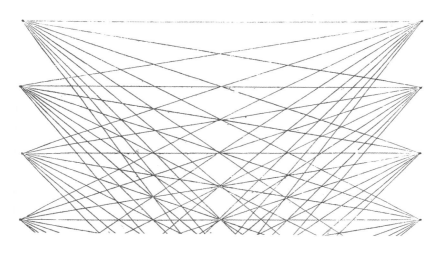

第 11 章
外星人密码和RSA算法

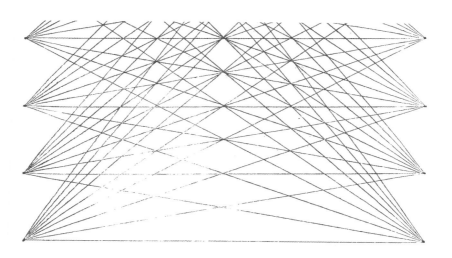

来自外星文明的加密信息可以算得上是一种终极形式的密码了。不管是外星人有意发送给我们的，还是我们截获的信息，都充满了神秘的色彩。在密码的历史上，曾多次出现过这类号称来自外星人的信息，但这其中有些信息非常简短。

火星人之光

1900年12月，世界各地的报纸在头条刊登了一条令人惊讶的消息——火星人正在向地球发送消息！这些报道称，火星人通过一些闪烁的光束向我们发来了信息，这种闪光持续的时间有时较长，有时较短。据称，这种“火星人之光”是由一位天文学家发现的。然而，当这位天文学家读到这条消息时，他本人却感到大吃一惊。

这位发现“火星人之光”的天文学家就是爱德华·C. 皮克林（Edward C. Pickering）教授，当时他是哈佛大学天文台台长。虽然皮克林确实观测到了一种有趣的闪光现象，但他从未说过这种闪光是火星人在给地球发送信号。在一篇名为《来自火星的信号，皮克林教授看到了火星上发出的明亮光束》的报道中，皮克林教授对这种闪光现象做出了解释。当然，如果不认真阅读这篇报道的详细内容，标题本身恐怕会加深我们的误解。然而，认真阅读过这篇报道的读者会发现，皮克林教授在该文中解释说，他于12月8日收到了一份洛厄尔天文台（位于亚利桑那州弗拉格斯塔夫）发给哈佛大学天文台的电报，这封电报的内容如下：

> 12月7日，A. E. 道格拉斯（A. E. Douglass）先生在观测火星的时候发现伊卡利亚海的北侧边界上出现了投射，且持续了70分钟。

这封电报里提到的“投射”是指光束。[1]收到上述电报以后，皮克林教授把这个消息转到了德国的基尔市，这条消息然后又从基尔市传到了整个欧洲。哈佛大学和德国的基尔市是传播天文学消息的两个枢纽。

消息传开后，法国巴黎的威尔弗里德·丰维埃尔（Wilfrid Fonvielle）将上述天文现象解读为火星人发给地球的信号。丰维埃尔将这种理论发表在了欧洲版《纽约先驱报》（*The New York Herald*）上，这条经过添油加醋的消息转了一圈后，最终又传回了美国。

关于火星光束的早期报道给性情古怪的电学天才尼古拉·特斯拉（Nikola Tesla）提供了进一步启发。特斯拉一直相信，地球人可以通过无线电信号与火星取得联系。他甚至还声称自己收到过来自火星智慧生命的无线电信号。

在各种谣言和猜测中，皮克林教授多次向媒体澄清自己从未说过收到火星人的信息。1901年3月，皮克林教授在《科学美国人》（*Scientific American*）杂志上发出感叹：“在欧洲，人们说我在和火星人通信，各种各样夸大事实的报道甚嚣尘上。不管那些光束到底是什么，我们都无法通过任何途径确认光束的性质。那些光束是不是智慧生命发出的，谁也说不准。这件事情目前根本无法解释。”[2]

虽然皮克林教授认为火星光束是完全无法解释的现象，但是不久之后，珀西瓦尔·洛厄尔（Percival Lowell）找到了一种能解释上述现象的理论。洛厄尔将其发表在了1901年12月的《美国哲学学会会刊》

（*Proceedings of the American Philosophical Society*）上，我们下面会谈到这篇文章的具体内容。

从前面的电报中我们已经知道，最先发现火星光束现象的是亚利桑那州的天文学家A. E. 道格拉斯教授。道格拉斯教授当时恐怕并未预料到这一发现后来会在公众中激起如此巨大的反响。他后来写道：

> 去年12月7日被描述为“来自火星人的消息”的现象事实上只是行星上的一团云被落日照亮了而已。那确实是一条来自火星的“消息”，它给我们提供了关于火星气候的情况，但那并不是来自智慧生物的消息。[3]

虽然天文学家一再声明火星光束并不是来自智慧生物的消息，有人却不愿意接受这种略显平淡的解释。1923年，查尔斯·福特（Charles Fort）在《新的土地》（*New Lands*）一书中仍把1900年的火星光束事件描写为智慧生命发来的信息。他写道：

> 这可能是一组间断性信息。也许在未来的某天，有人能把这些信息翻译出来。如果这是一颗行星通过图像的形式在向另一颗比它更强的行星致意的话，也许某一天这段信息会被视为整个地理系统中最伟大的演讲。请大家参考洛厄尔发表在《大众天文学》（*Popular Astronomy*）杂志第10–187期上的相关文章。这篇文章公布了一些数值，可能是某种密码，由长闪和短闪构成。洛厄尔将他认为长度正常的信号记录为1，然后将其他长度的信号分别记为2/3、$1\frac{1}{3}$、以及$1\frac{1}{2}$。如果在亚利桑那州的弗拉格斯塔夫还保存着1900年12月7日晚上的观察记录，那段长达70分钟的包含长闪和短闪的信号很可能是宇宙

> 中的一座孤岛上的智慧生物发给我们的。我们要么对这些信息视而不见，要么就该把地球上所有商博良的后代聚集到弗拉格斯塔夫，努力破译这段震耳欲聋的宇宙来信。[4]

英国著名科学杂志《自然》（*Nature*）在1902年5月1日登出了洛厄尔对火星光束现象的解释。2002年5月2日，该杂志又在“百年之前”栏目中重刊了洛厄尔的这篇论文。重印版中写道：

> 洛厄尔先生在他的结束语中说道，形成投影的地区——伊卡利亚海区域无疑是一片植被非常茂盛的区域。12月观察到的现象完全可以用以下理论进行解释：当时，在该区域的上空可能出现了一团云，这团云上升到距离地面13英里高的地方，然后向东北方向移动，移动速度大约是每小时27英里。这团云经过了埃里亚沙漠区域，然后消散不见了。[5]

也许，《自然》杂志之所以选择在100年后重印洛厄尔的文章，就是为了警告记者和科学家们不要妄下结论，更不要以非常确定的口吻来表达一些未经证实的观点。在100年后的今天，回首当年引起的轰动，再看看洛厄尔的上述结论，我们无疑可以对1900年的火星光束事件一笑了之。

长波上的密码

在查尔斯·福特的《新的土地》一书尚未出版之时，古列尔莫·马可尼（Guglielmo Marconi）的发现又引起了另一波关于外星密码的热潮。马可尼是诺贝尔物理学奖得主，有些人认为，他还是无线电波通信技术的发明者（但是许多特斯拉的支持者对这一点持有不同意见）。[6]

在马可尼掀起的这次热潮中，记者又一次扮演了推波助澜的角色。有些记者对马可尼的声明进行了错误的解读，还有一些记者甚至连马可尼的名字都不会拼。在一份名为《月度夜空地图》（*The Monthly Evening Sky Map*）的期刊上，登出了这样几段话[7]：

可能来自火星的信息震惊整个科学界

威廉·马可尼，无线电波通信技术的发明者，最近宣布，伦敦和纽约的无线电装置都记录下了一组神秘的信号。这组信号由点和划构成，就像电报中使用的莫尔斯电码一样。这组神秘的信号引起了世界范围内的广泛兴趣。有人认为，这是一组来自火星或者其他行星的信息。大部分科学家对上述说法持怀疑态度，他们认为这些信号很可能是由地球上的某种情况产生的，或者是由太阳产生的。但也有不少人认为，这些信号是由外星人发来的信息。

图 11-1　古列尔莫·马可尼（1874—1937）

古列尔莫·马可尼在《纽约新闻报》（*New York Journal*）上说："如果我十分确定地向大家宣布，一个或多个行星上的智慧生物正在试图向我们发送某种信息，那我一定是做出了一个大胆的猜测。但是，如果我坚决否认上述情况成立的可能性，那也是同样无用的。毕竟，对于这样一个深刻的问题，目前我们掌握的知识还是极不完备的。”他还说：“但可以肯定的一点是，地球上的无线电站会定期收到一些奇怪的信号。这些信号的意义和来源目前都不明确。”

这些信号像是莫尔斯电码

“在这个令人震惊的谜团中，真实的情况是，我们收到的信号似乎是由波长很长的电磁波造成的。因此，这些信号不可能是普通的杂散信号。”

“有时候，有人会想象这些信号似乎代表了莫尔斯电码中的某些字母。不管是在什么季节，都会有一些抱有上述信念的人偷偷地溜进我们的无线电站。”

“这些神秘的信号究竟是从哪里来的？能够回答这个问题的科学家将会成为不朽的伟大人物。”

“但是，与此同时，我们不应该忽视以下的可能性：这些信号有可能来自太阳。太阳上会发生电干扰现象，这一事实是众所周知的。”

“以下的事实也许能为有意研究这种现象的人提供一定的帮助：我们发现，只有在将波长设置在65英里以上时，我们才会收到这种信号。有时候，在发出长波20到30分钟以后，我们会听到行星或者行星间的声音。这种声音不会干扰通信，但是这种声音一般会持续一段时间。”

最常见的信号

“我们收到的最常见的信号具有一定的音乐性，这一点实在太令人好奇了。最常见的信号是3次短击，有人可能会把这种信号解读为莫尔斯电码中的字母“S”。但是，我们还收到过其他声音，也许代表着其他字母。”

“然而，我必须强调，在这些信号里找不到能连成单词的字母。

因此我们从未收到什么有意义的信息。”

“之前，由于战争的原因，我们无法对这种赫兹谜团进行详细的调查。但如今，我们这个机构打算对此进行彻底的探查。”

以上这篇文章已经比许多报道该事件的报道理智多了，这篇文章至少对长波信号现象提出了另一种可能的解释。后来，《月度夜空地图》上又登出了另一篇文章，[8]这篇文章比前一篇文章还要更加戏剧化：

马可尼说，火星在给我们发送信号

威廉·马可尼现在基本已能肯定自己确实收到了来自火星的无线电信号。这条消息来自J. H. C. 麦克贝思先生。麦克贝思先生是马可尼无线电有限公司驻伦敦的经理，他最近在曼哈顿的扶轮社发表了上述消息。大约在两年前，马可尼先生首次宣布自己收到了来自火星的消息。但是，科学家们却否认了马可尼先生的说法，他们认为，这些所谓的“消息”其实是大气扰动造成的。然而，马可尼先生通过进一步的测试发现，不管是否有其他干扰情况发生，他收到的信号都会定期中断。现在，马可尼先生和其他科学家已经宣布，这种信号来自另一颗行星，因为它的波长快达到地球上能量最大的无线电站所能产生的波长的10倍。据麦克贝思先生说，马可尼曾收到过波长估计为150 000米的信号。而我们地球上的无线电站所能产生的最大波长是17 000米。

麦克贝思先生表示，如果马可尼先生收到的信号确实来自火星，他相信这些未知信息的破译只是时间问题。在未来，一定会有一些富有创造性的密码破译天才能帮我们破译出这些来自火星的信息。届时，我们就可以建立起地球和火星之间的联系。

若干年以后，地球上的我们又收到了一些信号，有人再次认为这是来自外星人的信息。这一次，麦克贝思先生期待的“富有创造性的密码破译天才”终于出现了。

密码专家弗里德曼接受了挑战

如果有人要你翻译一段可能来自外星人的信息，你愿意吗？1924年，密码专家威廉·弗里德曼就收到了这样一项请求。

这个故事始于一个名叫戴维·佩克·托德（David Peck Todd）的人。托德是一名天文学教授，他在退休前曾经担任过阿默斯特学院天文台台长。1909年，托德提出，火星人可能正试图以无线电波的形式与人类取得联系。托德还认为，只要在高空中设置一个气球，就有可能侦测到火星人发来的信号。虽然后来托德并没能成功地实施这个在高空设置气球的实验构想，但是1924年8月，托德进行了另外一项测试。该月发生了火星冲的天文现象（即火星离地球距离最近的时刻），如果火星人真的在向地球发送信号，那么这将是一个接收信号的绝佳时机。

为了实验，托德说服美国陆军和海军在华盛顿特区颁布了无线电静默的命令，以尽可能减少无线电干扰。美国陆军和海军还在发生火星冲的时段命令军方的无线电接收站侦测来自火星的通信信号。

当天，托德选择接收波长为6 000米（频率为50千赫兹）的无线电信号。在29小时的实验窗口中，托德在6英寸（约15.2厘米）宽、30英尺（约9.1米）长的胶卷上记录了无线电信号。这种无线电信号记录装置是由

C. 弗朗西斯·詹金斯（C. Francis Jenkins）发明的。詹金斯将其称为“无线电照片信息连续传送机”。[9]对于该设备的工作原理，詹金斯做了如下解释：

> 当胶卷展开时，一种装置以每英尺50次的频率扫过胶卷。如果接收到声音信号，就会有一束光线打在胶卷上，并产生一条黑线。[10]

由于托德早就对火星来信的问题产生了兴趣，所以他对历史上地球收到的其他“火星信息”十分熟悉。托德说：“3年前，有报道称，马可尼声称收到了来自火星的信息。但是几天前又有报道称，马可尼表示他没空去收听可能来自火星的信息，并且认为接收火星信息是一个荒谬的想法。马可尼似乎改变了主意，也没有人知道第一次收到信号的时候马可尼究竟听到了什么。但是，我们有照片记录，因此不会产生‘某个人究竟听见了什么’的争论。我们的照片记录是一种永久性的记录，所有人都可以对我们的记录进行研究。”[11]

托德的实验是在华盛顿特区的詹金斯实验室内进行的。

詹金斯似乎并不太相信托德能收到来自火星的信息。詹金斯表示，他预测托德的实验只会得到一卷空白的胶卷。然而，当托德实验的胶卷被冲洗出来时，结果却和詹金斯的预期大不相同。一份报纸对托德的实验结果进行了以下报道：

> 胶卷上显示的“信息”是一系列排布规律的点和划。整卷胶卷从左到右都有，这些痕迹组合起来像一张张奇怪的脸。从胶卷的左侧到右侧，这一图案以固定的间隔不断重复。[12]

在图11–2中，读者可以看到冲洗出来的胶卷中的一段样本。

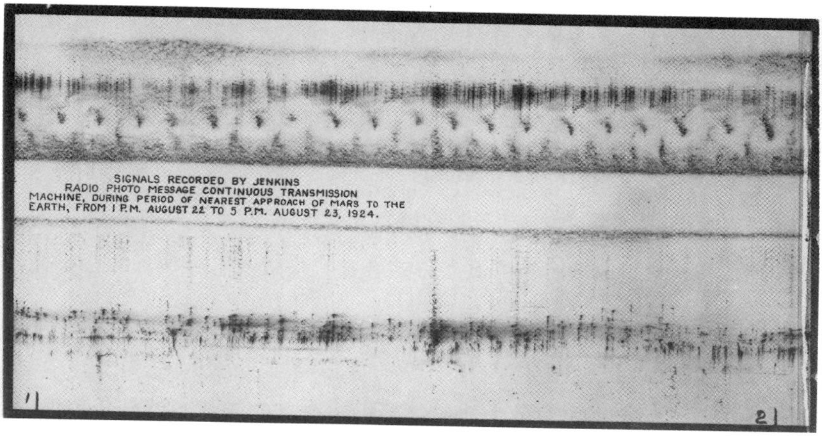

图 11-2 是来自火星的消息，还是另一个学习机会？

詹金斯看起来不像是喜欢搞恶作剧的人。事实上，詹金斯甚至不愿意将上述结果见报，因为他不希望公众认为这是个恶作剧。最终，为了满足公众的好奇心，詹金斯决定允许媒体登载胶片上的信息。但詹金斯强调，他参与该实验是出于科学上的好奇心，而不是因为他相信火星上有智慧生物。事实上，詹金斯表示：

> 我认为实验结果和火星没有关系。胶片上记录下的“声音”很可能是由于外差现象（heterodyning）或者无线电信号干扰而产生的。在胶卷上重复出现了一些像是人脸的图案，重复的频率大约是每半小时一次。我们无法解释这种奇怪的现象。[13]

而托德的态度则比詹金斯更开放（或者你可以说托德比詹金斯更容易深陷于一些无根据的臆想，这取决于你对此事采取什么态度了）。托德表示：

> 要是火星人真的在向地球发送消息，那么詹金斯的设备可能是我们接收这些消息的最佳工具了。如果火星人有某种机器（他们可能真的有），如果现在他们正在用这种机器向地球发送有关他们容貌的放大图像、风景、建筑、地形，还有其他诸如此类的东西，那么火星人选择的方式应是先把阳光信号转化成电信号，然后才向地球投射这些信号。所有这些一定会被我们这种奇异而独特的小机器捕捉到。[14]

最终，托德也未能对他的实验结果做出合理的解释。但是托德强调，留下这样的实验结果记录是非常有价值的。托德向弗里德曼提供了大约12英尺长的胶片记录，希望这位伟大的密码破译家能够从这些“密码”中找到一些有意义的信息。托德把胶片的另一部分交给了美国标准局无线电部门的负责人J. H. 德林杰（J. H. Dellinger）博士。

弗里德曼从未从这段信号中破译出任何结果。

然而，在进行科学研究的时候，有时研究者会在意外中得到一些教训。弗里德曼没有受过天文学或者物理学方面的训练，因此他没有发现问题是可以理解的。但是托德和詹金斯本来完全应该发现这个问题的。我不清楚托德和詹金斯为什么选择接收波长为6 000米的无线电信号，但是，在这个波段上根本不可能收到任何地球大气层以外发来的信息。因此，他们收到的信号很可能是某种地球上的信号源产生的干扰信号。然而，如果托德和詹金斯能够在不同的时间、从不同的方向、在不同的波长上重复上述实验的话（一般来说在不同条件下多次重复实验都是非常有意义的），他们有可能会发现一个全新的领域——射电天文学。

“信号”来自误差

在托德的实验结束之后，人类后来又发现了许多似乎很有趣的信号。这些信号产生的原因之一是电脑的内存会发生被电离的现象（即被带电的粒子击中），过多的电离作用有时会导致两个比特的翻转误差。因此，无论我们为了完成何种任务而将设备送入太空，这些设备收集到的数据都有可能因为上述误差而出现错误。虽然这些设备本身已经带有内置的修正算法，但是如果电离活动过多，算法就没有办法完全修正它们产生的错误了。于是，固态存储设备中就可能出现错误的数据。当这些错误的数据被传回地球时，我们有可能会将它们错误地解读为地外文明发给我们的信息。

对于有些种类的信号而言，上述误差很快就可以被发现。每一项任务或者每一种设备每年大约会发生两次这类可以被确认的误差 。然而有时候，科学家需要花费较长的时间才能发现数据中的误差，在这种情况下，公众往往会因为疑似外星信号的发现而兴奋一段时间。

更多外星信号！

在本部分中，我向读者介绍的都是较早的“外星信号”事件。对于今天的读者而言，我介绍的上述事件可能远不如1977年的“Wow信号”事件那么著名。如果读者有兴趣了解更多疑似来自外星文明的信号，可以阅读本书结尾的参考文献部分。受篇幅所限，关于外星信号的故事只能介绍到这里。接下来，我们再来看看由地球发向宇宙的信号。

角色的转换

到目前为止，人类不仅为接收地外信号而进行了认真的研究，也（极为谨慎地）向宇宙发出了一些我们自己的信号。向宇宙发送人类信号有一个很大的障碍，那就是军方经常对这类活动持强烈的反对态度。军方之所以反对向太空发送人类信号，是基于以下判断：如果我们真的能联系到某种地外文明，而且这种地外文明又有能力来到地球的话，他们的技术水平肯定远远领先于我们。因此，假如这种地外文明打算出于邪恶的目的入侵地球的话，我们将毫无还击之力。我们当然也有可能会联系上某个仁慈而友善的地外文明，但是万一对方没有那么和善的话，他们就可以对地球为所欲为，而以我们的技术水平是完全无法保护自己的。由于一些我个人无法理解的奇怪原因，有些人总是假设技术水平远超人类的外星文明一定会是一群热爱和平的智慧生物。我个人完全不认同这种观点。技术的进步从来没有改变过人性中最阴暗的部分。事实上，许多技术进步都被我们用在了一些非和平用途上，在这些技术进步的帮助下，一些人拥有了能更轻松、更大规模地屠杀其他人的能力。在有记载的几千年人类历史中，我们的技术水平确实极大地提高了，然而随着技术的进步，在每一场战争中死亡的人数也在不断地提高。在我的想象中，外星人在这一方面和我们不会有太大的差别。

不管怎么样，人类还是向太空发出了一些信息。这些信息都比较短，并且我们非常谨慎地采取了各种措施，把外星人收到这些信息的可能性降到了极低的水平。接下来，我要向读者介绍人类向地外发送信息的具体方法。

以下是一段1974年地球发向外太空的信息：

00000010101010000000000001010000010100000001001000
10001000100101100101010101010101010010010000000000
00000000000000000000000000011000000000000000000011
01000000000000000000011010000000000000000000001010100
00000000000000000111110000000000000000000000000000
00011000011100011000011000100000000000001100100001
10100011000110000110101111101111101111101111100000
00000000000000000000010000000000000000000100000000000
00000000000000000010000000000000000000001111110000000 0
00000111110000000000000000000000000110000110000111 00
01100010000000100000000010000110100001100011100110
10111110111110111110111110000000000000000000000000
01000000110000000001000000000001100000000000000000010
00001100000000001111110000011000000111110000000000
11000000000000010000000001000000000100000100000011 00
00000100000001100001100000010000000000110001000011
00000000000000011001100000000000000110001000011 0000
00000110000110000001000000010000001000000000100 0001

0000000110000000001000100000000011000000001000100000
00001000000001000001000000001000000001000000001000000000
000001100000000000110000000001100000000001000111010110
0000000000010000000100000000000000000100000011111000000
00000010000101110100101101100000010011100100111111
10111000011100000110111000000000010100000111011001 0
00000101000000111111001000000010100000110000000100000
110110000000000000000000000000000000000001110000010
000000000000011101010001010101010100111000000000010
10101000000000000000001010000000000000000011111000000
000000000111111111000000000000000111000000001110000000
001100000000000011000000011010000000000101100000 1100
1100000000110011000010001010000001010001000010001001
00010010001000000000100010100010000000000000010000100
00100000000000001000000000100000000000000001001010000
000000011110011111101001111000

设计上述信号的人相信，如果某种外星人能接收上述信号的话，他们会发现这段信号共有1 679个比特。而1 679是两个质数——23和73的乘积，即1 679 = 23 × 73。设计信号的人还认为，发现了上述事实的外星人可能会把上述信号组成一个长方形，这个长方形共有73行，每一行有23个比特：

00000010101010000000000
00101000001010000000100

10001000100010010110010

10101010101010100100100

00000000000000000000000

00000000000011000000000

00000000001101000000000

00000000001101000000000

00000000010101000000000

00000000011111000000000

00000000000000000000000

11000011100011000011000

10000000000000110010000

11010001100011000011010

11111011111011111011111

00000000000000000000000

00010000000000000000010

00000000000000000000000

00001000000000000000001

11111000000000000011111

00000000000000000000000

11000011000011100011000

10000000100000000010000

11010000110001110011010

11111011111011111011111

00000000000000000000000

00010000000110000000010

00000000001100000000000

00001000001100000000001

11111000001100000011111

00000000001100000000000

00100000000100000000100

00010000001100000001000

00001100001100000010000

00000011000100001100000

00000000001100110000000

00000011000100001100000

00001100001100000010000

00010000001000000001000

00100000001100000000100

01000000001100000000100

01000000000100000001000

00100000001000000010000

00010000000000001100000

00001100000000110000000

00100011101011000000000

00100000001000000000000

00100000111110000000000

00100000101110100101101 1

00000010011100100111111

10111000001110000011011 1

00000000010100000111011

00100000010100000111111

00100000010100000110000

00100000110110000000000

00000000000000000000000

00111000001000000000000

00111010100010101010101

00111000000000101010100

00000000000000101000000

00000000111110000000000

00000011111111100000000

00001110000000111000000

00011000000000001100000

00110100000000010110000

01100110000000110011000

01000101000001010001000

01000100100010010001000

00000100010100010000000

00000100001000010000000

00000001001010000000000

01111001111101001111000

接下来，设计者期待外星人能把每个0替换成一个空方块，而把每个1替换成一个有颜色的方块。这样外星人就会看到图11–3中的图案：

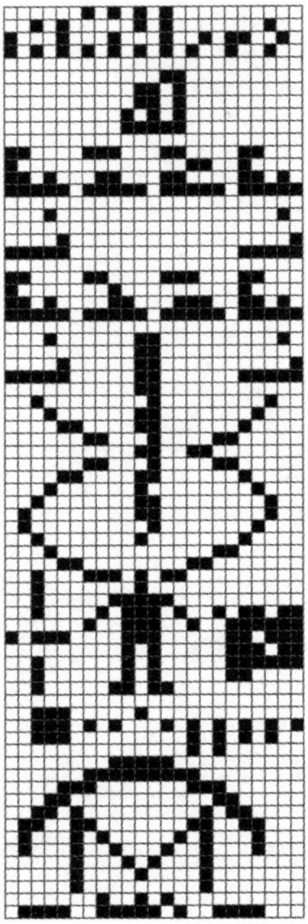

图 11–3　重建出的图像

你能看懂这幅图里隐藏的信息吗？设计上述信号的人希望，有足够技术水平来接收上述信号的外星人不仅能够破解人类构思出来的谜题，从而得到图 11–3 中的图案，还能理解上述图案中试图传递的信息。

这幅图试图传递的信息有以下这些（按照从上到下的顺序）：首先，展示了从 1 到 10 的二进制数字；然后，通过元素对应的数字列出 DNA（脱氧核糖核酸）的组成元素：1、6、7、8、15 分别代表氢、碳、氮、氧、磷；接着是核苷酸的分子式、DNA 的双螺旋结构，以及组成人类 DNA 的核苷酸数目；再接下来是一个简笔画图形，意在告诉外星人人类长什么样；人形简笔画的旁边则列出一个成年男性人类的平均身高，以及地球上的总人口数目；接下来是太阳系的简图，从图中可以看出太阳系各行星的相对大小，以及地球的位置；最后，我们向外星人展示了发出这段信号的阿雷西沃射电望远镜，并提供了该望远镜的口径信息！

老实说，我认为没有任何一种外星智慧生物能从图11–3中提取出如此多的信息。如果真想和外星人取得联系的话，一种更有诚意的做法应该是用更大的质数来传递一幅解析度更高的图片，这样才能更清楚地表达我们想表达的信息，而不是弄出这幅看起来像“雅达利2600”游戏界面一样的图像①。另一种选择是，我们可以将几千幅图片发向地外，形成一个“图解词典”。[15]

上述这条“阿雷西沃信息”只向地外发送过一次，时间不足3分钟。它的发送目标是距离地球 21 000~25 000光年的球状星团M13。球状星团M13是一个移动的目标，因此，当这段信息走完这段2万多光年的路途时，球状星团M13肯定已经不在那里了。所以，在这段信号到达比M13还要远得多的另一个星系之前，“阿雷西沃信息”应该不可能被任何地外文明接收到。我在前文中已经解释过，人类实际上是非常谨慎的，我们并不想真的联系上任何外星人。“阿雷西沃信息”只是一种宣传上的噱头而已。

“阿雷西沃信息”是在2 380兆赫兹的频率上发送的。就向地外文明发送信息的目的而言，任何给足的频率都有其特定的优势和劣势。但是，我认为，如果要向地外发送信息的话，我们应该考虑使用“扩频”（spread spectrum）技术——这种技术在情报界已经应用多年。所谓“扩频”是指把信号的频谱打散到很宽的一系列概率上去。“扩频”技术有以下这些优势：第一，如果敌方想要用某特定概率的大功率发射器来干扰我方的信号，只会干扰该频率上的很少一部分信息。也就是说，扩频技术能够提高信息的抗干扰能力。第二，把信号的频谱拓得很宽，可以使信号的功率比

① “雅达利2600”，英文为Atari 2600，是20世纪70年代流行的一种电子游戏机。——译者注

背景噪声还要低。敌方转动收音机的旋钮，听到的却只有静态噪声而已，谁又能想到在噪声之下居然悄悄传送着有意义的信息呢？这简直就是无线电界的“隐写术”！在地球上的秘密无线电通信领域中，扩频技术发挥了很大作用。这种技术可能同样适合用于非秘密的地外通信。只在一个频段上监听信息就好比面对一整片树林却只看其中一棵树。那么，能不能同时在几乎所有频段上发送信息呢？

随着时代的变化和技术的发展，人类接收地外信息的方式会发生怎样的变化？向太空发送人类信息的途径又会如何演进呢？这是两个值得我们关注的十分有趣的问题。在“阿雷西沃信息”发出近25年以后，人类又向外太空发出了另一条消息。这条消息是1999年5月由一台乌克兰的射电望远镜发出。截至2014年，人类又向外太空发送了7条新信息，这其中有一系列信息是2001年由一组俄罗斯的青少年与天文学家亚历山大·扎伊采夫（Aleksandr Zaitsev）合作发出的。各国军方仍对向太空发送人类信息的行为持反对态度，但是在今天这个世界中，想要向地外文明发送信息，也有许多方法可用。

阿雷西沃射电望远镜发出的那条简单的二进制信息是由弗兰克·德雷克（Frank Drake）、卡尔·萨根（Carl Sagan），以及其他几个人共同创作出来的。事实上，这条信息的创作过程与密码学有好几点联系。其中之一是：许多年以后，美国国家安全局邀请卡尔·萨根加入他们的科技咨询委员会。这是一个非常有趣的职位，但是成为美国国家安全局的一员就意味着萨根在写作方面会受到许多限制。如果萨根接受了上述职位，那么他在出版任何内容之前都必须经过国家安全局的审批程序，以保证他不会把任何加密信息无意识地泄露出去。为了避免受到上述限制，萨根拒绝了美国国家安全局的邀请。

我在研究如何用因数分解来把一串二进制数转化成一幅图像的过程

中，无意中发现了“阿雷西沃信息”和密码学的另一重联系。世界级的长笛演奏家兰布罗斯·D. 卡利马霍斯（Lambros D. Callimahos，他同时也是一名世界级的密码分析师，还在国家安全局的名人堂中有一席之地）曾在1966年3月刊的《科技纵览》（*IEEE Spectrum*）上讨论过如何发送与“阿雷西沃信息”类似的信息。那么发送“阿雷西沃信息”的主意是不是卡利马霍斯第一个想出来的呢？似乎并不是，因为他在这篇论文里告诉我们，发送“阿雷西沃信息”的主意是另一个人想出来的。他写道：

> 1961年，在西弗吉尼亚州的格林浅滩举行了一次会议，会议的主题是讨论与其他行星展开通信的可能性。会后，一位名为伯纳德·M. 奥利弗（Bernard M. Oliver）的与会者根据光栅原理创造了一条假想的信息。这条信息包含1 271个二进制数，或者说1 271个比特。读者可以在图1中看到这条信息。因为1 271只有两个质因数——31和41，所以我们可以很自然地把这段信息写成一个41行的长方形，每一行有31个比特，或者一个31行的长方形，每一行有41个比特。后一种写法产生的图案较不随机，因此我们可以判断后一种写法才是正确的写法。[16]

奥利弗把上述想法写进了一篇论文，这篇论文于1963年发表在一本以星际交流为主题的论文集中，这本论文集中既收录了旧论文，也有新发表的论文。奥利弗在这篇论文中解释了他的上述想法，他写道：“利用这种原理，国家射电天文台的弗兰克·德雷克创造了一条仿制的信息，并将这条信息寄给了所有参加上述会议的人（14）。”[17]

根据这些资料中的说法，“阿雷西沃信息”的创作者似乎是德雷克。奥利弗的论文中提到的“上述会议”就是卡利马霍斯所说的那次会议。但是奥利弗还给出了一些关于这次会议的更加详细的信息，即这次会议是

1961年11月在国家射电天文台召开的。[18]更加有趣的是奥利弗论文中的参考文献14（上述引文中也包含了引用标记）。在奥利弗的这篇论文的末尾，注明参考文献14是“海豚社成员之间的私下交流”。[19]

事实上，海豚社是一个半秘密的组织，这个组织的目标是与地外文明建立通信交流。海豚社组织的若干名成员都参加了1961年举行的上述学术会议，包括：该会议的组织者J. P. T. 皮尔曼（J. P. T. Pearman），天文学家卡尔·萨根、弗兰克·德雷克、黄授书、奥托·施特鲁韦（Otto Struve），发明家伯纳德·奥利弗（Bernard Oliver），神经科学家约翰·利利（John Lilly），演化生物学家J. B. S. 霍尔丹（J. B. S. Haldane），物理学家菲利普·莫里森（Philip Morrison），500强公司“微波联合公司”的创始人达纳·阿奇利（Dana Atchley），还有化学家梅尔文·卡尔文（Melvin Calvin，他在参加这次会议的时候获得了诺贝尔奖）。约翰·利利曾经研究过海豚的通信方式，海豚社因此而得名。在书面的材料中，我并没有找到太多关于海豚社组织的信息。如果有人能够找到该组织的内部通信，他一定可以以这些内部通信为素材写出一篇很有趣的论文来。海豚社组织是否还考虑过用其他方法来构建一些容易解读的信息呢？

价值20万美元的因数分解挑战

如果很久以后，在某个距离地球非常遥远的星系中真的有某种智慧生物收到了人类发出的“阿雷西沃信息”，那么这些外星人首先必须对1 679这个数字进行因数分解，才可能得到图11–3中的图像。对1 679这个数字进行因数分解是很容易的，因为1 679的两个质因数——23和73是两个较小的质数。但是，如果一个数的质因数有数百位长，对这个数进行因数分解就变成了一项非常困难的工作。许多年来，RSA实验室一直设重奖来鼓励人们参与因数分解的挑战，任何人只要能找出某数的质因数，并解释自己是如何找到这些质因数的，就能获得相当丰厚的奖金，奖金最高可达200 000美元。这次比赛的挑战数如下[20]：

RSA-2048

状态：未分解

25195908475657893494027183240048398571429282126204
03202777713783604366202070759555626401852588078440
69182906412495150821892985591491761845028084891200
72844992687392807287776735971418347270261896375014
97182469116507761337985909570009733045974880842840
17974291006424586918171951187461215151726546322822
16869987549182422433637259085141865462043576798423
38718477444792073993423658482382428119816381501067
48104516603773060562016196762561338441436038339044
14952634432190114657544454178424020924616515723350
77870774981712577246796292638635637328991215483143
81678998850404453640235273819513786365643912120103
97122822120720357

如果你想把上述数字输入电脑，看看电脑能不能帮你解开这个价值200 000美元的难题，那么你首先要注意不要输入错了——要是本书里有一个笔误，或者你复制这个挑战数的网站上有一个小错误，那可就糟糕了！为了防止这种情况的发生，我给大家提供一个验算的方法——上面这个大数共有617位，各个数位上的数字总和是2 738。

这个挑战数被称为“RSA–2048”，因为如果把这个数写成二进制数，它的总长度是2 048个比特。

目前，这笔200 000美元的奖金已经过期，但是上数的因数分解问题仍是一个悬而未决的挑战，没有任何人能够解开它。虽然这个问题的答案只是两个质因数，而不是一段明文信息，但我认为上述挑战非常符合本书的主题：一来这是一个未解之谜，二来这个挑战与密码学也有着很强的联系。那么，因数分解与密码学之间究竟有着怎样的联系呢？为什么一个公司要悬赏如此丰厚的一笔奖金来邀请人们对一个大数进行因数分解呢？

一种有300多年历史的密码系统

我们要讲的这个故事，从某种角度来看始于1977年。这一年，麻省理工学院的3位教授——罗纳德·里韦斯特（Ronald Rivest）、阿迪·沙米尔（Adi Shamir）和伦纳德·阿德尔曼（Leonard Adleman）共同发表了一篇论文。这篇论文做到了一件许多人认为不可能的事情：他们发现了一种方法，让发信人和收信人不需要事先沟通决定一个共同的私有密钥，就可以通过加密的信息进行安全的通信。这个系统被称为“RSA加密系统”，其中R、S、A分别是上述三位教授的姓氏的首字母。

虽然这种算法是在1977年被首次提出的，但是实际上它的数学基础早就已经存在了。要看到这个故事真正的开头，我们必须将时间倒回1640年——在这一年，数学家皮埃尔·德·费马（Pierre de Fermat）做出了一项重要的工作。由于某种原因，费马当时在研究这样一个问题：如果对一个数进行乘方，然后再用一个质数去除这个数，并取上述运算的余数，会得到怎样的结果？上述运算被称为“乘方模除p”，其中p就是用作除数的质数。

为了更加清楚地向读者解释这种运算，下面我举一个具体的例子。我们可以简单地计算出7的3次方，然后用11来除这个数，并取余数，也就是$7^3 \pmod{11}$。模除运算通常用符号mod来表示。由于$7^3 = 343$，用343除以11得31余2。因此以上运算的最终结果是2。也就是说$7^3 \pmod{11} = 2$。

当然，数学家是不会满足于仅仅进行简单的运算的。数学家希望能在这种运算中找出规律，并且证明这种规律对所有数都成立。于是，费马研究了各种“乘方模除p”运算的例子，试图找出其中的共同规律。

费马得到了以下结果：

$1^1 \pmod{2} = 1$	$1^2 \pmod{3} = 1$	$1^4 \pmod{5} = 1$	$1^6 \pmod{7} = 1$
	$2^2 \pmod{3} = 1$	$2^4 \pmod{5} = 1$	$2^6 \pmod{7} = 1$
		$3^4 \pmod{5} = 1$	$3^6 \pmod{7} = 1$
		$4^4 \pmod{5} = 1$	$4^6 \pmod{7} = 1$
			$5^6 \pmod{7} = 1$
			$6^6 \pmod{7} = 1$

你能够从上面的例子中找到规律吗？你是否可以预测出：对于小于11的数字，需要进行几次乘方才能让模除11（即mod 11）的结果等于1？

费马发现，对于任意质数p和一个小于p的正整数a，以下数学关系都成立：$a^{p-1} = 1 \pmod{p}$。今天，我们将上述结论称为“费马小定理”

（Fermat's little theorem），它是纯数学领域的一个定理。从某种角度来看，纯数学可以被定义为“只让数学家感到兴奋的数学”。如果某条数学结果能让非数学家也感到兴奋，那么这种结果多半就属于“应用数学”的范畴。

费马的上述定理是一个纯数学定理，但是这个定理却可以产生一个实际的应用。如果我们把$a^{p-1} = 1 \pmod p$这个等式两边同时乘以a[①]，就得到：$a^p = a \pmod p$。我很快就会向读者解释如何用这个公式来设计一种加密系统。但是，首先我需要解释如何将一条文字信息转化成一个数。

将一条文字信息转化成一个数的方式有很多，为了简单起见，这里我们将每个字母用其在字母表中对应的数字代替。那么，当我们看到17这个数字的时候，这个数字究竟是代表字母表中的第1个和第7个字母AG，还是代表第17个字母Q呢？为了消除这种歧义，我们规定字母表中的前9个字母也用两位数表示，也就是说，我们要在这9个字母对应的数字前面分别加上一个0，即：A=01，B=02，C=03…，到了J=10以后，我们就可以正常地书写接下来的数字了。根据上面这种对应关系，字母AG写作0107，而不是17。有了这套对应关系，我们就可以把任何文字信息转化成一个数字了。当然，对于一段很长的文字信息来说，我们将会得到一个非常大的数字，但是没关系，过一会儿我们再讨论如何处理过大的数字。不管我们最后得到的数字有多大，我们都可以用一个比该数更大的质数p来对这个数字进行“乘方模除p”的运算。

首先，我们将一段文字信息转化成一个数字，把这个数字称为a。接着，我们取a的e次幂，e代表加密指数，在这一步中，e必须小于p。再接下来，我们计算a的e次幂模p，这样就会得到一个完全不同的数字，这个数字和我们的原始数字（即文字信息对应的大数）没有任何明显的联系。

---

① 同余定理保证同余式两边同时乘以同一个数，同余式仍成立。——编者注

模除后得到的这个数字就是我们可以发送的密文信息。收到这个密文数字的人可以取这个数的d次幂，d代表解密指数，只要保证e和d的乘积是p，根据费马小定理，密文数字的d次幂就应该是a。一旦收信人得到数字a，他就可以根据`A=01，B=02，C=03…`的对应关系推出原始的明文文字信息。在上述系统中，通过$(\bmod p)$可以破译出原始的明文信息。

以上所有运算可以用算式表示为：$(a^e)^d = a \pmod p$。然而，这种加密方法是不成立的！读者可以再读一遍前面的内容，看看你是否能够找到这个方法的问题。

我们无法得出$(a^e)^d = a \pmod p$，是因为$(a^e)^d = a^{ed}$，如果我们要保证$a^{ed} = a \pmod p$，就必须保证$ed = p$。但是，p不可能是e和d的乘积，因为p是一个质数，不可能有除了1和自身以外的因数。

为了解决上述问题，我们可以尝试用一个非质数来取代质数p，但是，这样做又会导致另一个问题。比如，让我们来看看以下的这个算式：

$$2^{10} \pmod{10} = 4$$

我们希望上式的得数是2，然而如果n不是质数的话，我们就无法保证得到$a^n = a \pmod n$。

因为以上的原因，虽然费马小定理“几乎”可以帮我们设计出一种很好的加密系统，但终究还是不行。不过，对于费马来说，这些都不重要，因为费马对加密系统根本毫无兴趣。费马从来就不想发明什么加密系统，他只是想要研究纯数学问题而已。他可能从未意识到他的研究成果可以被用来设计加密系统。

故事在1760年左右迎来了转机。这次，故事的主角是伟大的瑞士数学家莱昂哈德·欧拉（Leonhard Euler）。欧拉知道费马小定理的内容，并且他一直在试图寻找一种方式让模除n对非质数n也能成立。在前文中，我只是简单地告诉读者“乘方模除n”的运算对非质数n不成立。但是，欧拉希望对费马的做法进行一些修改，让费马的结论对非质数n也成立。欧拉也从未想过这种纯数学定理能产生实际的应用，和费马一样，欧拉只是因为热爱这个游戏才进行研究的——对于像费马和欧拉这样的数学家而言，世界上最有趣、最伟大的游戏就是纯数学！

欧拉的研究得到了以下结论：如果n不是质数，他仍然可以找到一个指数使费马提出的数学关系成立，但是只有选择某些特定的底数才可能做到这一点。也就是说，进行乘方运算的数必须具有某些特定的性质才行。如果我们将底数写作a的话（就像刚刚一样），我们必须保证a和模数n没有大于1的公因数。比如，$a = 4$，$n = 9$是可以的。a的正因数是1、2、4，而n的正因数是1、3、9。a和n的最大公因数是1，所以$a = 4$，$n = 9$是可以的。而如果$a = 7, n = 21$则是不行的，因为7和21有一个大于1的公因数——7。

如果两个数的最大公因数是1，我们就说这两个数是“互质”的。两个数是否互质与这两个数本身是否是质数无关。比如，4和9是一对互质的数字，但4和9本身都不是质数。

因此，欧拉的研究结果告诉我们：如果a和n是一对互质数，那么就存在一个数字能让a的幂模n等于1。但是，欧拉的结果比费马的结果更加复杂一些。在费马的结果中，a的幂是$n-1$，而在欧拉的结果中，a的幂是n的函数。这个函数用语言表达起来比用算式表达更加清楚——这个函数就是小于n并且与n互质的正整数的个数。欧拉用希腊字母ϕ来表达这个函数。

比如，如果$n = 6$，那么小于6的正整数就是1、2、3、4、5。然而，1、2、3、4、5这几个数并不都与数字6互质。比如，数字2和数字6有大于1的公因数2。同样，数字4和数字6也有大于1的公因数。因此，在1、2、3、4、5这5个数字中，只有1、3、5与6互质。因为小于6并且与6互质的正整数共有3个，因此$\phi(6)=3$。

下面我们再举一个比6大的例子，即$n=15$。小于15的正整数是1、2、3、4、5、6、7、8、9、10、11、12、13、14。但是，在上述这些数中，只有与15互质的数才符合我们的要求。5和10都不符合我们的要求，因为这两个数字都与15有公因数5。3、6、9、12也不符合我们的要求，因为这4个数字都与15有公因数3。因此，在这14个数字中，只有1、2、4、7、8、11、13、14与15互质。因为小于15并且与15互质的正整数共有以上8个，所以$\phi(15)=8$。

欧拉的结果可以总结为以下算式：

如果a和n互质，那么$a^{\phi(n)}=1 \pmod n$。

对于n是质数的特殊情况，我们有$\phi(n)=n-1$，这是因为对于一个质数而言，所有小于这个质数的正整数都与这个质数互质。因此，如果我们把欧拉的上述定理中的n改为质数p，上述定理就变成了费马小定理。也就是说，费马小定理是欧拉定理的一种特殊情况。因此，我们说欧拉定理是"欧拉对费马小定理的推广"。

欧拉并没有用他证明的欧拉定理来设计加密系统。我们在前文中已经说过，欧拉和费马一样只对纯数学感兴趣。然而，在欧拉定理被发现200多年以后，麻省理工学院的3位教授发现，欧拉定理在密码学领域中有着非常神奇的应用。

在之前的段落中，我们曾经把费马小定理的算式两边同时乘以a。现在，我们把欧拉定理的算式两边也同时乘以a，这样我们就得到：$a^{\phi(n)+1}=a \pmod n$。

在前文中我已经解释过，用费马小定理的结果无法设计出我们想要的加密系统，因为我们选择的指数必须保证密文能被转化回明文a，然而这个指数必须是一个质数，因此我们无法对这个质数进行因数分解。有了欧拉定理以后，上述问题就不存在了。虽然存在一些数值n能使$\phi(n)+1$为质数，但是我们并不需要担心这个问题。以下几步可以解释我们为什么不需要担心这个问题。首先，我们在等式$a^{\phi(n)+1}=a(\mathrm{mod}\ n)$的两边同时乘以$a^{\phi(n)}$，这样就得到以下等式：$a^{2\phi(n)+1}=a^{\phi(n)+1}(\mathrm{mod}\ n)$。接着，我们改变以上等式右边的写法。由于$a^{2\phi(n)+1}=a^{\phi(n)}a^1(\mathrm{mod}\ n)$，而$a^{\phi(n)}=1(\mathrm{mod}\ n)$，所以以上等式的右边可以简化为$a^{2\phi(n)+1}=1a^1(\mathrm{mod}\ n)$。由于我们没必要把1写出来，所以上述等式就变成了：$a^{2\phi(n)+1}=a(\mathrm{mod}\ n)$。

上面的步骤看起来很复杂，但是其实我们只不过是把等式的两边同时乘以$a^{\phi(n)}$，然后再进行一些化简罢了。假如我们再次把等式的两边同时乘以$a^{\phi(n)}$，然后化简，就会得到：$a^{3\phi(n)+1}=a(\mathrm{mod}\ n)$。我们还可以继续重复上述步骤：选择你最喜欢的正数$k$，我们可以得到以下结果：$a^{k\phi(n)+1}=a(\mathrm{mod}\ n)$。

因此，虽然$\phi(n)+1$可能是质数，但是对于所有的正整数k而言，$k\phi(n)+1$不可能都是质数。只要保证当k取某个值时我们能对$k\phi(n)+1$进行因数分解，我们就能找到正数e和d，并分别用这两个数来加密和解密信息。

然而，我们并不想通过对$k\phi(n)+1$进行因数分解来找出e和d，因为因数分解这项工作太困难了！幸运的是，一种古老的算法能够帮助我们解决这个问题，这种算法叫作“欧几里得算法”。是的，这种算法真的有那么古老！欧几里得算法可以追溯到古希腊，通过这种算法，我们可以很快地找到e和d。我们只要选择一个与$\phi(n)$互质的数e，然后把$\phi(n)$和e都输入这种算法，几步以后，我们就会得到d了。

在前文中我已经说过，麻省理工学院的3位教授发明了一种加密系统，发信人和收信人不需要事先沟通决定一个共同的私有密钥，就可以通过这种加密系统进行安全的通信。读到这里，读者可能仍然很难看出为什么可以根据欧拉定理设计出这样一种加密系统，毕竟发信人和收信人似乎仍然需要通过事先的沟通来决定e、d和n的数值。但是实际情况并不是这样的。如果我希望接收某个素未谋面的人发来的加密信息，并且希望我能破译这段加密信息，实际上我可以很容易做到上述两点。我先选定e、d和n的数值，这一步我可以独立完成，而不需要通知对方。接着，我把e和n的数值放到我自己的网站上，这就是我的"公钥"，它们是公开的，任何人都可以在我的网站上找到。下面，想向我发送加密信息的人可以把他要发送的信息转化为数字a，然后计算出$a^e \pmod n$——这就是他要发给我的密文。收到密文以后，我求出密文的数字的d次方，然后模除p，就得到明文a。得到a以后，我就可以根据字母和数字的对应关系翻译出原始的文字信息了。在上述过程中，d的数值是我对外保密的私有信息，也就是我的"私钥"。

世界上的每一个人都可以利用上述系统进行加密通信，可以选择的数字很多。我们可以像出版黄页电话本一样出版公钥本（不管是印刷版的还是电子版的），把每个人的公钥都列在上面。只要私钥没有泄露，我们就可以安全地进行加密通信。

但是，关于上述加密系统还有一个重要的细节需要讨论，正是这个细节把这种公钥加密系统和之前提到的设立200 000美元奖金的因数分解大赛联系在一起。在选择公钥的时候，我们必须精心挑选n的数值。具体地说，你必须保证其他人很难计算出$\phi(n)$。因为如果其他人能计算出$\phi(n)$，他就可以把$\phi(n)$和e（e的数值是公钥的一部分）一起代入欧几里得算法，从而快速得到d的数值。而如果解密指数d被别人知道了，他就可以用d来破译所有别人发给我的密文，于是我的加密信息就不再保密了。

要让$\phi(n)$很难被计算出来，首先n必须是一个大数，但仅此一点还不够。如果我们能对大数n进行因数分解，就可以通过一个简单的公式算出$\phi(n)$的值。因此n不仅必须是一个大数，还必须是一个很难做因数分解的大数。制造这样一个大数的最简单的方法是取两个很大的质数的乘积。使用上述加密系统的人首先要选择两个很大的质数p和q，这两个质数都必须至少有100位。把这两个质数相乘就得到了n。为了让素未谋面的人能通过这个系统给你发送加密的信息，你必须把n的数值对所有人公开。

因此，RSA实验室才会重金悬赏能对这类大数（即两个大质数的乘积）做因数分解的人。RSA实验室出售的加密产品的理论基础就是两个大质数的乘积很难被因数分解，因此，他们希望所有可能购买他们的加密产品的顾客都能看到，大质数乘积的因数分解真的是一项难度极高的任务。RSA实验室向任何能找出该数的质因数的人提供高额奖金，不仅是在向客户展示他们对产品的信心，也是在吸引公众的注意力、借机宣传他们的产品。

然而，创造这个系统的人当然必须知道$\phi(n)$的数值。虽然对于其他任何人而言，$\phi(n)$的数值极难计算出来，但是对于这个系统的创造者而言，这却是一件很简单的事情，因为系统的创造者知道p和q这两个质数的值，因此只要将这两个质数相乘就可以得到$n = pq$。除了能计算出n的数值以外，系统的创造者还可以使用$\phi(n) = (p-1)(q-1)$这个简单的公式来得到他需要的数字。

为了让读者能够对RSA加密算法有一个基本的了解，接下来我还需要介绍几个其他的知识点。

第一，虽然我们模除的对象n是一个很大的数，但是我们不能用这个大数来一次性加密很长的信息。如果需要加密一段很长的信息，那么加密者需要首先把这段信息分成若干段，然后再对这些较短的信息分别进行加密处理。

第二，事实上我们甚至都不是用第一点中的方法来加密原始信息的。RSA加密算法是一个非常好的加密系统，但是这个系统的加密速度并不快。用户永远是缺乏耐心的，因此，在实际的RSA加密产品中，通常是将RSA加密系统与另一种速度快得多的加密系统结合在一起使用。后一种系统并不能保证发信人和收信人事先不交换密钥就能进行安全的加密通信。

整个系统通常是通过以下的方式来完成加密通信的：发信人首先制造出一个随机的密钥，再使用这个随机的密钥，通过一个加密速度很快的系统来加密他想要传输的信息。收信人并不知道这个随机的密钥是什么，因此发信人必须把这个随机的密钥和密文信息一起发给收信人。然而，如果发信人同时发出密文和密钥，那么任何收到这个信息的人就都可以读懂这段信息了。为了防止这种情况的发生，发信人首先用收信人的公钥来加密他想发送的随机密钥，然后再将经公钥加密后的随机密钥和密文信息一同发给收信人。这种加密方式被称为混合加密系统。混合加密系统结合了上述两种加密系统的优点。

数字签名

在我们曾经介绍过的蝉3301挑战密码中，经常出现包含数字签名的信息。在第9章中，我只简要地告诉读者，是现代的密码技术使得数字签名成为可能，但我没有向读者介绍数字签名技术的具体细节。现在，既然读者已经掌握了RSA加密系统的基本知识，那么数字签名的原理也就很容易解释清楚了。

在本章的上一部分，我已经向读者解释过：任何人都可以利用我的公钥向我发送加密信息，而只有我能够读懂这些信息，因为只有我知道我自己的私钥——即解密指数d。事实上，我们还可以把上述过程反转过来，也就是说我可以用我的私钥d来加密信息。经过这种加密的信息虽然看起来像是密码，事实上却没有任何安全性。因为任何人都可以使用我的公钥

e和n来破译我加密的信息。但是，由于只有我一个人知道d的数值，所以我是唯一有能力创造出这些加密信息的人。在以上这个系统中，除我以外的任何人都无法冒用我的身份来发送加密信息，因此我相当于用有一种“数字签名”证明了我的身份。

事实上，里韦斯特、沙米尔和阿德尔曼也很重视数字签名功能。他们三人首次发表论文提出这种新的加密系统的时候（这个系统很快就被命名为“RSA加密系统”），给这篇论文起的题目就叫作《关于数字签名》（*On Digital Signatures*）。[21]

以上，我向读者介绍了公共密钥加密系统的两种用途：一是加密出一段只有一个人能够读懂的信息；二是对一段信息进行数字签名，每个人都能够读懂这段信息，并且能够确认发信人的身份是真实可靠的。但是我要强调的是，RSA加密系统并不是每次只能完成上述其中一项任务，事实上，我们可以同时做到以上两点。如果我希望向你发送一条加密信息，并用数字签名来向你证明我的身份，那么我只需要同时使用你的公钥和我的私钥就可以了。

RSA加密系统是一种非常好的加密系统，但是这个系统的所有优越性都建立在一个前提条件之上，那就是：对一个大数做因数分解是非常困难的！那么，对大数做因数分解究竟有多困难呢？让我们来看一些相关的例子。

因数分解挑战

假如我们可以通过某种方法快速地对大数进行因数分解的话，我们就可以瞬间破解RSA加密系统。对于较小的数字而言，人类的头脑就可以轻松地完成因数分解的任务。比如，如果需要进行因数分解的数是15，那么一个小学生也能很快算出 $15 = 3 \times 5$。然而，如果待分解的数是4 697 296 523的话，这个小学生所掌握的因数分解技巧就几乎帮不上什么忙了。上题的正确答案是 $4\ 697\ 296\ 523 = 37\ 657 \times 124\ 739$。数学家和计算机科学家已经发明了一些因数分解的算法，可以轻松地分解像4 697 296 523这样的数字。但是对于非常大的数字而言，即使是世界上最快的计算机配上世界上最好的算法，也无法完成因数分解的任务。因此，前文中出现的大数RSA–2048至今也没有人能分解出来。

虽然RSA–2048的挑战至今无人能破，但是长度稍短的一些挑战数已经被解开了。RSA–640就是一个已经被成功解开的挑战数。RSA–640是这样一个数字：

```
3107418240490043721350750035888567930037346022842727545720161948823206440518081504556346829671723286782437916272838033415471073108501919548529007337724822783525742386454014691736602477652346609
```

一个德国团队于2005年成功地对RSA–640进行了因数分解。[22] RSA实验室向这个团队颁发了20 000美元的奖金，虽然这笔奖金没有RSA–2048的奖金那么高，但也是一个相当不错的奖励了！这个德国团队将对RSA–640进行因数分解的任务分配给8台计算机，每台计算机都拥有2.2兆赫兹的AMD皓龙（Opteron）中央处理器。然而即使是这样，这些计算机仍然花了5个月的时间才找出RSA–640的质因数。

2009年，另一个团队成功完成了对RSA–768进行因数分解的任务。

这个团队是由来自法国、德国、日本、荷兰、瑞士以及美国的研究者共同组成的。这项因数分解任务花费的计算时间“相当于一个单核2.2兆赫兹的AMD皓龙中央处理器大约2 000年的计算时间”。[23]这个团队当然不是用一个单核2.2兆赫兹的AMD皓龙中央处理器来完成这项任务的，他们将因数分解的任务分配给了“数百台计算机”。然而即便如此，这项任务仍然花费了这个团队近两年半的时间。

成功分解RSA–768的团队本来应该获得50 000美元的奖金，可惜的是2009年时这笔奖金已经过期了，因此上述团队并没有领到这笔50 000美元的奖金。

看来，上述这种“分而治之”（divide-and-conquer）的算法可以成功地解决一些困难的因数分解问题。但是，里韦斯特后来又设计出了另一项挑战，他认为这是用“分而治之”算法无法解决的挑战。里韦斯特写道：

> 试图解决这类问题的人通常会利用并行计算或者分布计算的方式来缩短计算所需要的时间。但是，我设计这项挑战的目的就是让上述方法发挥不了作用。要解决我设计的这个挑战，必须进行“具有串行本质”的计算。[24]

几年之前，里韦斯特和沙米尔，以及戴维·A. 瓦格纳（David A. Wagner）一起提出这个想法的时候，里韦斯特以一种更生动的方式解释了这个问题：

> 要完成我们这项挑战，就像生孩子一样：就算让两个女人同时怀孕，也不可能只用4.5个月就生出一个孩子来。[25]

时间胶囊密码

里韦斯特给他提出的这个挑战起了一个很特殊的名字——时间胶囊密码问题，因为他的设计意图就是：任何破解者都必须花35年的时间才能完成这个挑战。之所以选择35年，是因为1999年正好是麻省理工学院计算机科学实验室（LCS）成立35周年。在提出这个问题的同时，里韦斯特还制造了一个实体的时间胶囊，名叫"LCS创新时间胶囊"。这个时间胶囊是一个用铅制成的高达1米的容器。根据里韦斯特的计划，如果某个破解团队能从问题提出的1999年就开始不断地进行破解计算，那么2033年正好是他们解开这个问题的时间，实体的时间胶囊也将同时被打开。而如果在2033年之前有人能够通过其他方法解开这个问题的话，上述时间胶囊也可以被提前打开。

除了上述时间胶囊密码问题以外，里韦斯特还提供了另一个比较简单的问题，相当于一个迷你版的时间胶囊密码问题。通过研究这个迷你版的问题，试图破译上述时间胶囊密码问题的人可以更清楚地理解出题人的意图。介绍完这个迷你版的问题以后，里韦斯特才给出了真正待破解的时间胶囊密码问题。

在这个时间胶囊密码问题中，首先有一个数，用字母z来表示。z代表的是一段加密后的信息。我们需要用到另一个数字（数值密钥）才能破译上述加密信息，我们用字母w来表示这个数值密钥。我打算过一会再向读者详细解释破译信息的方法。现在，我们首先来研究如何找到w的问题。事实上，寻找w的运算十分简单，这种运算是：

$$w = 2^{2^t} \pmod{n}$$

里韦斯特已经向所有破解者提供了上述算式中的t和n的值。

在迷你版的时间胶囊密码问题中，$t = 10$，而$n = 253$。注意$253 = 11 \times 23$，也就是说253是两个质数的乘积。

代入t的值，我们就可以进行以下计算：

$$w = 2^{2^t} = 2^{2^{10}} = 2^{1\,024} = \cdots$$

上面这个数字太大了，我的计算器根本没法完成这种运算。我们也没有办法用科学记数法来写下上述运算的结果，因为我们需要知道这个数字中的每一位上的具体数字。别忘了，接下来的一步是用$n = 253$来除上面的得数，并取余数。我们当然可以写一个计算机程序，从数字2开始，不断地做乘2的运算，一共做1 024次，记录下得数的每一位具体数字，然后再用这个得数除以253，看看余数是多少。但是用上述方法来进行这种运算效率实在太低了。事实上，我们可以通过另一种方法来得到上述运算的答案，这种方法只有10个步骤，用不着1 024个步骤。而且，在这种方法下，不仅计算机可以轻松地完成上述运算，甚至用纸笔也可以完成（如果你足够仔细的话）。

这种方法的名字叫作“反复平方法”。顾名思义，我们要做的就是不断地进行平方运算。首先，我们从某个特定的数字开始（在我们的例子中，这个数字是2），然后取这个数的平方。接着，我们再对上一步的得数取平方。然后，对这一步的结果再取平方，以此类推。我们得到：

$2^2 = 4$,

$4^2 = 16$,

$16^2 = 256 = 3 \pmod{253}$.

不要忘了，我们要计算的是幂模除253的结果。

$3^2 = 9$,

$9^2 = 81$,

$81^2 = 6\,561 = 236 \pmod{253}$,

$236^2 = 55\,696 = 36 \pmod{253}$,

$36^2 = 1\,296 = 31 \pmod{253}$,

$31^2 = 961 = 202 \pmod{253}$,

$202^2 = 40\,804 = 71 \pmod{253}$.

到这一步，我们就可以停止了！上一步说明：$2^{1\,024} \pmod{253} = 71$。

我们究竟是如何得到上述结果的？为什么这种方法只需要10步，而不是1 024步？

为了回答这个问题，我们需要把上述操作中的几个步骤放在一起看。在我们做完4次平方运算以后，我得到的结果是$(((2^2)^2)^2)^2$。当我们对一个数的幂取平方的时候，我们事实上是在对指数做乘法。因此，我们得到：$(((2^2)^2)^2)^2 = 2^{16}$。也就是说，进行4次平方运算以后，我们得到了2的16次幂（指数的4个2的乘积等于16）。

根据上述原理，只要我们对2进行10次平方运算，我们最终得到的幂次就是10个2的乘积，也就是1 024。因此，反复平方运算可以让我们在10步内得到上述问题的答案，而不需要进行1 024步运算。

反复平方法还有另一个好处，那就是：如果你想要的最终得数是上述结果模253（或者模除任何一个数字），那么你可以在任何一步模除这个数字，然后继续将模除后的结果平方。不管你在哪一步进行模除运算，最终你仍然能得到正确答案。因此，我们可以利用这种性质，在进行反复平方运算的时候一有机会就模除253，这样可以避免得数变得过大，使得运算更加方便。因此，反复平方法不仅运算速度很快，而且几乎可以用任何一种计算器完成。即使是对很大的数字，我们也没必要特别写一段程序来完成上述运算。

通过反复平方法，我们已经得到 $w = 71$。接下来，我们来看一看在里韦斯特的例子中，如何根据 w 的值解出明文信息。

在里韦斯特给出的这个小例子中，已知 z 的值是“13(hex)”。括号中的“hex”是hexadecimal的缩写，意为十六进制。在通常的十进制计数系统中，我们使用0、1、2、3、4、5、6、7、8、9这10个数字符号，而在十六进制计数系统中，我们使用16个数字符号，这16个数字符号分别是0、1、2、3、4、5、6、7、8、9、A、B、C、D、E、F。

在表11–1中列出了我们常用的十进制数以及这些数对应的十六进制数。

表 11–1　十进制数及其对应的十六进制数

十进制数	十六进制数	十进制数	十六进制数	十进制数	十六进制数
1	1	11	B	30	1E
2	2	12	C	40	28
3	3	13	D	50	32
4	4	14	E	60	3C
5	5	15	F	70	46
6	6	16	10	80	50
7	7	17	11	90	5A
8	8	18	12	100	64
9	9	19	13	500	1F4
10	A	20	14	1 000	3E8

十六进制计数系统在计算机科学领域十分常见，在其他领域则很少出现。

至此，我们已经知道 $w = 71$（十进制），$z = 13$（十六进制）。在里韦斯特这个例子中，接下来的一步是把上述两个数字以某种方式结合在一起。但是，在进行这一步之前，我们首先必须将71转化成十六进制数，这样我们就能以同样的计数系统来表示w和z的值了。

十进制数71所对应的十六进制数是47。因此我们得到：$w = 47$（十六进制），$z = 13$（十六进制）。

接下来的步骤是对上述两个数字进行“XOR”运算。下面我将向读者解释如何进行该运算，并提供一种完成它的捷径。有了这种捷径以后，读者就不再需要理会该运算的细节了。首先，我们需要把十六进制数转换成二进制（即以2为基数的计数系统）数。十六进制数与二进制数的对应关系如下表所示：

0 = 0000	4 = 0100	8 = 1000	C = 1100
1 = 0001	5 = 0101	9 = 1001	D = 1101
2 = 0010	6 = 0110	A = 1010	E = 1110
3 = 0011	7 = 0111	B = 1011	F = 1111

因此，十六进制数47转化为二进制数是0100 0111，而十六进制数13转化为二进制数是0001 0011。我将每4个数字分成一组，并且在组与组之间加入空格，这并不是必需的，只是为了让读者能够更清楚地看懂我所做的替换处理。接下来，我们把以上的两个二进制数对齐，对齐的方式和做普通的十进制加法时一样。再接下来，我们将每一纵列中的两个数加起来。但是，做二进制加法的方式和十进制加法稍有不同。在二进制的加法运算中，$0 + 0 = 0$，$0 + 1 = 1$，$1 + 0 = 1$（读者应该可以猜到这一点），但是$1 + 1 = 0$。虽然这种规则看起来也许有点儿奇怪，但这实际上是一种非常有用的加法规则，在许多不同的情况下都适用。

XOR是“exclusive-or”的缩写，意为“异或”。与“异或”这个概念相对的概念是“同或”（inclusive-or，简写为OR）。在我们日常英文中，我们几乎不区分异或和同或。逻辑学家经常用“真”（True）与“假”（False）来代替1和0，这样做可以帮助我们理解OR和XOR之间的区别。在英语中，我们可以用“OR”一词造这样一个句子：“he’s strong if he can bench press 300 lbs or squat 400 lbs”（如果他能够卧推300磅或者蹲举400磅，那么他就很强壮了）。显然，如果他既能够卧推300磅，又能够蹲举400磅，我们当然会承认他是很强壮的。这就是一个同或（OR）的例子。而异或（XOR）和同或（OR）的区别是，异或不允许以上两个条件同时为真。在餐馆里点菜的时候，我们会遇到异或的例子。如果服务员问我们：“Would you like french fries or a potato with your meal? ”（你想要炸薯条还是一个土豆作为配餐？）我们显然不能回答“是的”来表示我们两种都想要，这样只会让服务员皱眉头而已。在这个例子中，“真”XOR“真”（我两种都想要）=“假”（服务员不会两种都给我），也就是1 + 1 = 0。将XOR运算的法则用于上述两个二进制数的加法运算，我们就得到：

```
    0100 0111
XOR 0001 0011
-------------
    0101 0100
```

接下来，我们再根据上表中的对应关系把这个二进制数转化成十六进制数，0101 0100（二进制）= 54（十六进制）。也就是说，在十六进制的计数系统中，上述加法运算可以写为：

```
    47
XOR 13
------
    54
```

以上就是对十六进制数做XOR运算的方法，为了让读者能够更方便地直接进行十六进制数的XOR运算，我们可以直接使用表11–2中的结果。有了表11–2，我们就不用先把十六进制数转化为二进制数，再把运算所得的二进制数重新转化为十六进制数了。下面，我们利用表11–2重复上述XOR运算。读者应该可以看出，有了表11–2以后，XOR运算的过程变得容易多了。

表 11–2　十六进制数的XOR运算

XOR	0	1	2	3	4	5	6	7	8	9	A	B	C	D	E	F
0	0	1	2	3	4	5	6	7	8	9	A	B	C	D	E	F
1	1	0	3	2	5	4	7	6	9	8	B	A	D	C	F	E
2	2	3	0	1	6	7	4	5	A	B	8	9	E	F	C	D
3	3	2	1	0	7	6	5	4	B	A	9	8	F	E	D	C
4	4	5	6	7	0	1	2	3	C	D	E	F	8	9	A	B
5	5	4	7	6	1	0	3	2	D	C	F	E	9	8	B	A
6	6	7	4	5	2	3	0	1	E	F	C	D	A	B	8	9
7	7	6	5	4	3	2	1	0	F	E	D	C	B	A	9	8
8	8	9	A	B	C	D	E	F	0	1	2	3	4	5	6	7
9	9	8	B	A	D	C	F	E	1	0	3	2	5	4	7	6
A	A	B	8	9	E	F	C	D	2	3	0	1	6	7	4	5
B	B	A	9	8	F	E	D	C	3	2	1	0	7	6	5	4
C	C	D	E	F	8	9	A	B	4	5	6	7	0	1	2	3
D	D	C	F	E	9	8	B	A	5	4	7	6	1	0	3	2
E	E	F	C	D	A	B	8	9	6	7	4	5	2	3	0	1
F	F	E	D	C	B	A	9	8	7	6	5	4	3	2	1	0

根据上表，我们可以看出7 XOR 3 = 4，4 XOR 1 = 5，因此我们可以直接得到：

```
    47
XOR 13
------
    54
```

在里韦斯特的例子中，最后一个步骤是在ASCII码表中查找对应的十六进制数。[26]由于十六进制数在计算机科学领域中非常常见，所以大部分ASCII码表都会同时提供每个字符对应的十进制数和十六进制数，参见表11–3和表11–4。

根据表11–3，我们可以看出54（十六进制）对应的字母是T。因此，这个例子的明文信息就是T。虽然明文信息的内容并不十分振奋人心，但里韦斯特提供的这个例子向我们解释了时间胶囊密码的原理。

由于里韦斯特特别指出破译者应该查找的是“扩展ASCII码表”，所以我怀疑时间胶囊密码中应该包含表11–4中的一个或多个字符（即十进制数128~255对应的字符）。

至此，我们已经了解了里韦斯特提出的时间胶囊密码的性质，下面我们来看这段密码的具体内容。

==

LCS35 时间胶囊密码谜题。

由罗伯特·L. 里韦斯特于 1999 年 4 月 2 日创建。

该谜题的参数（全部用十进制数表示是）：

表 11-3　扩展ASCII码表（较小数值）①

Dec Hex	名称	字符	Dec Hex	字符	Dec Hex	字符	Dec Hex	字符
0 0	空字符	NUL	32 20	Space	64 40	@	96 60	`
1 1	标题开始	SOH	33 21	!	65 41	A	97 61	a
2 2	正文开始	STX	34 22	“	66 42	B	98 62	b
3 3	正文结束	ETX	35 23	#	67 43	C	99 63	c
4 4	传输结束	EOT	36 24	$	68 44	D	100 64	d
5 5	请求	ENQ	37 25	%	69 45	E	101 65	e
6 6	确认	ACK	38 26	&	70 46	F	102 66	f
7 7	响铃	BEL	39 27	‘	71 47	G	103 67	g
8 8	退格	BS	40 28	(	72 48	H	104 68	h
9 9	水平制表符	HT	41 29	)	73 49	I	105 69	i
10 0A	换行键	LF	42 2A	*	74 4A	J	106 6A	j
11 0B	垂直制表符	VT	43 2B	+	75 4B	K	107 6B	k
12 0C	换页键	FF	44 2C	,	76 4C	L	108 6C	l
13 0D	回车键	CR	45 2D	–	77 4D	M	109 6D	m
14 0E	不用切换	SO	46 2E	.	78 4E	N	110 6E	n
15 0F	启用切换	SI	47 2F	/	79 4F	O	111 6F	o
16 10	数据链路转义	DLE	48 30	0	80 50	P	112 70	p
17 11	设备控制 1	DC1	49 31	1	81 51	Q	113 71	q
18 12	设备控制 2	DC2	50 32	2	82 52	R	114 72	r
19 13	设备控制 3	DC3	51 33	3	83 53	S	115 73	s
20 14	设备控制 4	DC4	52 34	4	84 54	T	116 74	t
21 15	否认	NAK	53 35	5	85 55	U	117 75	u
22 16	同步空闲	SYN	54 36	6	86 56	V	118 76	v
23 17	传输块结束	ETB	55 37	7	87 57	W	119 77	w
24 18	取消	CAN	56 38	8	88 58	X	120 78	x
25 19	媒介质结束	EM	57 39	9	89 59	Y	121 79	y
26 1A	替换	SUB	58 3A	:	90 5A	Z	122 7A	z
27 1B	转义	ESC	59 3B	;	91 5B	[	123 7B	{
28 1C	文件分隔符	FS	60 3C	<	92 5C	\	124 7C	|
29 1D	分组符	GS	61 3D	=	93 5D	]	125 7D	}
30 1E	记录分隔符	RS	62 3E	>	94 5E	^	126 7E	~
31 1F	单元分隔符	US	63 3F	?	95 5F	_	127 7F	DEL

① 表头中“Dec”表示十进制数，“Hex”表示十六进制数。——编者注

表 11-4 扩展ASCII码表（较大数值）[a]

Dec Hex	字符	Dec Hex	字符	Dec Hex	字符	Dec Hex	字符
128 80	Ç	160 A0	á	192 C0	└	224 E0	α
129 81	ü	161 A1	í	193 C1	┴	225 E1	β
130 82	é	162 A2	ó	194 C2	┬	226 E2	Γ
131 83	â	163 A3	ú	195 C3	├	227 E3	π
132 84	ä	164 A4	ñ	196 C4	─	228 E4	Σ
133 85	à	165 A5	Ñ	197 C5	┼	229 E5	σ
134 86	å	166 A6	ª	198 C6	╞	230 E6	μ
135 87	ç	167 A7	°	199 C7	╟	231 E7	τ
136 88	ê	168 A8	¿	200 C8	╚	232 E8	φ
137 89	ë	169 A9	⌐	201 C9	╔	233 E9	θ
138 8A	è	170 AA	¬	202 CA	╩	234 EA	Ω
139 8B	ï	171 AB	½	203 CB	╦	235 EB	δ
140 8C	î	172 AC	¼	204 CC	╠	236 EC	∞
141 8D	ì	173 AD	¡	205 CD	═	237 ED	ϕ
142 8E	Ä	174 AE	«	206 CE	╬	238 EE	ε
143 8F	Å	175 AF	»	207 CF	╧	239 EF	∩
144 90	É	176 B0	░	208 D0	╨	240 F0	≡
145 91	æ	177 B1	▒	209 D1	╤	241 F1	±
146 92	Æ	178 B2	▓	210 D2	╥	242 F2	⩾
147 93	ô	179 B3	│	211 D3	╙	243 F3	⩽
148 94	ö	180 B4	┤	212 D4	╘	244 F4	⌠
149 95	ò	181 B5	╡	213 D5	╒	245 F5	⌡
150 96	û	182 B6	╢	214 D6	╓	246 F6	÷
151 97	ù	183 B7	╖	215 D7	╫	247 F7	≈
152 98	ÿ	184 B8	╕	216 D8	╪	248 F8	°
153 99	Ö	185 B9	╣	217 D9	┘	249 F9	·
154 9A	Ü	186 BA	║	218 DA	┌	250 FA	·
155 9B	¢	187 BB	╗	219 DB	█	251 FB	√
156 9C	£	188 BC	╝	220 DC	▄	252 FC	^{n}
157 9D	¥	189 BD	╜	221 DD	▌	253 FD	^{2}
158 9E	₧	190 BE	╛	222 DE	▐	254 FE	■
159 9F	ƒ	191 BF	┐	223 DF	▀	255 FF	

a 本表中的某些符号会因为电脑和字体的不同而有不同的表达方式。

```
n = 631446608307288889379935712613129233236329881833
    084137558899077270195712892488554730844605575320
    651361834662884894808866350036848039658817136198
    766052189726781016228055747539383830826175971321
    892666861177695452639157012069093997368008972127
    446466642331918780683055206795125307008202024124
    623398241073775370512734449416950118097524189066
    796385875485631980550727370990439711973361466670
    154390536015254337398252457931357531765364633198
    906465140213398526580034199190398219284471021246
    488745938885358207031808428902320971090703239693
    491996277899532332018406452247646396635593736700
    9369212758092086293198727008292431243681
```

和前面的例子一样，n是两个质数的乘积。但是在真正的时间胶囊密码谜题中，要找出这两个质数是非常困难的！

```
t = 79685186856218
```

里韦斯特之所以选择上述t值，是为了让破译时间胶囊密码谜题所需的时间恰好等于35年。那么，他估计的计算时间是正确的吗？在对计算时间进行估计的时候，里韦斯特已经考虑了摩尔定律。根据摩尔定律，在1999—2012年期间，计算机内部芯片的速度会增加到原来的13倍，而在此后的年份中会再增加到原来的5倍。在这里值得强调的一点是，在里韦斯特的估计中，他假设破译者一直使用当时世界上运算速度最快的计算机，并且每年都会更换一种更新、更快的机型。

```
z = 427338526681239414707099486152541907807623930474
    842759553127699575212802021361367225451651600353
    733949495680760238284875258690199022379638588291
    839885522498545851997481849074579523880422628363
    751913235562086585480775061024927773968205036369
    669785002263076319003533000450157772067087172252
    728016627835400463807389033342175518988780339070
    669313124967596962087173533318107116757443584187
    074039849389081123568362582652760250029401090870
    231288509578454981440888629750522601069337564316
    940360631375375394366442662022050529545706707758
    321979377282989361374561414204719371297211725179
    2879310395477535810302267611143659071382
```

要解出这道密码谜题，要先计算w = 2^(2^t)(mod n).
然后对上述运算的结果用z做XOR运算。

由于w和z对应的数字串长度不同，于是就产生了一个有些令人迷惑的问题：究竟应该如何把这两个数字对齐进行XOR运算？对此，里韦斯特的解释是：应该把这两个数字的右侧对齐。左侧多余的数值无法与其他数值配对，这些多余的数值代表一个数字，里韦斯特将它称为b。里韦斯特指出，破译者应该把这个数字代入以下算式：5^b (mod $2^{1\,024}$)。就算b的数值很大，计算机也可以很快完成上述任务。上述算式的得数应该稍小于n的一个质因数。因此，通过以上的计算步骤，破译者就能够找到这个质因数，并据此对n进行因数分解。

里韦斯特提供的上述细节非常重要。因为有了这一细节，我们就可以快速地验证一个解答是否正确。试想，如果在未来的某一天，一位破译者称解出了里韦斯特的时间胶囊密码，而万一此时唯一知道正确解法的人（里韦斯特本人）已经死亡或者不记得密码的原文信息，那么麻省理工学院计算机科学实验室的主管人员该怎么做呢？里韦斯特生于1947年，我们很难保证密码被破解的时候他仍然在世。

显然，计算机科学实验室的主管不可能要求破译者当着他的面把所有破译工作重复一遍，以保证所有计算过程准确无误。因为，重复上述过程虽然不至于要再花费35年的时间，但肯定也是相当费时的。

然而，由于破译了时间胶囊密码就得到了对n进行因数分解的结果，所以，只要破译者将n的两个质因数交给计算机科学实验室的主管，就可以充分证明自己已经解开了里韦斯特的谜题。验证者只需要把这两个数重新相乘（计算机可以快速地完成这项任务），并确认乘积是否为n就可以了。至于密码本身，只要有上文所述的因数分解结果，就可以通过快速的计算取得密码的密钥。计算机科学实验室的主管可以利用破译者提供的因数分解结果快速高效地破译出密码的明文。

然而，对大数n进行因数分解并不是解开时间胶囊密码的捷径。事实上，这是一项极为困难的任务，因为里韦斯特挑选的大数n和上文提到的RSA–2048长度相当。因此，要对大数n进行因数分解，所耗费的时间理论上应该比直接破译时间胶囊密码所需的时间更长。

里韦斯特写道：

> 如果任何人认为自己已经解出了我的谜题，请将密码对应的明文信息和大数n的因数分解结果以及相关的破解笔记一同交给计算机科

> 学实验室的主管。在确认解法正确无误以后，计算机科学实验室的主管会举行一个特别的庆祝仪式，并在仪式上打开时间胶囊的封条。如果到2033年9月还没有任何人提交正确的解法，计算机科学实验室的主管会在实验室成立70周年的庆典仪式上（或者其他合适的仪式上）打开时间胶囊的封条。如果届时计算机科学实验室的主管无法前来，那么麻省理工学院的校长会指定另外的机构或个人完成上述程序。[27]

不管是对大数进行因数分解的挑战，还是里韦斯特提出的时间胶囊密码挑战，都是为了测试密码破译领域中最先进的技术和成果而设计的。除了本章中介绍的这些挑战以外，还有许多其他挑战也是为了上述目的而被创造出来的。

未来的期待

在我写作本书的过程中，又不断有新的未解密码出现在公众的视野中。这其中一些密码已有几十年的历史，而另外一些密码的历史甚至长达数百年。此外，新的挑战密码也在不断地涌现出来。如果要详细地研究每一条新出现的未解之谜，恐怕要花费一个人的毕生时间。然而，研究这些未解之谜实在是一项非常有趣的工作！

因此，万一你已经解出了本书中介绍过的所有密码，请不要觉得失落。记住，在本书之外，还存在许多其他的未解密码，你永远不必担心因缺乏挑战而变得无聊。也许再过几年，你还会看到本书的续篇，我将会从本书未提及的密码中挑选出较精彩的介绍给你们。

致 谢

在写作本书早期，我得到了美国国家安全局的支持。2011—2012 年间，我很荣幸地在美国国家安全局的密码历史中心任该机构的学者。宾夕法尼亚约克学院给了我一学期的学术休假，这为本书的完成提供了不小的帮助。国家密码博物馆图书馆管理员勒内·斯坦给了我巨大的帮助。克劳斯·施梅为我提供了许多案例，这都是我以前没有遇到的。施梅的部分英文版论文可以在《密码》（*Cryptologia*）杂志上找到，但是他的大部分作品目前只有德文版，包括他所著的优秀书籍，以及他的博客（博客网址为 http://scienceblogs.de/klausis-krypto-kolumne/author/kschmeh/）。我推荐懂德语的读者可以参考施梅的这些作品。

此外，我还要感谢以下这些人为我提供的帮助。他们包括：卡洛斯·阿尔瓦拉多、戈登·L. 安德森、珍妮·安德森、克里斯·贝内特、保拉·贝拉尔、贝齐·布卢门塔尔、保罗·博纳沃利亚、科琳·博伊尔、劳伦·布卡、凯伦·卡特、克里斯·克里斯滕森、保罗·M. 克莱门斯、约翰·W. 道森、拉尔夫·厄斯金、休·费尔柴尔德、盖里·费尔特斯、玛丽·安·福尔特、斯图尔特·弗里德、托尼·加夫尼、约翰·盖尔霍克、乔斯林·戈德温、帕特里克·格兰霍姆、卡罗琳·格雷夫斯-布朗、蒂娜·汉普森、洛根·哈里斯、戴维·哈奇、布赖恩·海因霍尔德、露西·休斯、利亚姆·赫德、梅根·卡娜贝、季米特里·卡伦特尼科夫、维基·卡恩、凯文·奈特、奥利弗·克内策尔、拜奈代克·朗、唐纳德·P. 勒瓦瑟、罗伯特·勒旺德、格雷格·林克、西蒙·马丁、阿德里安娜·梅厄、贝娅塔·迈杰希、安妮·玛丽·门塔、丹特·莫勒、布伦特·莫里斯、戴维·奥兰恰克、詹姆斯·拉姆、斯拉瓦纳·雷迪、吉姆·利兹、德克·赖默南兹、埃莉诺·罗布森、摩西·鲁宾、戈登·鲁格、凯茜安·萨拉诺娃、戴维·桑德斯、布鲁斯·施奈尔、克拉林·斯皮斯、约瑟夫·斯普劳斯基、克莱本·汤普森、凯尔·R. 特里普利特、吉姆·塔克、迈克尔·蒂姆、戴维·维斯科、亨特·威林厄姆、勒内·赞德伯根，以及露西·周。

如果我遗漏了某些人的名字，我在此表示歉意。我发自内心地感激所有人给我的帮助！

注　释

第 1 章　一位国王的探求

1 E-mail received by the author on November 21, 2014. Also see http://en.wikipedia.org/wiki/Jerusalem_artichoke.

2 Newbold included the section I labeled "cosmological" as part of the biological section. See Newbold, W. R., "The Voynich Roger Bacon Manuscript." *Transactions of the College of Physicians and Surgeons of Philadelphia* (1921): 431–74 (the relevant pages here are 461–63). Read April 20, 1921. Also see Newbold, W. R., and Kent, R. G., *The Cipher of Roger Bacon*, 1928, pp. 44 and 46. Brumbaugh labels the sections as 1. Botanical, 2. Astrological, 3. Medical (perhaps dealing with human reproduction), 4. Medicinal, and 5. Appendix or Postscript. See Brumbaugh, Robert S., "Science in Cipher," pp. 109–41, in Brumbaugh, Robert S. (ed.), *The Most Mysterious Manuscript*, 1978, p. 110 cited here.

3 Six are suggested, for example, in Reddy, Sravana, and Kevin Knight, "What We Know about the Voynich Manuscript," in *Proceedings of the 5th ACL-HLT Workshop on Language Technology for Cultural Heritage, Social Sciences, and Humanities*, pp. 78–86, Portland, OR, June 24, 2011. Available online at http://www.aclweb.org/anthology/W11-1511 and https://www.aclweb.org/anthology/W/W11/W11-15.pdf.

4 Taken from Act 3, Scene 2 of William Shakespeare's play *Julius Caesar*. This line is delivered by Marc Antony.

5 The quote was taken from Evans, R. J. W., *Rudolf II and His World: A Study in Intellectual History 1576–1612* (Oxford: Clarendon Press, 1973), 196.

6 Code penguin, http://tools.codepenguin.com/patterns/help#.

7 Evans, *Rudolf II and His World*, 58.

8 Evans, *Rudolf II and His World*, 198.

9 Evans, *Rudolf II and His World*, 230. Also see Mukherjee, Ashoke, "Giordano Bruno, An Ever-Burning Flame of Commitment to Truth and Reason," *Breakthrough* 8, no. 4 (2000): 1–13, available online at http://breakthrough-india.org/archives/bruno.pdf.

10 Evans, *Rudolf II and His World*, 279.

11 Stolfi, Jorge, Voynich Manuscript stuff, 2005, http://www.dcc.unicamp.br/stolfi/voynich/.

12 The graph is taken from Reddy and Knight, "What We Know," p. 4. Available online at http://www.isi.edu/natural-language/people/voynich-11.pdf.

13 Reddy and Knight, "What We Know," 4, footnote.

14 This translates to "On the Secret Workings of Art and Nature, and on the Vanity of Magic." Newbold, "The Voynich Roger Bacon Manuscript," p. 456 cited here.

15 Kennedy, Gerry, and Rob Churchill, *The Voynich Manuscript: The Unsolved Riddle of an Extraordinary Book Which Has Defied Interpretation for Centuries*, British hardcover edition, Orion Books, London, 2004, p. 20, which, in turn, took it from Bacon, Roger, *Compendium Studii Philosophiae*, 1271 or 1272.

16 Some sources give a death date two years later. See, for example, Kennedy and Churchill, *The Voynich Manuscript*, p. 21. At http://www.nndb.com/people/582/000114240/ a death date of 1294 is conjectured, although uncertainty is expressed. Also see Newbold, "The Voynich Roger Bacon Manuscript," 445–46, for the 1292 death date.

17 Lieutenant Colonel H. W. L. Hime, Royal Artillery, "Roger Bacon and Gunpowder," pp. 321–36 in *Roger Bacon Essays Contributed by Various Writers on the Occasion of the Commemoration of the Seventh Centenary of His Birth*, collected and edited by A. G. Little, Clarendon Press, Oxford, 1914, available online at http://archive.org/stream/rogerbaconessays00litt/rogerbaconessays00litt_djvu.txt. Also see Goldstone, Lawrence, and Nancy Goldstone, *The Friar and the Cipher: Roger Bacon and the Unsolved Mystery of the Most Unusual Manuscript in the World* (New York: Doubleday, 2005), 107–8.

18 Thorndike's argument does not get into the details of the cipher itself. See Thorndike, Lynn, "Roger Bacon and Gunpowder," *Science*, New Series 42 , no. 1092 (December 3, 1915): 799–800. This piece was reprinted as Appendix II (pp. 688–91) of Thorndike's *A History of Magic and Experimental Science during the First Thirteen Centuries of our Era*, Vol. II (New York: Columbia University Press, 1923), which is available online at https://archive.org/details/historyofmagicex02thor. The controversy is also mentioned in Goldstone and Goldstone, *The Friar and the Cipher*, 251.

19 Newbold, "The Voynich Roger Bacon Manuscript," p. 457 cited here. Newbold noted that this isn't what Ethicus actually did, at least not in the ciphers of his that are known today.

20 Newbold, "The Voynich Roger Bacon Manuscript," p. 456 cited here.

21 Newbold, "The Voynich Roger Bacon Manuscript," p. 457 cited here.

22 Roberts, R. J., and A. G. Watson (eds.), *John Dee's Library Catalogue* (London: The Bibliographical Society, 1990).

23 Kennedy and Churchill, *The Voynich Manuscript*, 2004, p. 66, citing Prinke's online article. Images comparing known samples of Dee's numbers to those on the manuscript pages are provided on p. 67 of Kennedy and Churchill. Also see Pelling, Nick, *The Curse of the Voynich: The Secret History of the World's Most Mysterious Manuscript* (Surbiton, Surrey, UK: Compelling Press, 2006), 11–13.

24 Goldstone and Goldstone, *The Friar and the Cipher*, pp. 187–88. When Sir Francis Walsingham received intercepted messages penned by the imprisoned Mary, Queen of Scots, and conspirators planning to set her free and assassinate Queen Elizabeth, he likely had much confidence that the plaintexts could be recovered. It was his cryptanalyst Thomas Phelippes who succeeded in breaking them, and the rest is history. Simon Singh's *The Code Book* (New York: Doubleday, 1999) does an excellent job of recounting this tale and the consequences for Mary

(see chapter 1). Thomas Phelippes was also at a meeting of Walsingham and Dee. See Goldstone and Goldstone, *The Friar and the Cipher,* 204–5.

25 Fell-Smith, Charlotte, *John Dee (1527–1608)*, (London: Constable & Co. Ltd., 1909), 311. Available online at https://archive.org/details/cu31924028928327.

26 Kelley was blackmailed by Sir Francis Walsingham into becoming a spy.

27 Brumbaugh, Robert S., editor and contributor, *The Most Mysterious Manuscript: The Voynich "Roger Bacon" Cipher Manuscript* (Carbondale, IL: Southern Illinois University Press, 1978), 131.

28 I learned of this in Pelling, *The Curse of the Voynich*, p. 158, where credit for the discovery of the numbering system is given to Allan Wechsler and Ivan Derzhanski, who came upon it independently.

29 Shelfmark APUG 557, fol. 353.

30 Translation taken from http://www.voynich.net/neal/barschius_translation.html, which incorrectly dates it as 1637, the date of a previous letter that we do not have. The original Latin may be found at http://www.voynich.nu/letters.html#gb39, along with other letters connected with the manuscript.

31 The text of this letter appears in Goldstone and Goldstone, *The Friar and the Cipher*, pp. 237–39 and elsewhere.

32 Identified as "Dr Raphael Missowsky, Attorney General to Ferdinand III" in Kennedy and Churchill, *The Voynich Manuscript*, 73.

33 Kircher, Athanasius, *The Volcanos: Or Burning and Fire-Vomiting Mountains, Famous in the World* (London: J. Darby,1669): 35.

34 Goldstone and Goldstone, *The Friar and the Cipher*, 239.

35 Goldstone and Goldstone, *The Friar and the Cipher*, 239–40.

36 Voynich, Wilfrid M., "A Preliminary Sketch of the History of the Roger Bacon Cipher Manuscript." *Transactions of the College of Physicians of Philadelphia* 43, ser. 3 (1921): 415–30. Read April 20, 1921.

37 Voynich, "A Preliminary Sketch."

38 Voynich, "A Preliminary Sketch."

39 Voynich, "A Preliminary Sketch."

40 Voynich, "A Preliminary Sketch."

41 Voynich, "A Preliminary Sketch."

42 Sowerby, E. Millicent, *Rare People and Rare Books* (London: Constable & Co., Ltd, 1967).

43 Sowerby, *Rare People and Rare Books*.

44 Taratuta, Evgeniya, "Our Friend Ethel Lilian Boole/Voynich." Moscow, 1964, translated from the Russian by Séamus Ó Coigligh with additional notes, 2008.

45 Orioli, Giuseppe, *Adventures of a Bookseller* (New York: Robert M. McBride & Co., 1938).

46 Sowerby, *Rare People and Rare Books*.

47 Taratuta, "Our Friend Ethel Lilian Boole/Voynich."

48 Manly, John M., "The Most Mysterious Manuscript in the World." *Harper's Monthly Magazine*, 143 (July 1921): 186–97.

49 Newbold, "The Voynich Roger Bacon Manuscript."

50 Newbold, "The Voynich Roger Bacon Manuscript."

51 Newbold, "The Voynich Roger Bacon Manuscript."
52 Newbold, "The Voynich Roger Bacon Manuscript."
53 Newbold, "The Voynich Roger Bacon Manuscript."
54 Newbold, "The Voynich Roger Bacon Manuscript."
55 Newbold, "The Voynich Roger Bacon Manuscript."
56 Newbold, "The Voynich Roger Bacon Manuscript."
57 Newbold, "The Voynich Roger Bacon Manuscript."
58 Newbold, "The Voynich Roger Bacon Manuscript."
59 Newbold, "The Voynich Roger Bacon Manuscript."
60 Newbold, "The Voynich Roger Bacon Manuscript."
61 Newbold, William Romaine. *The Cipher of Roger Bacon,* edited with foreword and notes by Roland Grubb Kent. (Philadelphia: University of Pennsylvania Press; London, H. Milford, Oxford University Press, 1928).
62 Newbold, "The Voynich Roger Bacon Manuscript."
63 Bird, J. Malcolm, "The Roger Bacon Manuscript Investigations into Its History, and the Efforts to Decipher It." *Scientific American Monthly*, June 1921, 492–96.
64 Anonymous, "The Roger Bacon Manuscript, What It Looks Like, and a Discussion of the Probabilities of Decipherment." *Scientific American*, May 28, 1921, 432, 439–40.
65 Anonymous, "The Roger Bacon Manuscript."
66 Thorndike, Lynn, "Roger Bacon." *The American Historical Review* 34, no. 2 (January 1929): 317–19.
67 Manly, John Matthews., "Roger Bacon and the Voynich Manuscript." *Speculum* 6 (July 1931): 345–91.
68 Manly, "Roger Bacon and the Voynich Manuscript."
69 Manly, "Roger Bacon and the Voynich Manuscript."
70 Manly, "Roger Bacon and the Voynich Manuscript."
71 Manly, "Roger Bacon and the Voynich Manuscript."
72 This name is hard to make out in the letter and may not have been reproduced correctly here.
73 "The Quirinal" refers to the Italian government of the time, so-called in reference to Quirinal Palace, which was used by kings, as well as being one of the homes of the presidents who followed them.
74 Also known as Peter and Pieter.
75 Kraus, H. P., *A Rare Book Saga* (New York: G. P. Putnam's Sons, 1978), 222.
76 Thanks to Klaus Schmeh for pointing many of these out. He wrote, "Regarding the Voynich manuscript alone, more than 15 'solutions' are known to me, and all have turned out to be false." See Schmeh, Klaus, "The Pathology of Cryptology—A Current Survey," *Cryptologia*, 36, no. 1 (January 2012): 14–45. By the end of 2015, he was aware of more than thirty false solutions.
77 Strong, Leonell C., "Anthony Askham, the Author of the Voynich Manuscript." *Science*, New Series 101, no. 2633 (June 15, 1945): 608–9.
78 Goldstone and Goldstone, *The Friar and the Cipher*, p. 276. The decipherment also included a "recipe for an herbal contraceptive."
79 Brumbaugh, R. S. "The Solution of the Voynich 'Roger Bacon' Cipher." *Yale University Library Gazette* 49, no. 4 (April 1975): 347–55. Brumbaugh later claimed that he found the solution

in 1972, but he didn't explain why no results were published until 1974. See Brumbaugh, R. S., "The Voynich 'Roger Bacon' Cipher Manuscript: Deciphered Maps of Stars." *Journal of the Warburg and Courtauld Institutes* 39 (1976): 139–50.

80 Brumbaugh, "The Solution of the Voynich 'Roger Bacon' Cipher."

81 Child, James R. "The Voynich Manuscript Revisited." *NSA Technical Journal* 21, no. 3 (Summer 1976): 1–4.

82 Child, J. R., "Again, the Voynich Manuscript." 2007. http://web.archive.org/web/20090616205410/http://voynichmanuscript.net/voynichpaper.pdf.

83 Goldstone and Goldstone, *The Friar and the Cipher*, 285.

84 Finn, James E., "The Voynich Manuscript. Extraterrestrial Contact during the Middle Ages?" January 14, 2001. http://www.bibliotecapleyades.net/ciencia/esp_ciencia_manuscrito02.htm. Accessed March 23, 2015.

85 Zbigniew Banasik's Manchu theory can be found online at http://www.ic.unicamp.br/~stolfi/voynich/04-05-20-manchu-theo/ and at the links contained therein.

86 Kennedy and Churchill, *The Voynich Manuscript*, 242.

87 Geheimnisvollstes Manuskript der Welt entschlüsselt, press release, 2005. Available online at http://www.ms408.de/downloads/Pressemitteilung.pdf.

88 http://www.ms408.de/context_d.htm.

89 http://www.edithsherwood.com/voynich_decoded/.

90 https://web.archive.org/web/20090609075537/http://www.voynich.nl/.

91 http://www.ciphermysteries.com/2009/11/12/richard-rogers-voynich-theory.

92 Watson, Leon, "Prophet of God" claims mysterious manuscript's code has been cracked, December 3, 2011, http://www.dailymail.co.uk/news/article-2069481/Prophet-God-claims-mysterious-manuscripts-code-cracked.html.

93 Watson, "Prophet of God."

94 Watson, "Prophet of God."

95 https://stephenbax.net/?page_id=11.

96 http://www.livescience.com/43542-voynich-manuscript-10-words-cracked.html.

97 http://theworldsfairest.com/writtenintongues/Written.in.Tongues-The.Voynich.Manuscript.Solved.pdf.

98 http://www.proza.ru/2015/01/11/2343.

99 Foster, Caxton C., "A Comparison of Vowel Identification Methods," *Cryptologia* 16, no. 3 (July 1992): 282–86.

100 An executable version of this algorithm can be downloaded from http://sun1.bham.ac.uk/G.Landini/evmt/evmt.htm. It's called VFQ and was written by Jacques Guy. The details of the algorithm are taken here from G. T. Sassoon, "The Application of Sukhotin's Algorithm to Certain Non-English Languages," *Cryptologia* 16, no. 2 (April 1992): 165–73.

101 This matrix is symmetric— although it mistakenly isn't in the example given in G. T. Sassoon, "The Application of Sukhotin's Algorithm to Certain Non-English Languages," *Cryptologia* 16, no. 2 (April 1992) 165–73.

102 Jacques Guy, "Statistical Properties of Two Folios of the Voynich Manuscript," *Cryptologia* 15, no. 3 (1991). The previous issue of *Cryptologia* (Jacques Guy, "Voynich Revisited" (letter to the editor), *Cryptologia* 15, no. 2, (April 1991)) contains a letter to the editor from Guy in which

he claims that Sukhotin's algorithm resulted in only four vowels, but he did qualify that "not having kept my notes, [I] am quoting mostly from memory."

103 This course was actually offered by Sanjoy Mahajan at MIT in 2008. See http://ocw.mit.edu/courses/mathematics/18-098-street-fighting-mathematics-january-iap-2008/.

104 Guy, "Voynich Revisited" (letter to the editor), p. 162 quoted here.

105 Values taken from William Ralph Bennett, Jr., *Scientific and Engineering Problem-Solving with the Computer*, (Englewood Cliffs, NJ: Prentice-Hall, Inc., 1976), 140, 193, and 194. Missing values in the third column simply hadn't been computed at the time this book was published.

106 Gerry Kennedy and Rob Churchill, *The Voynich Manuscript: The Unsolved Riddle of an Extraordinary Book Which Has Defied Interpretation for Centuries*, British hardcover edition, (London: Orion Books, 2004), 142. Pictures of seven different handwritings used in the manuscript appear on pp. 144–45. Currier originally estimated somewhere between six and eight different scribes.

107 John F. Clabby, *Brigadier John Tiltman: A Giant among Cryptanalysts* (Fort Meade, MD: Center for Cryptologic History, National Security Agency, 2007), 34.

108 Andreas Schinner, "The Voynich Manuscript: Evidence of the Hoax Hypothesis," *Cryptologia* 31, no. 2 (April 2007): 95–107.

109 D'Imperio, M. E. *The Voynich Manuscript—An Elegant Enigma* (Fort Meade, MD: National Security Agency, 1976). Reprinted by Aegean Park Press, Laguna Hills, CA, 1978.

110 Clark also published a biography of Edison that same year.

111 https://www.nsa.gov/public_info/declass/friedman_documents/.

112 W. F. Friedman and E. S. Friedman, "Acrostics, Anagrams, and Chaucer," *Philological Quarterly* 38 (1959): 1–20.

113 C. A. Zimansky, "William F. Friedman and the Voynich Manuscript," *Philological Quarterly* 49, no. 4 (1970): 433–42.

114 Lawrence Goldstone and Nancy Goldstone, *The Friar and the Cipher: Roger Bacon and the Unsolved Mystery of the Most Unusual Manuscript in the World* (New York: Doubleday, 2005), 284.

115 E-mail from David Hatch to the author, received July 6, 2015.

116 http://www.ciphermysteries.com/.

117 G. W. L. Hodgins, "2011 Forensic Investigation of the Voynich Manuscript," presented at the Voynich 100 Conference, Frascati, Italy, May 2011. Also reported on in A. J. T. Jull, "Some Interesting and Exotic Applications of Carbon-14 Dating by Accelerator Mass Spectrometry," *10th International Conference on Clustering Aspects of Nuclear Structure and Dynamics, Journal of Physics: Conference Series* 436 (2013): 012083.

118 Sravana Reddy and Kevin Knight, "What We Know about the Voynich Manuscript," *Proceedings of the 5th ACL-HLT Workshop on Language Technology for Cultural Heritage, Social Sciences, and Humanities*, Portland, OR (June 24, 2011), (Madison, WI: Omnipress, Inc.), 78–86.

119 David Kahn, *The Codebreakers* (New York: Macmillan, 1967), 863.

第 2 章　古董上的密码

1 Thomas Hoving, *False Impression: The Hunt for Big-Time Art Fakes* (New York: Simon & Schuster, 1996), 52–53.

2 Mary-Ann Russon, "Tiny Egyptian Mummy Confirmed by CT Scan to be Baby's Remains," *International Business Times*, May 8, 2014.

3 W. M. Flinders Petrie, *Ten Years' Digging in Egypt 1881–1891*, 2nd ed. (London: The Religious Tract Society, 1893), 124–25.

4 W. M. Flinders Petrie, *Illahun Kahun and Gurob, 1889–90* (London: David Nutt, 1891).

5 Alan R. Schulman, "The Ossimo Scarab Reconsidered," *Journal of the American Research Center in Egypt* 12 (1975): 15–18, p. 16 quoted here.

6 John Coleman Darnell, *The Enigmatic Netherworld Books of the Solar-Osirian Unity, Cryptographic Compositions in the Tombs of Tutankhamun, Ramesses VI and Ramesses IX*, (Göttingen, Germany: Academic Press Fribourg, Vandenhoek & Ruprecht, 2004), 1–2.

7 Étienne Drioton, "Les principes de la cryptographie égyptienne," *Comptes rendus des séances de l'Académie des Inscriptions et Belles-Lettres*, 97e année, no. 3 (1953): 355–64.

8 John Coleman Darnell, *The Enigmatic Netherworld Books of the Solar-Osirian Unity, Cryptographic Compositions in the Tombs of Tutankhamun, Ramesses VI and Ramesses IX*, (Göttingen, Germany: Academic Press Fribourg, Vandenhoek & Ruprecht, 2004), 5.

9 Alan R. Schulman, "The Ossimo Scarab Reconsidered," *Journal of the American Research Center in Egypt* 12 (1975): 15–18, p. 16 quoted here.

10 Plutarch's *Life of Lysander*, sec. 19.

11 Albert C. Leighton, "Secret Communication among the Greeks and Romans," *Technology and Culture* 10, no. 2 (April 1969): 139–54, p. 154 cited here. Leighton's citation for the letter was Ausonius' *Epistles,* 28.

12 Leighton, "Secret Communication," 153, footnote 61, cited here.

13 A dissenting view is offered by Henry R. Immerwahr in *Attic Script, A Survey* (New York: Oxford University Press, 1990). On p. 15, he wrote, "Before the time of Draco Athens was a cultural backwater—not a suitable location for the composition of the Homeric poems, as some have thought." And on p. 20 he wrote, "Corinth clearly took the lead in the custom of applying inscriptions to vases." He views the work in Athens as imitative of the work in Corinth.

14 Immerwahr, *Attic Script*, 45.

15 John Boardman, *Athenian Black Figure Vases* (New York: Oxford University Press, 1974), 201.

16 Dan Vergano, "Amazon Warriors' Names Revealed Amid 'Gibberish' on Ancient Greek Vases," *National Geographic* (September 22, 2014), http://news.nationalgeographic.com/news/2014/09/140923-amazon-greek-vase-translations-science/.

17 Umberto Eco, *Foucault's Pendulum*, trans. William Weaver (San Diego, New York, London: Harcourt Brace Jovanovich, 1989), 24.

18 *Big Bang Theory* transcripts, https://bigbangtrans.wordpress.com/series-4-episode-09-the-boyfriend-complexity/.

19 Adrienne Mayor, John Colarusso, and David Saunders, "Making Sense of Nonsense Inscriptions Associated with Amazons and Scythians on Athenian Vases," *Hesperia: The Journal of the American School of Classical Studies at Athens* 83, no. 3 (July–September 2014): 447–93, p. 466 cited here.

20 Adrienne Mayor, John Colarusso, and David Saunders, "Making Sense of Nonsense Inscriptions Associated with Amazons and Scythians on Athenian Vases," *Hesperia: The Journal of the American School of Classical Studies at Athens* 83, no. 3 (July–September 2014): 447–93, pp. 486–87 cited here.

21 Dan Vergano, "Amazon Warriors' Names Revealed Amid 'Gibberish' on Ancient Greek Vases," *National Geographic*, September 22, 2014.

22 Adrienne Mayor, *The Amazons: Lives and Legends of Warrior Women across the Ancient World* (Princeton, NJ: Princeton University Press, 2014), 11.

23 E-mail from Adrienne Mayor to the author, received June 25, 2015.

24 Adrienne Mayor, John Colarusso, and David Saunders, "Making Sense of Nonsense Inscriptions Associated with Amazons and Scythians on Athenian Vases," Version 2.0, Princeton/Stanford Working Papers in Classics, July 2012, 27–28.

25 E-mail from Adrienne Mayor to the author, received June 25, 2015.

26 E-mail from Adrienne Mayor to the author, received June 25, 2015.

27 Adrienne Mayor, John Colarusso, and David Saunders, "Making Sense of Nonsense Inscriptions Associated with Amazons and Scythians on Athenian Vases," *Hesperia: The Journal of the American School of Classical Studies at Athens* 83, no. 3 (July–September 2014): 447–93, p. 489 cited here.

28 Adrienne Mayor, John Colarusso, and David Saunders, "Making Sense of Nonsense Inscriptions Associated with Amazons and Scythians on Athenian Vases," *Hesperia: The Journal of the American School of Classical Studies at Athens* 83, no. 3 (July–September 2014): 447–93, p. 465 cited here.

29 E-mail from David Saunders to the author, received June 26, 2015.

30 Ole Franksen, *Mr. Babbage's Secret: The Tale of a Cypher and APL* (Englewood Cliffs, NJ: Prentice-Hall, 1985), 86.

31 Stig Eliasson, "When 'Chance' Explanations Turn Tenuous: Basque Verbal Morphology, Argument Marking, and an Allegedly Nonsensical Danish Runic Inscription."

32 Eliasson, "When 'Chance' Explanations Turn Tenuous."

33 Michael P. Barnes, *Runes: A Handbook* (Suffolk, U.K.: Boydell Press, 2012), 167.

34 Barnes, *Runes*, 202.

35 Barnes, *Runes*, 24.

第3章　多拉贝拉密码

1 This is actually a series of six marches titled *Pomp and Circumstance Military Marches*, of which the *Trio* section, "Land of Hope and Glory," of the first is played at many American graduations.

2 Edward Elgar, *My Friends Pictured Within* (London: Novello and Company Ltd, 1946).

3 John C. Tibbetts, "John C. Tibbetts Interviews Eric Sams," available online at http://www.ericsams.org/index.php/interview-by-john-c-tibbetts.

4 Ron Rosenbaum, "A Visit with an Avenging Angel," *The Shakespeare Wars* (New York: Random House, 2008), 66–75, available online at http://www.ericsams.org/index.php/ron-rosenbaum-the-avenging-angel.

5 John C. Tibbetts, "John C. Tibbetts Interviews Eric Sams," available online at http://www.ericsams.org/index.php/interview-by-john-c-tibbetts.

6 Solution #04, revised as #36, posted at http://unsolvedproblems.org/index_files/Solutions.htm.

7 E-mail received by author on June 19, 2015.

8 Found in an odd book by one of the people who claimed a solution to the Voynich manuscript: Joseph Martin Feely, *Electrograms from Elysium: A Study on the Probabilities in Postmortuary Communication through the Electronics of Telepathy and Extrasensory Perception, Including the Code of Anagrams in the Purported Sender's Name* (New York, 1954), 246.

9 See https://www.cqu.edu.au/about-us/staff-directory/profiles/higher-education-division/higher-education-division-honadjunct-appoint/robertst and http://unsolvedproblems.org/S12x.pdf.
10 Jerrold Northrup Moore, *Edward Elgar: A Creative Life* (Oxford, U.K.: Oxford University Press, 1984), 114. The passage also appears online at http://www.benzedrine.ch/dorabella.html.
11 John Holt Schooling, "Secrets in Cipher IV. From the Time of George II. to the Present Day," *The Pall Mall Magazine* 8, no. 36 (April 1896): 608–18, p. 618 cited here.
12 http://www.ciphermysteries.com/2013/10/09/elgars-little-cipher.
13 Robert J. Buckley, *Sir Edward Elgar* (London: Ballantyne, Hanson & Co., 1905).
14 The Elgar Birthplace Museum, L189.
15 The Elgar Birthplace Museum, L296.
16 This was short for "Darling Chuck," where Chuck was Elgar's daughter Carice.
17 Mrs. Richard Powell, *Edward Elgar: Memories of a Variation*, 4th ed., rev. and ed. Claud Powell, addendum Jerrold Northrop Moore (Aldershot, Hants, U.K.: Scolar Press, 1994), 36.
18 Powell, *Edward Elgar*, 4th ed., 90–91.
19 Powell, *Edward Elgar*, 4th ed., 93.
20 R.B.T. was Richard Baxter Townshend. Elgar's variation for Townshend ridiculed Townshend's inability to control his voice in the amateur theatre productions he performed in.
21 Powell, *Edward Elgar*, 4th ed., 132–33.
22 Ian Parrott, *Elgar* (London: J. M. Dent and Sons Ltd., and New York: Farrar, Straus and Giroux, Inc., 1971), 122.
23 Kevin Jones, "The Puzzling Mr. Elgar," *New Scientist* 184, no. 2479 (December 25, 2004): 56.
24 Mrs. Richard Powell, *Edward Elgar: Memories of a Variation* (London: Oxford University Press, 1937), 35.
25 Rosa Burley and Frank Carruthers, *Edward Elgar: The Record of a Friendship* (London: Barrie and Jenkins, 1972), 132.
26 Elgar Birthplace Museum, L1104.
27 Martin Bird, ed., *Darling Chuck, The Carice Letters*, Edward Elgar: Collected Correspondence, ser. II, vol. 1, Elgar Works (Chippenham, Wiltshire, U.K.: Antony Rowe Ltd., 2014). Examples from pp. 145, 159, 160, 162, 163, 164, 164, 164, 168, 182, 192, 201, 204, 237, 272, 381, 381, and 381, respectively.

第 4 章 黄道十二宫杀手密码

1 Robert Graysmith, *Zodiac* (New York: St. Martin's/Marek, 1986), 30.
2 Robert Graysmith, *Zodiac* (New York: St. Martin's/Marek, 1986), 32–33.
3 Images from Robert Graysmith, *Zodiac* (New York: St. Martin's/Marek, 1986), 50–51.
4 Robert Graysmith, *Zodiac* (New York: St. Martin's/Marek, 1986), 58 for some "solutions."
5 Robert Graysmith, *Zodiac* (New York: St. Martin's/Marek, 1986), 67.
6 Robert Graysmith, *Zodiac* (New York: St. Martin's/Marek, 1986), 69.
7 Robert Graysmith, *Zodiac* (New York: St. Martin's/Marek, 1986), 71.
8 Robert Graysmith, *Zodiac* (New York: St. Martin's/Marek, 1986), 75.
9 Robert Graysmith, *Zodiac* (New York: St. Martin's/Marek, 1986), 83.
10 Robert Graysmith, *Zodiac* (New York: St. Martin's/Marek, 1986), 85 and 91.

11 Robert Graysmith, *Zodiac* (New York: St. Martin's/Marek, 1986), 91.
12 Robert Graysmith, *Zodiac* (New York: St. Martin's/Marek, 1986), 102.
13 Robert Graysmith, *Zodiac* (New York: St. Martin's/Marek, 1986), 107.
14 Michael D. Kelleher and David van Nuys, *"This Is the Zodiac Speaking": Into the Mind of a Serial Killer* (Westport, CT: Praeger, 2001).
15 Robert Graysmith, *Zodiac* (New York: St. Martin's/Marek, 1986), 122.
16 Robert Graysmith, *Zodiac* (New York: St. Martin's/Marek, 1986), 243.
17 Robert Graysmith, *Zodiac* (New York: St. Martin's/Marek, 1986), 136–37.
18 Robert Graysmith, *Zodiac Unmasked: The Identity of America's Most Elusive Serial Killer Revealed* (New York: Berkley Books, 2002), 128.
19 Robert Graysmith, *Zodiac* (New York: St. Martin's/Marek, 1986), 148.
20 Gareth Penn, "The Calculus of Evil," *MENSA Bulletin*, no. 288 (July/August 1985): 5–8, 25, p. 5 quoted here.
21 Gareth Penn, "The Calculus of Evil," *MENSA Bulletin*, no. 288 (July/August 1985), 5–8, 25, p. 5 quoted here.
22 Gareth Penn, "The Calculus of Evil," *MENSA Bulletin*, no. 288 (July/August 1985), 5–8, 25, p. 25 quoted here. Also see Gareth Penn, *Times 17: The Amazing Story of the Zodiac Murders in California and Massachusetts 1966–1981* (New York: Foxglove Press, 1987), 26.
23 http://www.zodiackillerfacts.com/Ray%20Grant.htm.
24 Fletcher Pratt, *Secret and Urgent* (New York: Bobbs-Merrill, 1939), 258. This is a book used by Harden!
25 See the killer's letter of January 29, 1974, for the claim of thirty-seven victims. Available online at http://www.zodiackillerfacts.com/gallery/thumbnails.php?album=19.
26 Robert Graysmith, *Zodiac* (New York: St. Martin's/Marek, 1986), 308–11.
27 I refer the interested reader to Kelleher and Van Nuys, *"This Is the Zodiac Speaking"* again. Their argument against Zodiac being connected with Bates is persuasive.
28 Robert Graysmith, *Zodiac* (New York: St. Martin's/Marek, 1986), 308.
29 http://www.imdb.com/title/tt0469999/.
30 http://www.imdb.com/title/tt0371739/.
31 http://www.imdb.com/title/tt0443706/.

第 5 章　罪案密码

1 Cheri L. Farnsworth, *Adirondack Enigma: The Depraved Intellect and Mysterious Life of North Country Wife Killer Henry Debosnys* (Charleston, SC: The History Press, 2010).
2 Farnsworth, *Adirondack Enigma*, p. 26, quoting a reporter who visited Henry in jail.
3 Farnsworth, *Adirondack Enigma*, 69.
4 Farnsworth, *Adirondack Enigma*, 27.
5 Farnsworth, *Adirondack Enigma*, 69.
6 Farnsworth, *Adirondack Enigma*, 69–71.
7 Farnsworth, *Adirondack Enigma*, 71–72.
8 Farnsworth, *Adirondack Enigma*, 72.
9 Farnsworth, *Adirondack Enigma*, 72.
10 Farnsworth, *Adirondack Enigma*, 72.

11 Farnsworth, *Adirondack Enigma*, 72–73.
12 Farnsworth, *Adirondack Enigma*, 73.
13 Farnsworth, *Adirondack Enigma*, 73.
14 Farnsworth, *Adirondack Enigma*, 73.
15 Farnsworth, *Adirondack Enigma*, 73.
16 Farnsworth, *Adirondack Enigma*, 43.
17 Farnsworth, *Adirondack Enigma*, 76.
18 Farnsworth, *Adirondack Enigma*, 76.
19 Farnsworth, *Adirondack Enigma*, 77–78.
20 Farnsworth, *Adirondack Enigma*, 79–80.
21 Farnsworth, *Adirondack Enigma*, 80.
22 Farnsworth, *Adirondack Enigma*, 80–81.
23 *Philadelphia Times*, August 11, 1882.
24 Farnsworth, *Adirondack Enigma*, 27 and 63.
25 See http://www.isi.edu/~knight/ to get an idea of the lively, creative, eccentric way he views the world.
26 E-mail received by the author on May 8, 2015.
27 Farnsworth, *Adirondack Enigma*, 107.
28 American Cryptogram Association, http://cryptogram.org/.
29 Taken from Tom Mahon and James J. Gillogly, *Decoding the IRA* (Cork, Ireland: Mercier Press, 2008), 266, fig. 27, pt. 9.
30 Tom Mahon and James J. Gillogly, *Decoding the IRA* (Cork, Ireland: Mercier Press, 2008), 48.
31 Mahon and Gillogly, 33.
32 Mahon and Gillogly, 53.
33 Brainy Quotes, John Walsh quotes, http://www.brainyquote.com/quotes/authors/j/john_walsh.html.
34 *America's Most Wanted*, http://en.wikipedia.org/wiki/America's_Most_Wanted#Profiling_missing_persons. The man was captured in April 2008.
35 That would give us 156 distinct symbols instead of the 155 we actually observed. That's why I used the word "approximately."
36 Nick Pelling, Cipher Mysteries blog, http://www.ciphermysteries.com/2014/05/19/cracking-the-scorpion-ciphers.
37 Successories website, John Walsh Quotes, http://www.successories.com/iquote/author/6223/john-walsh-quotes/1.
38 IMDb, *America's Most Wanted*, http://www.imdb.com/title/tt0094415/quotes.
39 It was actually February 2, 2004. The bodies were found on the third.
40 Tom Farmer, "Suicide Note Leaves Motive for Saugus Killings a Mystery," *The Boston Herald*, February 5, 2004, p. 14.
41 Mac Daniel and Diane Allen, "Killings, Suicide Baffle Authorities. Husband Left Note; Motive a Mystery," *The Boston Globe*, February 5, 2004, 3rd ed., p. B1.
42 Posted January 30, 2006, 11:16 a.m.
43 Posted January 30, 2006, 11:42 a.m.
44 Posted January 31, 2006, 11:58 a.m.

45 E-mail received by the author on August 3, 2015.

46 Posted February 18, 2006, 1:29 p.m.

47 Posted February 19, 2006, 7:02 a.m.

48 Successories, John Walsh Quotes, http://www.successories.com/iquote/author/6223/john-walsh-quotes/8.

49 Greg Olsen and Rebecca Morris, *If I Can't Have You: Susan Powell, Her Mysterious Disappearance, and the Murder of Her Children* (New York: St. Martin's Press, 2014), 102. Also see p. 228.

50 Anonymous, "Police Close Susan Powell Case, Offer New Details," *USA Today*, May 20, 2013, available online at http://www.usatoday.com/story/news/nation/2013/05/20/susan-powell-case/2344681/.

51 David Lohr, "Susan Powell's Parents Search For Clues in Father-in-Law's Former Home," *Huffington Post*, May 27, 2014, available online at http://www.huffingtonpost.com/2014/05/27/susan-powell-missing-clues_n_5398147.html.

52 Communication with the author.

53 Ingo Wagner, "Verschlüsselt und nicht knackbar, Polizei scheitert an Festplatten des 'Masken manns,' " *Focus Magazine*, January 23, 2012. Available online at http://www.focus.de/panorama/welt/verschluesselt-und-nicht-knackbar-polizei-scheitert-an-festplatten-des-maskenmanns_aid_705621.html. Quoted portion translated by the author.

第6章 受害者密码

1 Anthony Boucher, "QL696.C9." *Ellery Queen's Mystery Magazine*, May 1943.

2 Eye Doctor Guide.com, http://www.eyedoctorguide.com/Eye-Color/hazel-eyes-eye-color.html.

3 "Inquest into the Death of a Body Located at Somerton on 1.12.48." [refers to Dec. 1, 1948], 1949, p. 69. Hands can tell us a lot about a person. In this instance, it was also noted that the hands were "hard, but were not rough from performing manual work."

4 Gerald (Gerry) Michael Feltus, *The Unknown Man: A Suspicious Death at Somerton Park* (Richmond, Australia: Hyde Park Press, 2010), 143.

5 Feltus, *The Unknown Man*, 39–40.

6 Kerry Greenwood, *Tamam Shud: The Somerton Man Mystery* (Sydney, Australia: NewSouth Publishing, University of New South Wales Press Ltd., 2012), 28.

7 Descriptions vary slightly.

8 "Inquest into the Death of a Body," 79.

9 All of the marks were on the pocket label of the same pair of pants.

10 "Inquest into the Death of a Body," 81.

11 Feltus claims that the paper was rolled up. Greenwood described it as folded.

12 "Inquest into the Death of a Body," 37.

13 "Inquest into the Death of a Body," 39.

14 "Inquest into the Death of a Body," 39.

15 "Inquest into the Death of a Body," 41.

16 "Inquest into the Death of a Body," 45.

17 "Inquest into the Death of a Body," 47.

18 "Inquest into the Death of a Body," 81.

19 "Inquest into the Death of a Body," 85.

20 "Inquest into the Death of a Body," 85.

21 "Inquest into the Death of a Body," 85.

22 "Inquest into the Death of a Body," 95.

23 "Inquest into the Death of a Body," 97.

24 "Inquest into the Death of a Body," 97.

25 "Inquest into the Death of a Body," 97.

26 "Inquest into the Death of a Body," 9. Note: There are two Clelands involved with this case, Thomas and John. They were cousins.

27 "Inquest into the Death of a Body," 9.

28 "Inquest into the Death of a Body," 7.

29 Greenwood, *Tamam Shud*, 80.

30 The suitcase can be seen in the 1978 documentary at http://www.youtube.com/watch?v=nnPqlYPQ9lY, but it was destroyed in 1986.

31 "Inquest into the Death of a Body," 73.

32 Feltus, *The Unknown Man*, 17.

33 Feltus, *The Unknown Man*, 171.

34 He wished to remain anonymous. His name is given as Ronald Francis in the books by Feltus and Greenwood, but this is a pseudonym. A list of common errors made by people writing about this case states, "Reports sometimes refer to the man who found the copy of *The Rubaiyat,* tossed in the back of his car, as a 'doctor' or 'chemist'. This is not certain. In fact both his name and real occupation have been withheld. Also the reason why his details are withheld is also suppressed. All we can assume is that he was some kind of professional." From https://www.eleceng.adelaide.edu.au/personal/dabbott/wiki/index.php/List_of_facts_on_the_Taman_Shud_Case_that_are_often_misreported.

35 *The Advertiser* (Adelaide, South Australia), "Police Test Book for Somerton Body Clue," July 26, 1949, p. 3.

36 Maybe. The 1949 inquest, p. 5, claims that the "Tamam Shud" paper came from a second edition. In any case, by "first edition," I mean the first from this publisher. It is not the first appearance of the book in print.

37 Some reports say "in the back of the book," whereas others say "on the back of the book." One official police document says that the back cover was missing.

38 If the hypothesis that follows is correct, this is exactly what we would expect. If the cipher consists of just initial letters, we should have a close match with initial frequencies, less of a match with overall frequencies (but not a *total* disconnect, since initial letters are included in determining overall frequencies), and a strong mismatch when comparing with terminal frequencies, since almost none of the initial letters are counted when compiling terminal frequencies—the only exceptions arise from the one-letter words A and I.

39 I redacted most of the editor's phone number. Please e-mail him if you have the solution!

40 Simon Singh, *The Codebook* (New York: Doubleday, 1999).

41 Feltus believes that he found the entire correct solution, but he doesn't reveal all of it in his book. Instead, he refers the reader to the back cover, which bears an image of the cipher. He also included a phone number alongside the cipher, but it's a fake. That is, it doesn't match the phone number that played an important role in the case.

42 Some proposed solutions, displayed on pages 119 and 120, are of the type I propose (taking cipher letters as the first letters of words), but there is no reason to favor them over many other potential solutions.

43 There are claims that more than one telephone number was present, but nothing authoritative enough to warrant inclusion here. This is another instance where we are greatly hampered because most of the police file is no longer available.

44 The term "nurse" may not have been technically accurate for the woman at this point in her life. See https://www.eleceng.adelaide.edu.au/personal/dabbott/wiki/index.php/List_of_facts_on_the_Taman_Shud_Case_that_are_often_misreported, item no. 4.

45 Boxall's copy was a 1924 Sydney edition.

46 Greenwood, *Tamam Shud*, 60.

47 "Inquest into the Death of a Body," 7.

48 "Inquest into the Death of a Body," 53.

49 Feltus, *The Unknown Man*, 151.

50 Feltus, *The Unknown Man*, 108.

51 Feltus, *The Unknown Man*, 178–79.

52 For more on early intelligence sharing with Australia, and a slightly different version of the story above, see Tom Johnson, *American Cryptology during the Cold War: 1945–1989, Book I: The Struggle for Centralization 1945–1960* (Ft. Meade, MD: Center for Cryptologic History, National Security Agency, 1995), 18–19. Discrepancies are minor, such as labeling the cryptanalyst as being from SIS (Signals Intelligence Service) instead of ASA. There were frequent name changes during these years. The evolution led to the National Security Agency (NSA) in 1952.

53 Wikipedia, http://en.wikipedia.org/wiki/Taman_Shud_Case, citing Phillips, John Harber, "So When That Angel of the Darker Drink," *Criminal Law Journal* 18, no. 2 (April 1994): 110.

54 Wikipedia, http://en.wikipedia.org/wiki/Taman_Shud_Case.

55 Victor Navasky, ed., *The Nation*, New York, Commentary, "Harry Dexter White," letter to the editor, April 1, 1988, available online at http://www.commentarymagazine.com/article/harry-dexter-white/.

56 On November 3, 1948, the day after Harry S. Truman defeated Thomas E. Dewey in the presidential election, the *Chicago Daily Tribune* ran the headline "Dewey Defeats Truman." Actually, the Dewey presidency has more sources to back it up than the digitalis-induced death of White. *The Journal of Commerce* ran the headline "Dewey Victory Seen as Mandate to Open New Era of Government-Business Harmony, Public Confidence" on November 3.

57 Trove digitized newspapers, http://trove.nla.gov.au/ndp/del/article/76017408. Thanks to James Ramm for pointing this article out to me!

58 Greenwood, *Tamam Shud*, 122–23.

59 https://web.archive.org/web/20110612061900/http://www.weirld.com/Paranormal/UFOs-Aliens/The-Strange-Case-of-the-Somerton-Man-Time-Traveller-Human/Alien-Hybrid-or-Secret-Agent-Spy.html.

60 "Weirld" is not a typo but rather a portmanteau of "weird" and "world."

61 *America's Most Wanted*, http://en.wikipedia.org/wiki/America's_Most_Wanted#Profiling–missing–persons. The man was captured in April 2008.

62 *Unsolved Mysteries*, https://web.archive.org/web/20140422010517/http://unsolved.com/about.html.

63 Greenwood, *Tamam Shud*, 129.

64 Available online at http://www.youtube.com/watch?v=iy5s50F3uB8 ("Somerton Man" segment). This is the source of the quotes that follow.

65 Some secondary sources say "formaldehyde," but that is technically wrong, as it refers to a gas. However, "formaldehyde solution" is equivalent to formalin, which is a mix of formaldehyde gas and water.

66 Renato Castello, "New Twist in Somerton Man Mystery as Fresh Claims Emerge," *Sunday Mail* (SA), November 23, 2013.

67 Castello, "New twist."

68 Castello, "New twist."

69 Gerry Feltus, "Second Application to Exhume Refused - 4 Jun 2014," http://theunknownman.com/comments.html This opinion also appeared at Pete Bowes, "the dead will not arise ... Rau, his second ruling," June 26, 2014, http://tomsbytwo.com/2014/06/26/the-dead-will-not-arise-rau-his-second-ruling/.

70 Kynton Grace, "South Australia's *X-Files*: Part 2—The Somerton Man Mystery and the Secrets of Adelaide's Tunnels," *The Advertiser*, June 9, 2014.

71 "Dad Identifies Poisoned Youth as N.Y. Student," *Washington Times-Herald*, January 22, 1953, p. 3.

72 "Body Found at Airport, Coded Note on Abdomen," January 21, 1953.

73 "Code Note Taped to Body in Ditch," *The Evening Bulletin*, Philadelphia, 1, column 2, Sports Final edition, January 20, 1953.

74 "Cyanide Caused Death at Airport," *The Evening Bulletin*, Philadelphia, p.1, column 3, Postscript edition, January 21, 1953.

75 "Cyanide Caused Death at Airport."

76 "Mystery Man at Airport Is Found Dead of Poison," *The Philadelphia Inquirer* January 21, 1953, p. 1.

77 "Mystery Man Cyanide Clue Hints." Murder Most accounts just say "cyanide." Potassium cyanide is specified in "Man Poisoned, Paper in Code Taped to Body; FBI Called In," *Washington Star*, p. A-26, and in some other articles. It is "potassium or sodium syanide [sic]" in "Code Message Found Taped to Dead Man," *Washington Times-Herald*, January 21, 1953, p. 35.

78 "Mystery Man Cyanide Clue Hints Murder."

79 "Mystery Man Cyanide Clue Hints Murder."

80 "Code Note Taped to Body in Ditch," *The Evening Bulletin*, Philadelphia, January 20, 1953, p. 1.

81 "Body with Code Note Found Here," *Daily News*, Philadelphia, January 20, 1953, p. 2.

82 " 'Dulles' Name Taped to Body of 'Suicide,'" *Washington Post*.

83 "Code Message Found Taped to Dead Man," *Washington Times-Herald*, January 21, 1953, p. 35. The quote is attributed to Dr. Edward Burke of the police crime laboratory, in "Mystery Man Cyanide Clue Hints Murder."

84 We also have the quote "Police said they found no cyanide container near the body." in "Cyanide Caused Death at Airport."

85 "Test Tube Found at Site of Airport Ditch Death," *The Philadelphia Inquirer*, January 23, 1953, p. 28. Available online through http://fultonhistory.com/Fulton.html.

86 "'Dulles' Name Taped to Body of 'Suicide.'" Also stated in "Student Identified as Cyanide Victim," *New York Times*, January 22, 1953.

87 "Mystery Man Cyanide Clue Hints Murder."

88 "Code Note Taped to Body in Ditch," *The Evening Bulletin*, Philadelphia, January 20, 1953, p. 1. Described as eight inches by three inches in "Code Message Found Taped to Dead Man," *Washington Times-Herald*, January 21, 1953, p. 35. It was described as only five inches in "'Dulles' Name Taped to Body of 'Suicide.'"

89 E-mail from David Hatch to the author, October 3, 2012.

90 "Mystery Man at Airport Is Found Dead of Poison." Another source claims that there was no wallet! See "Code Message Found Taped to Dead Man."

91 "Body Found at Airport, Coded Note on Abdomen."

92 "Cyanide Caused Death at Airport." Some sources only mention "a car key." Were there four keys or just one?

93 "Code Message Found Taped to Dead Man."

94 "Codes Seen Key in Airport Death," January 22, 1953.

95 For false report, see "Codes Seen Key in Airport Death."

96 "Mystery Man Cyanide Clue Hints Murder."

97 Rubin's age was typically reported as eighteen, but some articles gave his age as nineteen. See "Body at Airport Is Identified as NYU Student," Philadelphia, January 21, 1953, and "Airport Death Called Murder by Ominsky," Philadelphia, January 21, 1953. In this second article, Rubin's age is later stated as eighteen.

98 "Youth Dabbled in Cryptography," *Philadelphia Evening Bulletin*, January 22, 1953.

99 In most accounts, it was forty-seven cents. For example, see "Code Note Taped to Body in Ditch." But it was forty-six cents in "Body with code note found here."

100 "Poison Victim Identified as N.Y. Youth," *The Philadelphia Inquirer*, January 22, 1953, p.1, illustrated on p. 3 and continued on p. 13, quote from p. 13.

101 "Poison Victim Identified as N.Y. Youth," quote from p. 13.

102 "FBI Investigates Possibilities of Murder in Mysterious Cyanide-Death of Student," *Schenectady Gazette*, January 22, 1953.

103 Jay Apt, "Note Baffles Decoders in Poison Death," *Daily News*, January 22, 1953, p. 2.

104 "FBI Investigates Possibilities of Murder in Mysterious Cyanide-Death of Student."

105 "Hazing Studied in Airport Death," *The Philadelphia Inquirer*, January 28, 1953, p. 5.

106 "Student Identified as Cyanide Victim," *New York Times*, January 22, 1953.

107 E-mail from Claire Ashley Wolford, May 2, 2012.

108 "Life on the Newsfronts of the World, Student Leaves a Coded Clue to His Death, a Soldier Plants a Flag and Flu Virus Sweeps the World," *Life*, February 2, 1953, p. 30.

109 "Youth Dabbled in Cryptography."

110 "Poison Victim Identified as N.Y. Youth," quote from p. 13.

111 "Mystery Shrouds Death by Cyanide," *The Philadelphia Inquirer*, March 25, 1953, p. 19.

112 "Mystery Shrouds Death by Cyanide."

113 "Mystery Shrouds Death by Cyanide." The names Edwin and Edward both appear in newspaper accounts.

114 "Body at Airport."

115 "Body at Airport."

116 "Gay rights campaigner calls for fresh inquiry into death of Alan Turing," *Manchester Evening News*, December 24, 2013, http://www.manchestereveningnews.co.uk/news/greater-manchester-news/codebreaker-alan-turing-gay-rights-6445286, accessed July 11, 2014.

117 Langen, Harry [*sic*—it should say Henry], *My Life Between the Lines*.

118 Langen, *My Life Between the Lines*.

119 I found this paper as a clipping in an uncatalogued portion of the National Cryptologic Museum's collection. It is not clear from the clipping which paper it came from!

120 Wolfram Alpha, http://www.wolframalpha.com/input/?i=average+english+word+length. You may find much larger estimates, but these simply average the lengths of all English words from some list. They do not weight the average by frequency. That is, the value above looks at actual usage.

121 Shane Anthony, "From 1999: Body found in field puzzles police," *St. Louis Post-Dispatch*, July 2, 1999, but Christpher Tritto implied that the sentence was about two years in his article.

122 According to Christopher Tritto. Shane Anthony reported that he served 11 months.

123 Christopher Tritto, "Code Dead: Do the encrypted writings of Ricky McCormick hold the key to his mysterious death?" *Riverfront Times* (St. Louis), June 14, 2012.

124 Tritto, "Code Dead." This article reports that "At least one gas-station employee" claims to have seen McCormick there that day. There's no elaboration concerning the uncertainty expressed in the number of witnesses.

125 Anthony, "From 1999." This article actually says "Highway 367," but it's a typo, one that has spread to the Wikipedia page for the McCormick cipher.

126 Still available online at the FBI site, http://www.fbi.gov/news/stories/2011/march/cryptanalysis_032911.

127 Jennifer Mann, "FBI wants public to help crack code in St. Charles County cold case," *St. Louis Post-Dispatch*, March 31, 2011.

128 Mann, "FBI wants public to help."

129 Tritto, "Code Dead."

130 Tritto, "Code Dead."

131 Tritto, "Code Dead."

第 7 章 来自坟墓的密码

1 R. H. Thouless, "A Test of Survival," *Proceedings of the Society for Psychical Research* 48 (July 1948): 253–63.

2 Thouless, "A Test of Survival."

3 Thouless, "A Test of Survival."

4 Thouless, "A Test of Survival."

5 Cambridge University Library, "The Society for Psychical Research [3000] collection," SPR.MS 63/2/113. Thanks to Klaus Schmeh for sharing this.

6 *Oxford English Dictionary*, 2nd ed. (Oxford, U.K.: Oxford University Press), http://public.oed.com/history-of-the-oed/dictionary-facts/.

7 Thouless, "A Test of Survival."

8 Thouless, "A Test of Survival."

9 Thouless, "A Test of Survival."

10 Thouless, "A Test of Survival."

11 Thouless, "A Test of Survival."

12 Thouless, "A Test of Survival."

13 Thouless, "A Test of Survival," p. 262.

14 There are several different ways to carry out "double Playfair" encipherment. This is just one example.

15 Robert H. Thouless, "Additional Notes on a Test of Survival," *Proceedings of the American Society for Psychical Research* 48 (1948): 342–43.

16 James J. Gillogly and Larry Harnisch, "Cryptograms from the Crypt," *Cryptologia* 20, no. 4 (October 1996): 325–29.

17 Gillogly and Harnisch, "Cryptograms from the Crypt."

18 He apparently didn't want much attention and published his paper under the name T. E. Wood. No point of contact was given in the paper. I found his full name in a document from Cambridge University Library, "The Society for Psychical Research [3000] collection," SPR.MS 63/1/1. Thanks to Klaus Schmeh for sharing this. A fuller biography also appears in this file and more from Wood follows. The biography might help suggest what languages he used.

19 Originally, he had a normal Vigenère cipher in mind, but he learned of its weakness before going to print. See Cambridge University Library, "The Society for Psychical Research [3000] collection," SPR.MS 63/2/4. Thanks to Klaus Schmeh for sharing this.

20 T. E. Wood, "A Further Test for Survival," *Proceedings of the Society for Psychical Research* 49, (1950): 105–6, p. 105 quoted here.

21 Wood, "A Further Test for Survival," p. 105 quoted here.

22 Cambridge University Library, "The Society for Psychical Research [3000] collection," SPR.MS 63/2/46. Thanks to Klaus Schmeh for sharing this.

23 Cambridge University Library, "The Society for Psychical Research [3000] collection," SPR.MS 63/2/97. Thanks to Klaus Schmeh for sharing this.

24 Frank C. Tribbe, "The Tribbe/Mulders Code," *Journal of the Academy of Religion and Psychical Research* 3, no. 1 (January 1980): 44–46.

25 Arthur S. Berger, *Aristocracy of the Dead: New Findings in Postmortem Survival* (London and Jefferson, NC: McFarland & Co., 1987), p. 152 cited here.

26 Berger, *Aristocracy of the Dead*, 165.

27 Berger, *Aristocracy of the Dead*, 163.

28 Berger, *Aristocracy of the Dead*, 165.

29 Arthur S. Berger, "Reincarnation by the Numbers: A Criticism and Suggestion," *Reincarnation: Fact or Fable?* Arthur S. Berger and Joyce Berger, eds. (London: The Aquarian Press, 1991), 221–33, p. 229 cited here.

30 Berger, "Reincarnation by the Numbers," 231.

31 For details of the lock, see p. 248 of Ian Stevenson, "The Combination Lock Test for Survival," *Journal of the American Society for Psychical Research* 62 (1968): 246–54.

32 See, for example, Matt Blaze, "Safecracking for the computer scientist," 2004 draft. Available online at http://www.crypto.com/papers/safelocks.pdf (accessed May 17, 2016).

33 Stevenson, "The Combination Lock Test for Survival."

34 Joseph Martin Feely, *Electrograms from Elysium: A Study on the Probabilities in Postmortuary Communication through the Electronics of Telepathy and Extrasensory Perception, Including the Code of Anagrams in the Purported Sender's Name* (New York, 1954), 202.

35 Feely, *Electrograms from Elysium*, 338–39.

36 Feely, *Electrograms from Elysium*, 236.

37 Feely, *Electrograms from Elysium*, 274.

38 Feely, *Electrograms from Elysium*, 240, 247.

39 Feely, *Electrograms from Elysium*, 329.

40 Feely, *Electrograms from Elysium*, 331.

41 Michael Levin, "Encryption Algorithms in Survival Evidence," *The Journal of Parapsychology* 58, no. 2 (June 1994): 189–95.

42 Levin, "Encryption Algorithms in Survival Evidence."

43 Levin, "Encryption Algorithms in Survival Evidence."

44 Levin was much more successful with his other papers, which earned a total of 7,214 citations as of this writing.

45 Susy Smith (pen name of Ethel Elizabeth Smith). Introduction by Gary E. R. Schwartz and Linda G. S. Russek. *The Afterlife Codes: Searching for Evidence of the Survival of the Human Soul* (Charlottesville, VA: Hampton Roads Publishing Company, Inc., 2000).

46 The Internet Archive Wayback Machine, https://archive.org/web/.

47 Afterlife Codes, https://web.archive.org/web/20010517033428/http://www.afterlifecodes.com/.

48 The linked page was archived from June 14, 2001.

49 It was just a MASC. See the section on Frank C. Tribbe earlier in this chapter or, for more detail, read Frank C. Tribbe, "The Tribbe/Mulders Code," *Journal of the Academy of Religion and Psychical Research* 3, no. 1 (January 1980): 44–46.

50 Elsewhere on the website, Smith wrote, "Naturally, we also thought of the possibility that someone might try to break my code by fraud, but we discounted that idea because we felt sure people weren't mean and unprincipled enough to do such a thing. But then came the '90s and an unfortunate moral decline. Thus we are having to advise those who register now not to offer rewards, which might be tempting. After considering the idea that I should apply this to myself as well, I have decided not to revoke my offer. It's such a part of my history now that a few more years won't make any difference. Except that although my body complains constantly with arthritis and such, my mind still enjoys hanging around keeping in touch. I could go on living until my funds are all used up. Not to worry. In that case Gary and Linda have vowed to provide the reward, should I be able to send my message after my demise. Fortunately also, the amount of $10,000 has been left with the University of Arizona Foundation for the Susy Smith Project by Robert Bigelow's National Institute for Discovery Science of Las Vegas. This, plus future contributions, should guarantee the continuation of our computer's code activity over the years."

51 Klaus Schmeh, *Nicht zu knacken* (Munich, Germany: Carl Hanser Verlag, 2012).

第 8 章　是否有绝对安全的密码？

1 In 2011, Steven Bellovin uncovered the fact that the one-time pad had, in fact, only been rediscovered by Mauborgne and Vernam. It was previously known to Frank Miller and was published by him in an 1882 commercial code book. For details, see Steven M. Bellovin, "Frank Miller: Inventor of the One-Time Pad," *Cryptologia* 35, no. 3 (July 2011): 203–22.

2 A line from the film *Fight Club*.

3 A line from the film *Saw*.

4 A line from the film *Freddy Got Fingered*.

5 It was re-presented by William F. Friedman, who had already solved it but delayed revealing his solution until the next issue to give readers a chance to attack it for themselves.

6 Taken here from Louis Kruh, "A 77-Year Old Challenge Cipher," *Cryptologia* 17, no. 2 (April 1993): 172–74.

7 Kruh, "A 77-Year Old Challenge Cipher," 172.

8 Louis Kruh, "Riverbank Laboratory Correspondence, 1919 (SRH-50)," *Cryptologia* 19, no. 3 (July 1995): 236–46.

9 A line from the film π.

10 Thanks to Moshe Rubin for helping me identify this individual.

11 TRIO (ACA pen name for Fenwick Wesencraft), "Solutions of the M-94 Test Messages," *The Cryptogram* 48, no. 8 (November–December 1982): 6–7.

12 Byrne promoted the system most famously in John F. Byrne, *Silent Years: An Autobiography with Memoirs of James Joyce and Our Ireland* (New York: Farrar, Straus and Young, 1953).

13 The error was pointed out in Jeff Calof, Jeff Hill, and Moshe Rubin, "Chaocipher Exhibit 5: History, Analysis, and Solution of *Cryptologia*'s 1990 Challenge," *Cryptologia* 38, no. 1 (January 2014): 1–25.

第 9 章　欲言又止的挑战密码

1 For a fuller account of this episode of government censorship, see Craig Bauer and Joel Burkholder, "From the Archives: Reading Stimson's Mail," *Cryptologia* 31, no. 2 (April 2007): 179–84.

2 Alexander d'Agapeyeff, *Codes and Ciphers* (London: Oxford University Press, 1939), front inside dust jacket.

3 d'Agapeyeff, *Codes and Ciphers*, 62–63.

4 J 77/2621/1445, Divorce Court File: 1445. Appellant: Josephine Christian Lilian Passy d'Agapeyeff. Respondent: Alexander d'Agapeyeff. Type: Wife's petition for divorce [WD], 1929, The National Archives, Kew, England.

5 Ralph Erskine and John Gallehawk located and photographed d'Agapeyeff's SOE file for me, but it didn't indicate why he was turned down. The reference is Special Operations Executive: Personnel Files (PF Series). Alexander d'Agapeyeff, Collection: Records of Special Operations Executive, 01 January 1939–31 December 1946, HS 9/9/5, The National Archives, Kew, England. Erskine thought it was doubtful that d'Agapeyeff was turned down because of the adultery. He pointed out, "Sir Stewart Menzies, head of the UK SIS, had at least 1 mistress, and was also divorced. Hinsley said privately that Godfrey, the Director of Naval Intelligence had a number of women." On the other hand, when interviewing potential new hires, NSA

polygraphers ask questions like "Have you ever cheated on your wife?" Beyond the trust issue, they don't want to hire people who may be subject to blackmail.

6 d'Agapeyeff, *Codes and Ciphers*, 131.

7 Image from d'Agapeyeff, *Codes and Ciphers*, 103.

8 The series begins with Gordon Rugg and Gavin Taylor, "A very British mystery: The case of the D'Agapeyeff Cipher," https://hydeandrugg.wordpress.com/2013/07/02/a-very-british-mystery-the-case-of-the-dagapeyeff-cipher/. As of this writing, part 6, in which Van Zandt's work is discussed, is unpublished. Rugg shared a draft version of it with me. I expect that a combined and revised version of these articles will see print in *Cryptologia*.

9 Gordon Rugg and Gavin Taylor, "A very British mystery, part 6: A possible solution? Probably not ... ," draft version.

10 Richard Feynman, *Surely You're Joking, Mr. Feynman* (New York: W. W. Norton and Co., 1985), 148–49.

11 Feynman's cipher, https://groups.google.com/forum/#!topic/sci.crypt/u1O1W1mL7Fw.

12 Feynman ciphers, https://groups.google.com/forum/#!searchin/sci.crypt/Jack$20C.$20Morrison$20/sci.crypt/RAxvau5mxJ4/VWLRaSbb-xMJ.

13 The last word didn't appear in its entirety in the deciphered text. That's why it was completed in brackets. http://www.vanderbilt.edu/usac/documents/BoTPresentation.pdf.

14 *Central Intelligence Agency Employee Bulletin*, June 1, 1988, released in response to a Freedom of Information Act (FOIA) request.

15 Remarks by William H. Webster, Director of Central Intelligence, at the dedication of the sculpture for the New Headquarters Building, November 5, 1990, released in response to a FOIA request.

16 Webster remarks.

17 Robert M. Andrews, "Sculpture with Code Poses Mystery at the CIA," *Los Angeles Times*, April 28, 1991, p. A34.

18 Part of a petition sent to the CIA Fine Arts Commission on August 2, 1989, released under a FOIA request. The sixteen signatures were primarily from the Office of Soviet Analysis (SOVA).

19 This method can be found at https://kryptosfan.wordpress.com/k3/k3-solution-3/.

20 John Markoff, "C.I.A.'s Artistic Enigma Yields All but Final Clue," *New York Times*, June 16, 1999, p. A24.

21 Markoff, "C.I.A.'s Artistic Enigma Yields All but Final Clue."

22 Markoff, "C.I.A.'s Artistic Enigma Yields All but Final Clue."

23 Markoff, "C.I.A.'s Artistic Enigma Yields All but Final Clue."

24 Lester S. Hill, "Cryptography in an Algebraic Alphabet," *The American Mathematical Monthly* 36, no. 6 (1929): 306–12.

25 These figures actually include non-invertible matrices. It was easier to just test all matrices than to eliminate those that weren't desirable from a cryptologic perspective.

26 Markoff, "C.I.A.'s Artistic Enigma Yields All but Final Clue."

27 Andrews, "Sculpture With Code Poses Mystery at the CIA."

28 Wikipedia, http://en.wikipedia.org/wiki/Cicada_3301.

29 Quote taken from *Anonymous: How Hackers Changed the World*, BBC documentary, https://www.youtube.com/watch?v=jwA4tSwmS-U.

30 This message can now be found at this site: https://www.reddit.com/r/a2e7j6ic78h0j/comments/o2rbx/ukbn_txltbz_nal_hh_uoxelmgox_wdvg_akw_hvu_ogl_rsm/.

31 You can hear the message at this site: https://www.youtube.com/watch?v=k24ZrFR2IUQ.

32 http://pgp.mit.edu:11371/pks/lookup?search=0x7A35090F&op=index.

33 See http://www.clevcode.org/cicada-3301/ for Eriksson's interpretation of the rest of this clue.

34 And still is, as of this writing. See http://www.williamgibsonbooks.com/source/agrippa.asp.

35 "Norway Terror Attacks Fast Facts," CNN Library, July 17, 2015.

36 "Norway Terror Attacks Fast Facts," CNN Library, July 17, 2015. But according to Bjoern Amland and Sarah DiLorenzo, "Lawyer: Norway Suspect Wanted a Revolution," July 24, 2011, he killed at least ninety-two and wounded ninety-seven others. It's unclear where this number for the deaths came from. Much later accounts don't support it. See, for example, Olivia Herstein, "Still Standing—As One," *Viking* 108, no. 12 (December 2011): 16–20, 22, 24. Wikipedia says that Breivik killed 77 and injured 241.

37 Amland and DiLorenzo, "Lawyer."

38 Dave Ramsden, *Unveiling the Mystic Ciphers: Thomas Anson and the Shepherd's Monument Inscription* (CreateSpace, 2014).

39 Archived at https://web.archive.org/web/20110919074847/http://jahbulon.com/.

40 https://web.archive.org/web/20121210124921/http://jahbulon.com/about/.

41 List of errors taken from Todd D. Mateer, "Cryptanalysis of Beale Cipher Number Two," *Cryptologia* 37, no. 3 (July 2013): 225–26.

42 Letter from M. E. Ohaver to John Ingles, August 8, 1933. This letter is discussed in Louis Kruh, "A Basic Probe of the Beale Cipher as a Bamboozlement," *Cryptologia* 6, no. 4 (October 1982): 378–82.

43 Wayne S. Chan, "Key Enclosed: Examining the Evidence for the Missing Key Letter of the Beale Cipher," *Cryptologia* 32, no. 1 (January 2008): 33–36.

44 Chan was unable to get any listings closer to 1832 for St. Louis.

45 Forrest Fenn, *The Thrill of the Chase, A Memoir* (Santa Fe, NM: One Horse Land & Cattle Co., 2010), 128.

46 Fenn, *The Thrill of the Chase*, 129–32.

47 Fenn, *The Thrill of the Chase*, 133.

48 Forrest Fenn, *Too Far to Walk* (Santa Fe, NM: One Horse Land & Cattle Co., 2013), 164.

49 Fenn, *Too Far to Walk*, 164.

50 Fenn, *Too Far to Walk*.

51 Jennifer Deutschmann, "Forrest Fenn: Hunt for Buried Treasure Is 'Out of Control,' " *Inquisitr*, April 28, 2015, available online at http://www.inquisitr.com/2050321/forrest-fenn/.

52 Fenn, *Too Far to Walk*, 86.

53 Margie Goldsmith, "Well Over $1 Million in Buried Treasure: Find It!" *Huffington Post*, February 18, 2011, updated February 27, 2013.

54 Fenn, *Too Far to Walk*, 265.

55 Deutschmann, "Forrest Fenn."

56 Athol Thomas, *Forgotten Eden: A View of the Seychelles Islands in the Indian Ocean* (London and Harlow, U.K.: Longmans, Green and Co. Ltd., 1968), 133–34.

57 Thomas, *Forgotten Eden*, 134.

58 E-mail sent to the author on July 13, 2015.

59 Klausis Krypto Kolumne, http://scienceblogs.de/klausis-krypto-kolumne/2016/03/27/wer-kann-diesen-grabstein-entschluesseln-und-wo-befindet-er-sich-ueberhaupt/.

第 10 章 未解的长密码

1 Klausis Krypto Kolumne, http://scienceblogs.de/klausis-krypto-kolumne/klaus-schmehs-list-of-encrypted-books/.

2 Benedek Láng, "Why Don't We Decipher an Outdated Cipher System? The Codex of Rohonc," *Cryptologia* 34, no. 2 (April 2010): 115–44.

3 The solutions were pointed out by Klaus Schmeh in "The Pathology of Cryptology—A Current Survey," *Cryptologia* 36, no. 1 (January 2012): 14–45, and are included in the references at the end of this book.

4 Lynda Roscoe Hartigan, *James Hampton: The Throne of the Third Heaven of the Nations' Millennium General Assembly* (Boston: Museum of Fine Arts, October 19 (1975?)–February 13, 1976). *This essay was originally published for a 1976 exhibition of James Hampton's work.*

5 Stephen Jay Gould, "James Hampton's Throne and the Dual Nature of Time," *Smithsonian Studies in American Art* 1, no. 1 (Spring 1987): 46–57.

6 Gould, "James Hampton's Throne."

7 Stephen Jay Gould, "Nonoverlapping Magisteria," *Natural History* 106 (March 1997): 16–22.

8 Gould, "James Hampton's Throne."

9 Gould, "James Hampton's Throne."

10 This is 00011 in Klaus Schmeh's list. See http://www.scm.keele.ac.uk/research/knowledge_modelling/km/people/gordon_rugg/cryptography/penitentia/index.html.

11 Keele University, School of Computing and Mathematics, http://www.keele.ac.uk/scm/staff/academic/drgordonrugg/.

12 E-mail from Gordon Rugg to the author on June 29, 2015.

13 IMDb, http://www.imdb.com/name/nm0235710/.

14 Search Visualizer, http://www.searchvisualizer.com/.

15 This is 00012 in Klaus Schmeh's list.

16 Keele University, Knowledge Modelling, http://www.scm.keele.ac.uk/research/knowledge_modelling/km/people/gordon_rugg/cryptography/ricardus_manuscript.html.

17 Sandra and Woo, http://www.sandraandwoo.com/2013/07/29/0500-the-book-of-woo/. This is number 00022 in Klaus Schmeh's list.

18 E-mail from Oliver Knörzer to the author, July 16, 2015.

19 E-mail from Oliver Knörzer to the author, July 16, 2015.

20 E-mail from Oliver Knörzer to the author, July 16, 2015. The wiki is at http://bookofwoo.foogod.com/wiki/Book_of_Woo_Wiki.

21 Book of Woo Wiki, http://bookofwoo.foogod.com/wiki/What_We_Know. The hint was given on July 21, 2014.

22 E-mail from Oliver Knörzer to the author, July 16, 2015.

23 E-mail from Oliver Knörzer to the author, July 16, 2015.

24 E-mail from Oliver Knörzer to the author, July 16, 2015.

25 Auguste Kerckhoffs, "La Cryptographie Militaire." *Journal des science militaires* (Paris: Baudoin, 1883), vol. IX, 5–83.

26 See imgur, http://imgur.com/a/8xnWx/all.

27 Klausis Krypto Kolumne, http://scienceblogs.de/klausis-krypto-kolumne/klaus-schmehs-list-of-encrypted-books/.

第 11 章 外星人密码和RSA算法

1 "Science Notes," *Scientific American* 84, no. 12 (March 23, 1901), 179.

2 "Science Notes," *Scientific American*.

3 "Annual Report of the Board of Regents of the Smithsonian Institution," 1900, p. 169 (taken from *Boston Transcript*, February 2, 1901). Available online at http://www.biodiversitylibrary.org/item/53390#page/291/mode/1up. Note: This is not to be confused with a different volume of almost identical title, but specifying "1900 Incl. Rept. US Natl Mus."

4 Charles Fort, *New Lands*, in The Books of Charles Fort, with an introduction by Tiffany Thayer (omnibus ed.) (New York: Henry Holt and Co., 1941), ch. 32, p. 494. Available online in single-volume edition at http://www.resologist.net/landsei.htm. See part 2, chapter 20.

5 "100 Years Ago," *Nature* 417, no. 6884 (May 2, 2002): 31.

6 Charles Fort also wrote about this episode. See Fort, *New Lands*, ch. 36, pp. 526–27. Available online in single-volume edition at http://www.resologist.net/landsei.htm. See part 2, chap. 24.

7 *The Monthly Evening Sky Map* XIV, no. 159 (March 1920).

8 *The Monthly Evening Sky Map* XV, no. 179 (November 1921). Available online at Google Books. More detail can be found in "Marconi Believes He Received Wireless Messages From Mars," *New York Tribune*, Sept. 2, 1921, p. 3.

9 "Weird 'Radio Signal' Film Deepens Mystery of Mars," *Washington Post*, August 27, 1924.

10 "Weird 'Radio Signal' Film."

11 "Weird 'Radio Signal' Film."

12 "Weird 'Radio Signal' Film."

13 "Weird 'Radio Signal' Film."

14 "Weird 'Radio Signal' Film."

15 This idea was mentioned by Ronald Bracewell in "What to Say to the Space Probe when it Arrives," *Horizon*, January 1977. Bracewell preferred the idea of sending animated cartoons because they can convey words that can't be made clear by single images.

16 Harold Wooster (moderator), Paul J. Garvin, Lambros D. Callimahos, John C. Lilly, William O. Davis, and Francis J. Heyden, "Communication with Extraterrestrial Intelligence," *IEEE Spectrum* 3, no. 3 (March 1966): 156 and 158.

17 Bernard M. Oliver, "Interstellar Communication," *Interstellar Communication;—A Collection of Reprints and Original Contributions*, A. G. W Cameron, ed. (New York: W.A. Benjamin, 1963), 294–305.

18 This conference is also where Frank Drake first put forth what's become known as the Drake equation.

19 Oliver, "Interstellar Communication," 305.

20 Reproduced from http://www.rsa.com/rsalabs/node.asp?id=2093.

21 R. L Rivest, A. Shamir, and L. Adleman, *On Digital Signatures and Public-Key Cryptosystems* (There was soon a title change to *A Method for Obtaining Digital Signatures and Public-key Cryptosystems*. The date is the same for both.), M.I.T. Laboratory for Computer Science Report

MIT/LCS/TM-82, April 1977. This report later appeared in *Communications of the ACM* 21, no. 2 (February 1978): 120–26, with the second title.

22 The team consisted of F. Bahr, M. Boehm, Jens Franke, and Thorsten Kleinjung.

23 Thorsten Kleinjung, Kazumaro Aoki, Jens Franke, Arjen K. Lenstra, Emmanuel Thomé, Joppe W. Bos, Pierrick Gaudry, Alexander Kruppa, Peter L. Montgomery, Dag Arne Osvik, Herman te Riele, Andrey Timofeev, and Paul Zimmermann, "Factorization of a 768-bit RSA modulus," version 1.4, February 18, 2010, 14, available online at http://eprint.iacr.org/2010/006.pdf.

24 Ronald L. Rivest, "Description of the LCS35 Time Capsule Crypto-Puzzle," April 4, 1999, available online at http://people.csail.mit.edu/rivest/lcs35-puzzle-description.txt.

25 Ronald L. Rivest, Adi Shamir, and David A. Wagner, "Time-lock puzzles and timed-release Crypto," Revised March 10, 1996, p. 2, available online at http://theory.lcs.mit.edu/~rivest/RivestShamirWagner-timelock.ps.

26 Rivest specified "ascii at 8 bits per character," meaning that extended ASCII is intended. The 8 bits offer 256 possibilities.

27 Ronal L. Rivest, "Desription of the LCS35 Time Capsule Crypto-Puzzle," http://people.csail.mit.edu/rivest/lcs35-puzzle-description.txt.

参考文献与延伸阅读

第 1 章

Jim Reeds's bibliography follows, with my additions (and some deletions). Permission to use Jim's work as a basis for my own bibliography was granted by Jim at the Cryptologic History Symposium in October 2011. Jim got some references from voynich@rand.org, Rene Zandbergen, Gabriel Landini, Denis Mardle, Dennis Stallings, and especially from Brian Smith.

手　稿

Items with WFF numbers are in the William F. Friedman collection of the George Marshall Library in Lexington, Virginia.

"The Voynich 'Roger Bacon' Cipher Manuscript." Beinecke Rare Book Library, Yale University: MS 408. Supplementary material in folders and boxes labelled A–N.

Positive photocopies of ff. 1–56 of Voynich MS. British Library MS Facs 461.

Positive photocopies of miscellaneous folios of Voynich MS, misc. VMS correspondence of R. Steele, 1921, and misc. VMS articles. (Contains: 68r1/68r2, 65v/66r, 78v/79r, 107v/108r, 108v/111r, 111v/112r, 112v/113r, 1r, and 116v. Correspondence includes letters to Steele from W. Voynich, W. Newbold, and from A. W. Pollard. Articles include a clipping from *Morning Post* newspaper, 26.9.21, "Astrological Anagrams: the Diary of Roger Bacon," clipping from *Daily Chronicle* newspaper, n.d., "Key to Cypher in historic MS./America's new light on Roger Bacon/600 years' mystery," pencilled draft of article by Steele, photostat of typed lecture notes by Newbold, copy of J. Manly's *Harper's* article, copy of Louis Cons February 4, 1922, article.) British Library MS Facs 439.

Brumbaugh, Robert S. "Voynich Newsletter." Various issues: February 1978, 6-p typescript. November 1978, 7-p typescript. January 1980, 7-p typescript.

Carter, Albert H. "Some Impressions of the Voynich Manuscript." Unpublished notes, September 10, 1946. WFF 1614.

Carton, Raoul. "The Cipher of Roger Bacon." 55-p typescript, translated from French by E. L. Voynich. In N.Y. Academy of Medicine Library. 1930.

Currier, Prescott. "Voynich MS Transcription Alphabet; Plans for Computer Studies; Transcribed Text of Herbal A and B Material; Notes and Observations." Unpublished communications to John H. Tiltman and M. D'Imperio. Darimascotta, ME.

D'Imperio, M. E. "Structure of Voynich Text Groups: A Statistical Model." 2-p typescript, 1978.

D'Imperio, M. E. "An Application of Cluster Analysis and Multiple Scaling to the Question of 'Hands' and 'Languages' in the Voynich Manuscript." January 28, 1992. [Confirms Currier's findings. In Gillogly collection. From abstract: "This paper is an extensively revised and updated version of an earlier paper in an in-house technical journal, dated 20 June 1978. It includes corrected and expanded data sets."]

D'Imperio, M. E. "Odd Repetitions or Near Repetitions in the Text." January 1992.

D'Imperio, M. E. "Some Ideas on the Construction of the Voynich Script." 3-p typescript, January 1992.

Firth, Robert. "Notes on the Voynich Manuscript." Numbered series of essays: 1–24, 1991–1995.

Friedman, William F. Two "First Study Group" transcription alphabet sheets. WFF 1609.1.

Friedman, William F. "First Study Group" transcription alphabet sheet. WFF 1609.2.

Friedman, William F., Mark Rhoads, et al. Minutes of the "Voynich Manuscript Research Group," 1944–45. National Cryptologic Museum, VF 10-8.

Friedman, William F., et al. Printout of transcription, onto 131 printout sheets, ca. 1944–46. WFF 1609.

Friedman, William F., et al. Printout of partial transcription, Second Study Group, ca. 1963. WFF 1609.3.

Friedman, William F., et al. Correspondence with RCA Corp. about activities of "Second Study Group," 1963. WFF 1609.4.

Friedman, William F., et al. Printout of partial transcription, ca. 1944–46 Unnumbered item in Friedman Collection.

Friedman, William F., et al. Printout of partial transcription, ca. 1944–46. Unnumbered item in NSA Historical Records Collection.

Guy, Jacques B. M. "The distribution of letters c and o in the Voynich Manuscript: Evidence for a real language?" April 1994.

Krischer, Jeffrey P. *The Voynich Manuscript*, Harvard University term paper, 1969.

Mervyn, Tim, unpublished manuscript. [This manuscript is mentioned briefly on p. ix and then in much more detail on pp. 216–20 of Kennedy, Gerry, and Rob Churchill, *The Voynich Manuscript: The Unsolved Riddle of an Extraordinary Book Which Has Defied Interpretation for Centuries*, British hardcover edition (London: Orion Books, 2004).]

Panofsky, Erwin. "Answers to Questions for Prof. E. Panofsky." Letter to William F. Friedman, March 19, 1954. Correspondence between Friedman, Panofsky, and J. v. Neumann. Letters from Richard Salomon to Erwin Panofsky and Gertrud Bing. WFF 1614.

Petersen, Theodore C. "Notes to Mr. Tiltman's [1951] Observations on the Voynich Cipher MS." Unpublished. April 23, 1953.

Petersen, Theodore C. Hand Transcript and Concordance of the Voynich Manuscript and Other Working Papers. In the Friedman Collection, George Marshall Library, Lexington, Virginia, dated 1966, but this is simply when the library acquired it (following Petersen's death).

Puckett, Frances M. Partial transcription of ff. 111v–114r, in WFF 1613.

Strong, Leonell C. Collection of letters, notebook entries, and worksheets. In a private collection.

Tiltman, John. "Interim Report on the Voynich MS." Personal communication to William F. Friedman, May 5, 1951. 2-p typescript. In NSA Historical Records Collection. Copy in WFF 1615.13.

Tiltman, John. Partial transcription, 1951. In NSA Historical Records Collection.

Tiltman, John. Biography of T. C. Petersen. WFF 1615.

Tiltman, John. "The Voynich MS" Script of an address presented at the Baltimore Bibliophiles. March 4, 1951. [The date looks wrong. Cited by D'Imperio. Almost certainly the 1967 Baltimore Bibliophiles address.]

Voynich, Wilfrid, Ethel Voynich, and A. M. Nill. Notes concerning the history of the cipher manuscript. Voynich Archives, Library of the Grollier Club of New York, 1917–196? [Possibly identical with Beinecke MS 408 B.]

纸质书籍与文章

Altick, Richard D. *The Scholar Adventurers.* (New York: The Free Press, 1966). [First ed. in 1950. Has discussion of VMS and Manly's refutation on pp. 200–6. A revised reprint appeared in 1987.]

Anonymous. "Art Works Worth $1,500,000 Arrive to Escape War." *Chicago Daily Tribune.* Oct. 9, 1915, p. 1, col. 2. [Exhibit of WMV's books at the Art Institute of Chicago, which includes "a work by Roger Bacon in cipher to which the key has never been discovered."]

Anonymous. "Antique Books Worth $500,000." *Chicago Sunday Tribune.* Oct. 10, 1915, sec. 2, p. 1, col. 5. [Exhibit of WMV's books at the Art Institute of Chicago. A few paragraphs describe the VMS. "The manuscript is from the hand of Roger Bacon . . . was bought by Emperor Rudolf . . . and at the end of the sixteenth century passed into the hands of King Ferdinand of Bohemia."]

Anonymous. "Review of Brumbaugh's Most Mysterious Manuscript." *Choice* 15 (October 1978), 1080.

Anonymous. "The Roger Bacon Manuscript." *Scientific American* 124 (May 7, 1921): 362.

Anonymous. "The Roger Bacon Manuscript: What It Looks Like, and a Discussion of the Possibilities of Decipherment." *Scientific American* 124, no. 22 (May 28, 1921): 432, 439, 440.

Ashbrook, Joseph. "Roger Bacon and the Voynich Manuscript." *Sky and Telescope* (April 1966): 218–19.

Barlow, Michael. "The Voynich Manuscript—By Voynich?" *Cryptologia* 10, no. 4 (October 1986): 210–16.

Barlow, Michael. "Voynich Solved?" (Review of Levitov), pp. 47–48 in "Reviews and Things Cryptologic." *Cryptologia* 12, no. 1 (January 1988): 37–51.

Barthélémy, Pierre. "L'indéchiffrable manuscrit Voynich résiste toujours au décryptage." *Le Monde,* December 20, 2000.

Barthélémy, Pierre (introduction only). *Le Code Voynich.* (Paris: Jean-Claude Gawsewitch, 2005). Color facsimile edition of the full text of the Voynich ms.

Bauer, Craig P. *Secret History: The Story of Cryptology* (Boca Raton, FL: Chapman and Hall/CRC, 2013).

Bax, Stephen. Voynich—a provisional, partial decoding of the Voynich script, https://www.youtube.com/watch?v=fpZD_3D8_WQ.

Bennett, William Ralph. *Scientific and Engineering Problem Solving with the Computer* (Englewood Cliffs, NJ: Prentice-Hall, 1976). [Contains a chapter on VMS.]

Bird, J. Malcom. "The Roger Bacon Manuscript: Investigation into its History, and the Efforts to Decipher It." *Scientific American Monthly* 3 (June 1921): 492–96.

Black, John. *Codex Gigas* (the Devil's Bible), the largest manuscript in the world, http://www.ancient-origins.net/myths-legends-europe/codex-gigas-devil-s-bible-largest-manuscript-world-001276, January 27, 2014.

Blunt, Wilfrid, and Sandra Raphael. *The Illustrated Herbal* (London: Thames and Hudson, in association with the Metropolitan Museum of Art, 1979). [Provides two-color reproductions of folios from the Voynich manuscript, along with many other pictures from medieval herbals, allowing comparison.]

Bolton, Henry Carrington. *The Follies of Science at the Court of Rudolph II 1576–1612* (Milwaukee: Pharmaceutical Review Publishing Co., 1904), available online at https://archive.org/details/folliesscienceа00boltgoog.

Boston Transcript. Review of Newbold's Cipher of Roger Bacon, June 30, 1928, p. 2.

Brooke, Tucker. "Doctor Mirabilis." *Yale Review* 19 (1929): 207–8. [Review of Newbold.]

Brumbaugh, Robert S. "Botany and the Voynich 'Roger Bacon' Manuscript Once More." *Speculum* 49 (1974): 546–48.

Brumbaugh, Robert S. "The Solution of the Voynich 'Roger Bacon' Cipher." *Yale Library Gazette* 49, no. 4 (April 1975): 347–55.

Brumbaugh, Robert S. "The Voynich 'Roger Bacon' Cipher Manuscript: Deciphered Maps of Stars." *Journal of the Warburg and Courtauld Institutes* 39 (1976): 139–50.

Brumbaugh, Robert S., ed. and contrib. *The Most Mysterious Manuscript.* (Carbondale, IL: Southern Illinois University Press, 1978, and London: Weidenfeld and Nicholson, 1977).

Bühler, Markus. *Der Fährtenleser 4: Das Voynich-Manuskript: Ein kryptobotanisches Rätsel* (German ed.), Feb. 4, 2009. This is a magazine for cryptozoology. The Voynich ms qualifies for inclusion as the supposed nonexistence of the plants it depicts makes it a case of cryptobotany, a related field.

Bull. Art Inst. Chicago IX (1915): 100.

Carton, Raoul. "Le Chiffre de Roger Bacon." *Révue d'Histoire de la Philosophie* 3 (1929): 31–66, 165–79.

Casanova, Antoine. *Méthodes d'analyse du langage crypte: une contribution a l'étude du manuscrit de Voynich: These pour obtenir le grade de Docteur de l'Universite Paris 8*, March 19, 1999.

Child, James R. "The Voynich Manuscript Revisited." *NSA Technical Journal* 21, no. 3 (Summer 1976): 1–4, available online at https://www.nsa.gov/public_info/_files/tech_journals/voynich_manuscript_revisited.pdf, p. 1.

Child, James R. "Again, The Voynich Manuscript," 2007, available online at http://web.archive.org/web/20090616205410/http://voynichmanuscript.net/voynichpaper.pdf.

Clabby, John F. *Brigadier John Tiltman: A Giant among Cryptanalysts* (Fort Meade, MD: Center for Cryptologic History, National Security Agency, 2007).

Clark, Ronald W. *The Man Who Broke Purple: The Life of Colonel William F. Friedman, Who Deciphered the Japanese Code in World War II* (Boston: Little Brown & Co., 1977).

Cons, Louis. "Un manuscrit mystérieux: Un traité scientifique du treizième siècle, attribué à Roger Bacon." *L'Illustration* 159, no. 4118 (February 4, 1922): 112.

Cons, Louis. "Newbold's Trail." *Saturday Review of Literature* 5 (October 27, 1928): 292.

Corrales, Scot. "The Books of the Damned: Fact or Fiction?" *FATE* (July 2000).

Currier, Captain Prescott H. "Some Important New Statistical Findings." *Proceedings of a Seminar Held on 30th November 1976 in Washington, D.C.*, available online at http://www.ic.unicamp.br/~stolfi/voynich/mirror/gillogly/currier.paper.

Daiger, Michael. "The World's Most Unusual Manuscript." *Occult* (January 1976).

"Devil's Bible." season 1, episode 5 of the National Geographic Channel series *The Truth Behind*, 2008, aired July 30, 2010. [On Codex Gigas].

Dilas, Jonathan. Das Voynich-Manuskript, http://www.jonathan-dilas.de/Texte/voynich.html.

D'Imperio, M. E., ed. *New Research on the Voynich Manuscript: Proceedings of a Seminar* (Washington, DC: Privately printed pamphlet, November 30, 1976).

Partial contents:

James Child. "A Linguistic Approach to the Voynich Text."

Capt. Prescott H. Currier. "Some Important New Statistical Findings."

Dr. Sydney Fairbanks. "Suggestions Toward a Decipherment of the 'Key.' "

M. E. D'Imperio. "The Solution Claim of Dr. Robert S. Brumbaugh."

Capt. Prescott Currier. "Further Details of New Statistical Findings."
Capt. Prescott Currier. "The Voynich Manuscript, Some Notes and Observations."
Stuart H. Buck. "What Constitutes Proof?"

D'Imperio, M. E. *The Voynich Manuscript—An Elegant Enigma* (Fort Meade, MD: National Security Agency, 1976). Reprinted by Aegean Park Press, Laguna Hills, CA, 1978.

D'Imperio, M. E. "The Voynich Manuscript: A Scholarly Mystery." *Manuscripts* 29, no. 2 (Spring 1977): 85–93 and vol. 29, no. 3 (Summer 1977): 161–73 and vol. 30, no. 1 (Winter 1978): 34–48. [Three-part article. Parts about physical history of the manuscript, about attempts at decipherment, and about Brumbaugh's, Currier's, and Child's work, respectively.]

D'Imperio, M. E. "An Application of Cluster Analysis and Multidimensional Scaling to the Question of 'Hands' and 'Languages' in the Voynich Manuscript." *National Security Agency Technical Journal* 23, no. 3 (Summer 1978): 59–75.

D'Imperio, M. E. "An Application of PTAH to the Voynich Manuscript." *National Security Agency Technical Journal* 24, no. 2 (Spring 1979): 65–91.

Drucker, Johanna. *The Alphabetic Labyrinth* (New York: Thames and Hudson, 1995).

Ephron, H. (Pseud. "DENDAI"). "A burning question in re the Voynich MS (slightly revised)." *The Cryptogram* 43 (March–April 1977): 22, 46–48; (May–June 1977): 49, 51–52, 72.

Evans, R. J. W. *Rudolf II and His World: A Study in Intellectual History 1576–1612* (Oxford, U.K.: Oxford University Press, 1973).

Feely, Joseph M. *Roger Bacon's Cipher: The Right Key Found* (Rochester, NY: self-published, 1943).

Fell-Smith, Charlotte. *John Dee (1527–1608)* (London: Constable & Company Ltd., 1909). Available online at https://archive.org/details/cu31924028928327.

Fischer, Elliot. "Language Redundancy and Cryptanalysis." *Cryptologia* 3, no. 4 (October 1979): 233–35.

Fleischaker, Julia. "Has a botanist solved the mystery of the Voynich Manuscript?" http://www.mhpbooks.com/has-a-botanist-solved-the-mystery-of-the-voynich-manuscript/. [On Dr. Arthur Tucker's "solution."]

Fletcher, John Edward. *A Study of the Life and Works of Athanasius Kircher, 'Germanus Incredibilis'* (Leiden, Netherlands, and Boston: Brill, 2011).

Foster, Caxton C., "A Comparison of Vowel Identification Methods." *Cryptologia* 16, no. 3 (July 1992): 282–86.

Friedman, Elizebeth S. " 'The Most Mysterious Manuscript' Still an Enigma." *The Washington Post*, August 5, 1962, pp. E1, E5.

Friedman, W. F., and E. S. Friedman. "Acrostics, Anagrams, and Chaucer." *Philological Quarterly* 38 (1959): 1–20.

Garland, Herbert. "The Mystery of the Roger Bacon Cipher MS." *Bookman's Journal and Print Collector* 5 (October 1921): 11–16.

Garland, Herbert. "A Literary Puzzle Solved?" *Illustrated London News* 160 (May 20, 1922): 740–42.

Garland, Herbert. "Notes on the Firm of W. M. Voynich." *Library World* 34 (April 1932): 225–28.

Gilson, Étienne. Review of Newbold's *The Cipher of Roger Bacon*. *Révue critique d'histoire et de littérature*, Paris 62 (August 1928): 378–83.

Goldstone, Lawrence, and Nancy Goldstone. *The Friar and the Cipher: Roger Bacon and the Unsolved Mystery of the Most Unusual Manuscript in the World* (New York: Doubleday, 2005).

Grossman, Lev. "When Words Fail: The Struggle to Decipher the World's Most Difficult Book." *Lingua Franca* 9, no. 3 (April 1999): 9–15.

Gutmann, J. *Athanasius Kircher (1602–1680) und das Schöpfungs—und Entwicklungsproblem* (Fulda, Germany: Druck Parzeller & Co., 1938). [This is a doctoral dissertation from the University of Würzburg that discusses Kircher's anticipation of evolution.]

Guy, Jacques. "Voynich Revisited." Letter to the editor. *Cryptologia* 15, no. 2 (April 1991): 161–66.

Guy, Jacques B. M. "Statistical Properties of Two Folios of the Voynich Manuscript." *Cryptologia* 15, no. 3 (July 1991): 207–18. [Applies Sukhotin's vowel-finding algorithm to VMS text.]

Guy, Jacques B. M. "On Levitov's Decipherment of the Voynich Manuscript." December 9, 1991. Available online at http://www.necronomi.com/magic/hermeticism/levitov+voynich.txt.

Guzman, Gregory S. Review of Brumbaugh's "Most Mysterious Manuscript." *Historian* 42 (November 1979): 120–21.

Harnisch, Larry (Pseud. "AR-MYR"). "The Voynich Manuscript." *The Cryptogram* 43 (May–June 1976): 45, 62–63; (July–August 1976): 69, 74–77.

Holzer, Hans. *The Alchemist: The Secret Magical Life of Rudolf von Habsburg* (New York: Stein and Day, 1974).

James, Peter J., and Nick Thorpe. *Ancient Inventions* (New York: Ballantine Books, 1994).

Jay, Mike. "Maze of Madness." *Fortean Times (UK)*, no. 130 (January 2000).

Johnson, Charles. Review of Newbold's "Cipher of Roger Bacon." *English Historical Review* 44 (October 1929): 677–78.

Joven, Enrique. *The Book of God and Physics: A Novel of the Voynich Mystery* (New York: William Morrow, 2009).

Kahn, David. "The Secret Book." *Newsday*, June 26, 1962.

Kahn, David. *The Codebreakers* (New York: Macmillan, 1967). [VMS discussed on pp. 863–72, 1120–21.]

Kennedy, Gerry, and Rob Churchill. *The Voynich Manuscript: The Unsolved Riddle of an Extraordinary Book Which Has Defied Interpretation for Centuries*, British hardcover ed. (London: Orion Books, 2004).

Kennedy, Gerry, and Rob Churchill. *The Voynich Manuscript: The Mysterious Code That Has Defied Interpretation for Centuries*, American ed. (Rochester, VT: Inner Traditions, 2006). [Note: The pagination differs from the British edition.]

Kent, Roland G. "Deciphers Roger Bacon Manuscripts." *Pennsylvania Gazette* 19 (May 27, 1921): 851–53.

Kircher, Athanasius. *Polygraphia Nova et Universalis ex Combinatoria Arte Detecta* (Rome: Varese, 1663).

Kircher, Athanasius. *Ars Magna Sciendi Sive Combinatoria* (Amsterdam: Johannes Janssonius a Waesberge, 1669).

Knobloch, Eberhard. "Renaissance Combinatorics." *Combinatorics: Ancient and Modern*, Robin Wilson and John J. Watkins, eds. (Oxford, U.K.: Oxford University Press, 2013). [Pp. 123–46 and 135–37 are relevant here for more on Kircher's combinatorial work.]

Knox, Sanka. "700-Year-Old Book for Sale; Contents, In Code, Still Mystery." *New York Times,* July 18, 1962, p. 27.

Knuth, Donald E. "Two Thousand Years of Combinatorics." *Combinatorics: Ancient and Modern*, Robin Wilson and John J. Watkins, eds. (Oxford, U.K.: Oxford University Press, 2013), 7–37.

Kraus, H. P. *Catalogue 100. Thirty-Five Manuscripts: Including the St. Blasien Psalter, the Llangattock Hours, the Gotha Missal, the Roger Bacon (Voynich) Cipher Ms.* (New York: H.P. Kraus, 1962).

Kraus, H. P. *A Rare Book Saga* (New York: G. P. Putnam's Sons, 1978).

Kullback, Solomon. *Statistical Methods in Cryptanalysis* (Laguna, CA: Aegean Park Press, 1976).

Landini, G., and R. Zandbergen. "A Well-kept Secret of Medieval Science: The Voynich Manuscript." *Aesculapius*, no. 18 (July 1998): 77–82.

Landmann, E. "Das sogenannte Voynich-Manuskript." *Magazin 2000plus*, Alte Kulturen Spezial, 2007.

"Law Report, May 10. High Court of Justice. King's Bench Division. Cathedral Library Thefts: Old Volume Traced. The Dean and Chapter of the Cathedral Church of Lincoln v. Voynich." *The Times*, London, May 11, 1916, p. 4.

Levitov, Leo. *Solution of the Voynich Manuscript: A Liturgical Manual for the Endura Rite of the Cathari Heresy, the Cult of Isis* (Laguna Hills, CA: Aegean Park Press, 1987).

Liebert, Herman W. "The Beinecke Library Accessions 1969." *Yale University Library Gazette* 44 (April 1970). [Describes Kraus's gift of VMS on pp. 192–93.]

Loeser, R. "Roger Bacon's Chiffremanuskript." *Die Umschau* 26 (1922): 115–17.

"Lovecraft and the Voynich Manuscript." *INFO Journal*, no. 48 (ca. 1984). [The International Fortean Organization's *INFO Journal*.]

Manly, John M. "Roger Bacon's Cypher Manuscript." *American Review of Reviews* 64 (July 1921): 105–6.

Manly, John M. "The Most Mysterious Manuscript in the World." *Harper's Monthly Magazine* 143 (July 1921): 186–97.

Manly, John M. "Roger Bacon and the Voynich Manuscript." *Speculum* 6 (July 1931): 345–91.

Marshall, Peter. *The Magic Circle of Rudolf II: Alchemy and Astrology in Renaissance Prague* (New York: Walker & Company, 2006).

McKaig, Betty. "The Voynich Manuscript—Cipher of the Secret Book." (Interview with Leonell Strong). *North County Independent*, Oct. 7, 1970. Reprinted courtesy Independent Newspapers, Inc., San Diego.

McKenna, Terence K. "Has the World's Most Mysterious Manuscript Been Read at Last?" *Gnosis Magazine* 7 (Summer 1988): 48–51.

McKenna, Terence K. *The Archaic Revival: Speculations on Psychedelic Mushrooms, the Amazon, Virtual Reality, UFOs, Evolution, Shamanism, the Rebirth of the Goddess, and the End of History* (San Francisco: HarperSanFrancisco, 1991).

McKeon, Richard. "Roger Bacon." *The Nation* 127 (August 29, 1928): 205–6.

Moses, Montrose J. "A Cinderella on Parchment: The Romance of the New 600 Year-Old Bacon Manuscript." *Hearst's International* (1921): 16–17, 75.

Nadis, Steve. "Look Who's Talking." *New Scientist* 179, no. 2403 (July 12, 2003).

New York Times. March 26, 1921, p. 6; March 27, 1921, sec 2, p. 1; April 21, 1921, p. 3; April 22, 1921, p. 13. [All on Newbold's findings.]

New York Times. "Roger Bacon's Formula Yields Copper Salts, Proving Newbold Secret Cipher Translation." December 2, 1926, p. 5, with follow-up articles December 3, p. 22, and December 12, sec. 20, p. 12.

New York Times. "Will Orders Sale of Bacon Cipher." April 15, 1930, p. 40.

Newbold, William R. "The Voynich Roger Bacon Manuscript." *Transactions of the College of Physicians and Surgeons of Philadelphia* (1921) 431–74. Read April 20, 1921.

Newbold, William R. "The Eagle and the Basket on the Chalice of Antioch." *American Journal of Archaeology* 29, no. 4 (October–December 1925): 357–80. [A paper on the Holy Grail.]

Newbold, William Romaine. *The Cipher of Roger Bacon,* edited with foreword and notes by Roland Grubb Kent. (Philadelphia: University of Pennsylvania Press; London, H. Milford, Oxford University Press, 1928).

Newsom, Eugene. *A Split in the Mystery Curtain*, 20-p pamphlet held at Central Arkansas Library System, Butler Center for Arkansas Studies, Little Rock, AR, 1994.

O'Neill, Hugh. "Botanical Remarks on the Voynich MS." *Speculum* 19 (1944): 126.

Orioli, Giuseppe. *Adventures of a Bookseller* (New York: Robert M. McBride & Co., 1938).

Pelling, Nick. *The Curse of the Voynich: The Secret History of the World's Most Mysterious Manuscript* (Surbiton, Surrey, U.K.: Compelling Press, October 2006).

Pesic, Peter. "François Viète, Father of Modern Cryptanalysis—Two New Manuscripts." *Cryptologia* 21, no. 1 (January 1997). [Viète may be the father of vowel recognition algorithms.]

Pollak, Michael. "Can't Read It? You Can Look at the Pictures." *New York Times*, September 16, 1999, p. G11. Available online at https://partners.nytimes.com/library/tech/99/09/circuits/articles/16voyn.html.

Poundstone, William. *Labyrinths of Reason* (New York: Doubleday, 1988).

Pratt, Fletcher. *Secret and Urgent* (New York: Bobbs Merrill, 1939). [Discussion of VMS on pp. 30–38.]

Prinke, Rafał T. in "Did John Dee *Really* Sell the Voynich MS to Rudolf II?" available online at http://main2.amu.edu.pl/~rafalp/WWW/HERM/VMS/dee.htm.

Publishers Weekly. "Kraus Marks Anniversary with Catalog of Treasures." 181 (June 25, 1962): 39–40.

Reddy, Sravana, and Kevin Knight. "What We Know About the Voynich Manuscript." *Proceedings of the 5th ACL–HLT Workshop on Language Technology for Cultural Heritage, Social Sciences, and Humanities*, Portland, OR (Madison, WI: Omnipress, Inc., 2011), 78–86. Available online at http://www.aclweb.org/anthology/W11-1511 and https://www.aclweb.org/anthology/W/W11/W11-15.pdf.

Reeds, James. "Entropy Calculations and Particular Methods of Cryptanalysis." *Cryptologia* 1, no. 3 (July 1977): 235–54.

Reeds, Jim. "William F. Friedman's Transcription of the Voynich Manuscript." *Cryptologia* 19, no. 1 (January 1995): 1–23.

Review of "The Cipher of Roger Bacon (Newbold)." *Quarterly Review of Biology* 3 (December 1928): 595–96.

Ricci, Seymour de. *Census of Medieval and Renaissance Mss in the United States and Canada*, 2 vols. (1937) (Kraus reprint, 1961) [VMS: vol. 2, pp. 1845–47.]

Roberts, R. J., and Andrew G. Watson, eds. *John Dee's Library Catalogue* (London: The Bibliographical Society, 1990). [Claims the folio numbers in the VMS are in John Dee's hand.]

Roitzsch, E. H. Peter. *Das Voynich-Manuskript: ein ungelöstes Rätsel der Vergangenheit* (Münster, Germany: MV-Verlag, 2008). There's also a second edition from 2010.

Rugg, Gordon. "An Elegant Hoax? A Possible Solution to the Voynich Manuscript." *Cryptologia* 28, no. 1 (January 2004): 31–46.

Rugg, Gordon. "The Mystery of the Voynich Manuscript." *Scientific American* (July 2004): 104–9.

Rugg, Gordon. *Blind Spot: Why We Fail to See the Solution Right in Front of Us* (New York: HarperCollins, 2013).

Ruysschaert, Jose. *Codices Vaticani Latini 11414–11709* (Rome: Biblioteca Apostolica Vaticana, 1959). [Describes the MSs acquired by the Vatican from the Collegium Romanum and mentions that

W. Voynich bought a number of them, which have been transferred to various American libraries, including the VMS.]

Salomon, Richard. Review of Manly's Critique of Newbold's Decipherment, *Bibliotek Warburg, Kulturwissenschaftliche Bibliographie zum Nachleben der Antike* 1 (1934): 96.

Sarton, George. Review of Manly's Critique of Newbold's Decipherment, in "Eleventh Critical Bibliography of the History and Philosophy of Science and of the History of Civilization. (To October 1921)." *Isis* 4, no. 2 (October 1921): 390–453, p. 404 relevant here.

Sarton, George. Review of Newbold's *Cipher of Roger Bacon, Isis* 11 (1928): 141–45.

Sassoon, George T. "The Application of Sukhotin's Algorithm to Certain Non-English Languages." *Cryptologia* 16, no. 2 (April 1992): 165–73.

Schaefer, Bradley E. "The Most Mysterious Astronomical Manuscript: Baffled Researchers Are Looking for Astronomical Clues to Help Decipher a Medieval Manuscript." *Sky & Telescope* 100, no. 5 (November 2000).

Schinner, Andreas. "The Voynich Manuscript: Evidence of the Hoax Hypothesis." *Cryptologia* 31, no. 2 (April 2007): 95–107.

Schuster, John. *Haunting Museums* (New York: Tom Doherty Associates, 2009).

Sebastian, Wencelas. "The Voynich Manuscript; Its History and Cipher." *Nos Cahiers,* 2 (1937): 47–69 (Montréal: Studium Franciscain).

Serafini, Luigi. *Codex Seraphinianus* (Milan, Italy: Franco Maria Ricci, 1981) (in two volumes). [There have also been later editions, at least one of which contains some new material.]

Seymour, Ian. "Thirteenth Century Magic Glass." *Astronomy Now* (June 1992): 59.

Shepherdson, Nancy. "Mystery Codes." *Boys' Life* 87, no. 11 (November 1997): 42.

Shuker, Karl P. N. *The Unexplained: An Illustrated Guide to the World's Natural and Paranormal Mysteries* (London: Carlton Books Ltd., 1996).

Smolka, J., and R. Zandbergen, eds. "Athanasius Kircher und seine erste Prager Korrespondenz." *Bohemia Jesuitica 1556–2006, Festschrift zum 450. Jahrestag des Ordens in Böhmen* (Prague, Czech Republic: V. Praze, 2010), 677–705.

Smyth, Frank. "A Script Full of Secrets." *The Unexplained* 6, no. 70 (1982): 1381–85.

Smyth, Frank. "The Uncrackable Code." *The Unexplained* 6, no. 71 (1982): 1418–20.

Smyth, Frank. "A Script Full of Secrets" and "The Uncrackable Code," reprinted in *Mysteries of Mind, Space & Time: The Unexplained* (Westport, CT: H. S. Stuttman, Inc., 1992), pp. 3062–69. [Originally published in *The Unexplained* in the U.K.]

Sowerby, E. Millicent. *Rare People and Rare Books* (London: Constable & Co. Ltd., 1967).

Stallings, Dennis, J. "Catharism, Levitov, and the Voynich Manuscript." October 10, 1998, available online at http://www.bibliotecapleyades.net/ciencia/esp_ciencia_manuscrito04.htm.

Steele, Robert. "Luru Vopo Vir Can Utriet." *Nature* 121 (February 11, 1928): 208–9. [About Bacon "gunpowder cipher," not VMS.]

Steele, Robert. "Science in Medieval Cipher." *Nature* 122 (October 13, 1928): 563–65.

Stojko, John. *Letters to God's Eye: The Voynich Manuscript for the First Time Deciphered and Translated into English* (New York: Vantage Press, 1978).

Stokley, James. "Did Roger Bacon Have a Telescope?" *Science News Letter* 14 (September 1, 1928): 125–26, 133–34.

Strong, Leonell C. "Anthony Askham, the Author of the Voynich Manuscript." *Science*, New Series 101, no. 2633 (June 15, 1945): 608–9.

Strong, Leonell C., and E. L. McCawley. "A Verification of a Hitherto Unknown Prescription of the 16th Century." *Bulletin of the History of Medicine* (Baltimore, Md.) 21 (November–December 1947): 898–904.

Sypher, F. J. *Eric Sams, Cryptography and the Voynich Manuscript*, a pamphlet (New York, 2011), available online at http://www.ericsams.org/index.php/a-portrait-of-f-j-sypher/278-eric-sams-cryptography-and-the-voynich-manuscript.

Taratuta, Evgeniya. *Our Friend Ethel Lilian Boole/Voynich* (Moscow: Izdatel'stvo, 1957, translated from the Russian by Séamus Ó Coigligh with additional notes, 2008), available online at http://www.corkcitylibraries.ie/media/SOCoiglighwebversion171.pdf.

Taylor, Frances Grandy. "The Mystery of Manuscript 408." *The Hartford Courant*, October 12, 1999, available online at http://articles.courant.com/1999-10-12/features/9910120627_1_manuscript-library-rare-documents-mystery.

Theroux, Michael. "Deciphering 'The Most Mysterious Manuscript in the World' The Final Word?" *Borderlands* 50 (1994): 36–43.

Thorndike, Lynn. "The "Bacon" Manuscript," letter to *Scientific American* 124 (June 25, 1921): 509.

Thorndike, Lynn. "Review of Newbold's Cipher of Roger Bacon." *American Historical Review* 34 (1929): 317–19.

Tiltman, John. "The Voynich Manuscript, 'The Most Mysterious Manuscript in the World.'" *NSA Technical Journal* 12 (July 1967): 41–85.

Times [Newspaper of London]. "Mr. W. M. Voynich." March 22, 1930, p. 17; March 25, 1930, p. 21; March 26, 1930, p. 18. [Obituary.]

Toresella, Sergio. "Gli erbari degli alchimisti." [Alchemical herbals.] *Arte farmaceutica e piante medicinali—erbari, vasi, strumenti e testi dalle raccolte liguri* [Pharmaceutical arts and medicinal plants—herbals, jars, instruments and texts of the Ligurian collections.] Liana Saginati, ed. (Pisa, Italy: Pacini Editore, 1996), 31–70. [Profusely illustrated. Fits the VMS into an "alchemical herbal" tradition.]

Von Schleinitz, Otto. "Die Bibliophilen W. M. Voynich." *Zeitschrift fur Bücherfreunde* 10 (1906–1907): 481–87. [Contains information about Voynich's life.]

Voynich, Wilfrid M. "A Preliminary Sketch of the History of the Roger Bacon Cipher Manuscript." *Transactions of the College of Physicians and Surgeons of Philadelphia* (1921): 415–30. Read April 20, 1921. Available online at https://archive.org/stream/s3transactionsstud43coll/s3transactionsstud43coll_djvu.txt.

"Voynich Manuscript. Botanical Clue, Evidence indicating Roger Bacon could not have written the Voynich manuscript ..." *Science News Letter* (July 29, 1944): 69.

"Voynich Manuscript Translated." *INFO Journal* no. 56 (ca. 1988). [The International Fortean Organization's *INFO Journal*.]

Way, Peter. *Codes and Ciphers* (London: Aldus, 1974).

Weekly World News [newspaper]. March 7, 2000, 4–7. [Interview with Mike Jay, author of VMS article in *Fortean Times*.]

Werner, Alfred. "The Most Mysterious Manuscript." *Horizon* 5 (January 1963): 4–9.

Westacott, Evalyn. *Roger Bacon in Life and Legend* (New York: Philosophical Library, 1953).

Wickware, Francis Sill. "The Secret Language of War." *Life* 19 (November 26, 1945): 63–70. [Only one sentence mentions the VMS, calling it "possibly the only unbreakable code," which provoked Strong to write an angry letter to the editor.]

Williams, Robert L. "A Note on the Voynich Manuscript." *Cryptologia* 23, no. 4 (October 1999): 305–9.

Wilson, Colin. *The Encyclopedia of Unsolved Mysteries* (Chicago: Contemporary Books, 1988).

"The World's Most Baffling Manuscript." *Parade Magazine*, February 21, 1982.

Wrixon, Fred B. *Codes, Ciphers, and Other Cryptic and Clandestine Communication* (New York: Black Dog and Leventhal, 1998).

Zandonella, Catherine. "Book of riddles." *New Scientist* 172, no. 2317 (November 17, 2001): 36–39.

Zimansky, Curt A. "William F. Friedman and the Voynich Manuscript." *Philological Quarterly* 49 (1970): 433–42.

网 站

Anonymous. http://hurontaria.baf.cz/CVM/a0.htm. [Many articles on the Voynich manuscript are linked from this page.]

Landini, G., and René Zandbergen. The European Voynich Manuscript Transcription Project Home Page, http://web.bham.ac.uk/G.Landini/evmt/evmt.htm.

Prinke, Rafał T. Some facts, thoughts and speculations related to the Voynich manuscript, http://main2.amu.edu.pl/~rafalp/WWW/HERM/VMS/vms.htm.

Voynich Manuscript Mailing List HQ, www.voynich.net. [This is the website for the mailing list begun by Jim Gillogly and Jim Reeds in 1991.]

Yale University. Beinecke Rare Book and Manuscript Library, General Collection of Rare Books and Manuscripts, Medieval and Renaissance Manuscripts, http://brbl-net.library.yale.edu/pre1600ms/docs/pre1600.ms408.htm. [This is Yale's website for the Voynich manuscript.]

Zandbergen, René. The Voynich Manuscript, www.voynich.nu.

第 2 章

关于古埃及

Black, Jeremy, Graham Cunningham, Eleanor Robson, and Gábor Zólyomi. *The Literature of Ancient Sumer* (Oxford, U.K.: Oxford University Press, 2004). The introduction by Robson is relevant here.

Bohleke, Briant. "Amenemopet Panehsi, Direct Successor of the Chief Treasurer Maya." *Journal of the American Research Center in Egypt* 39 (2002): 157–72.

Bosticco, S. "Scarabei egiziani della Necropoli die Pithecusa nell'Isola di Ischia." *La Parola del Passato* 54 (1957): 215–29.

Charles, Robert P. "Les scarabées Égyptiens et Égyptisants de Pyrga district de Larnaca (Chypre)." *ASAE* 58 (1964): 3–36.

Charles, Robert P. "Remarques sur une maxime religieuse à propos d'un scarabée égyptien à Kyrenia (Chypre)." In *Melanges de Carthage offerts à Charles Saumagne* (Louis Poinssot, Maurice Pinard, Paris, Libraire Orientaliste, Paul Geuthner 1964–1965), 11–20.

Darnell, John Coleman. *The Enigmatic Netherworld Books of the Solar-Osirian Unity, Cryptographic Compositions in the Tombs of Tutankhamun, Ramesses VI and Ramesses IX* (Fribourg, Germany:

Academic Press and Göttingen, Germany: Vandenhoek & Ruprecht, 2004). This book is a slight reworking of Darnell's doctoral dissertation, which is available through University Microfilms International.

Deveria, T. "L'écriture secrète dans les textes hiéroglyphiques des anciens Égyptiens." *Bibliothèque Égyptologique* 5 (1897): 49–90.

Drioton, Étienne. "Essai sur la cryptographie privée de la fin de la XVIIIe dynastie." *Revue d'Égyptologie* I (1933): 1–50.

Drioton, Étienne. "Une figuration cryptographique sur une stèle du Moyen Empire." *Revue d'Égyptologie* I (1933): 203–29.

Drioton, Étienne. "La cryptographie égyptienne." *Revue Lorraine d'Anthropologie* VI (1933–1934): 5–28.

Drioton, Étienne. "La cryptographie égyptienne." *Chronique d'Egypte* IX, (1934): 192–206.

Drioton, Étienne. "Les jeux d'écriture et les rébus de l'Egypte antique." *Rayon d'Egypte* VIII (1935): 173–75.

Drioton, Étienne. "Notes sur le cryptogramme de Montouemhêt." *Université libre—de Bruxelles, Annuaire de l'Institut de philologie et d'histoire orientales* III. Volume offert à Jean Capart, Bruxelles, 1935, 133–40.

Drioton, Étienne. "Les protocoles ornementaux d'Abydos." *Revue d'Égyptologie* II (1936): 1–20.

Drioton, Étienne. "Le cryptogramme de Montou de Médamoud." *Revue d'Égyptologie* II (1936): 22–33.

Drioton, Étienne. "Un rébus de l'Ancien Empire." *Mémoires de l'Institut français d'Archéologie orientale du Caire* LXVI (Cairo, Egypt: Mélanges Gaston Maspero, I, 1935–1938): 697–704.

Drioton, Étienne. "Note sur un cryptogramme récemment découvert à Athribis." *Les Annales du Service des Antiquités de l'Egypte* XXXVIII (1938): 109–16.

Drioton, Étienne. "Deux cryptogrammes de Senenmout." *Les Annales du Service des Antiquités de l'Egypte* XXXVIII (1938): 231–46.

Drioton, Étienne. "Senenmout cryptographe." *Atti del XIX Congresso internazionale degli Orientalisti* September 23–29, 1935, XIII (Rome, Tipografia del Senato, G. Bardi, 1938), 132–38.

Drioton, Étienne. "Cryptogrammes de la reine Nefertari." *Les Annales du Service des Antiquités de l'Egypte* XXXIX (1939): 133–44.

Drioton, Étienne. "Recueil de cryptographie monumentale." *Les Annales du Service des Antiquités de l'Egypte* XL (1940): 305–427.

Drioton, Étienne. *Recueil de cryptographie monumentale* (Cairo, Egypt: L'institut français d'archéologie orientale, 1940). This book is available online at http://www.cfeetk.cnrs.fr/fichiers/Documents/Ressources-PDF/documents/K1234-DRIOTON.pdf.

Drioton, Étienne. "L'écriture énigmatique du *Livre du Jour et de la Nuit*." A. Piankoff, *Le Livre du Jour et de La Nuit* (Cairo, Egypt: L'institut français d'archéologie orientale, 1942), 83–121.

Drioton, Étienne. "La cryptographie du Papyrus Sait 825." *Les Annales du Service des Antiquités de l'Egypte* XLI (1942): 199–234.

Drioton, Étienne. "A propos du cryptogramme de Montouemhêt." *Les Annales du Service des Antiquités de l'Egypte* XLII (1943): 177–81.

Drioton, Étienne. "Procédé acrophonique ou principe consonantal?" *Les Annales du Service des Antiquités de l'Egypte* XLIII (1943): 319–49.

Drioton, Étienne. "La cryptographie par perturbation." *Les Annales du Service des Antiquités de l'Egypte* XLVI (1944): 17–33.

Drioton, Étienne. "Notes diverses. 9. Le cynocéphale et l'écriture du nom de Thot. — 10. Chawabtiou à inscriptions cryptographiques." *Les Annales du Service des Antiquités de l'Egypte* XLV (1945): 17–29.

Drioton, Étienne. "Plaques bilingues de Ptolémée IV." *Discovery of the Famous Temple and Enclosure of Serapis at Alexandria*, Alan Rowe, *Supplément aux Annales du Service des Antiquités de l'Egypte* no. 2, (1946), 97–112.

Drioton, Étienne. "La cryptographie de la chapelle de Toutânkhamon." *The Journal of Egyptian Archaeology* XXXV (1949): 117–22.

Drioton, Étienne. "Inscription énigmatique du tombeau de Chéchanq III à Tanis." *Kêmi* XII (1952): 24–33.

Drioton, Étienne. "Les principes de la cryptographie égyptienne." *Comptes rendus des séances de l'Académie des Inscriptions et Belles-Lettres* 97e année, no. 3 (1953): 355–64, available online at http://www.persee.fr/web/revues/home/prescript/article/crai_0065-0536_1953_num_97_3_10159.

Fairman, H. W. "Notes on the Alphabetic Signs Employed in the Hieroglyphic Inscriptions in the Temple of Edfu." *ASAE* 43 (1943): 191–310.

Fairman, H. W. "An Introduction to the Study of Ptolemaic Signs and their Values." *BIFAO* 43 (1945): 52–138.

Graves-Brown, Carolyn. "A Mummified Baby?" Egypt Centre, Swansea, May 1, 2014, available online at http://egyptcentre.blogspot.co.uk/2014/05/a-mummified-baby.html. This is the most scholarly account available. Graves-Brown is the Egypt Centre curator.

Halévy, Joseph. "Observations critiques sur les prétendus Touraniens de la Babylonie." *Journal Asiatique* 3 (1874): 461–536. This puts forth the incorrect argument that Sumerian was a form of secret writing used by priests.

Hornung, E., and E. Staehelin. *Skarabäen und andere Siegelamulette aus Basler Sammlungen* (Mainz, Germany: P. von Zabern, 1976), 173–80. [For information on cryptographic scarabs and seals.]

Hoving, Thomas. *False Impression: The Hunt for Big-Time Art Fakes* (New York: Simon & Schuster, 1996).

Junker, H. *Über das Schriftsystem im Tempel der Hathor in Dendera* (Berlin: August Schaefer, 1903).

Jurman, Claus. "Ein Siegelring mit kryptographischer Inschrift in Bonn." *Ägypten und Levante XX/Egypt and the Levant* 20 (2010): 227–42.

Kahn, David. *The Codebreakers* (New York: Macmillan, 1967). This classic book on the history of cryptology includes not only a couple of pages on ancient Egyptian cryptology, but also an accessible account of how Egyptian hieroglyphs came to be deciphered using techniques similar to those of the cryptanalysts.

Leibevitch, J. "Un écho posthume du Chanoine Étienne Drioton." *BSFÉ* 36 (1963): 34–36.

Livius. "Mummified Fetus Found in Egyptian Sarcophagus." The History Blog, http://www.thehistoryblog.com/archives/30530.

Lorenzi, Rossella. "Mummified Fetus Found in Tiny Ancient Egyptian Sarcophagus." May 13, 2014, http://news.discovery.com/history/archaeology/mummified-fetus-found-in-tiny-ancient-egyptian-sarcophagus-140512-140512.htm.

Morenz, Ludwig D. "Tomb Inscriptions: The Case of the I Versus Autobiography in Ancient Egypt." *Human Affairs* 13 (2003): 179–96. On pp. 190–91, Morenz wrote

A stela from Naga ed Deir provides an interesting example of illiterate imitation of writing (Un of Cal N 3993). The execution of the monument is rather crude. Tile proportions of the sitting man are not well done. As in high culture the picture is provided with a framing line. On top of it we see a line with signs imitating hieroglyphs. They are probably inspired by the offering-formula. On this stelae writing is imitated. Unfortunately we know nothing more about this man. He remains even nameless. On the other hand he becomes a representative of a social group we know next to nothing about except ceramics. It is quite important to see that this stelae imitates the monuments in the style of high culture. It suggests a broader acceptance and imitation of this code of the elite. From Gebelein we have no such examples, but Un of Cal. (6-19911) comes fairly close. The layout of the inscription also seems rather confused but still is intelligible.

Morenz, Ludwig D. *Sinn und Spiel der Zeichen: Visuelle Poesie im Alten Ägypten* (Cologne, Germany: Böhlau Verlag, 2008).

Petrie, W. M. Flinders. *Illahun Kahun and Gurob. 1889–90* (London: David Nutt, 1891), 27–28, available online at http://www.lib.uchicago.edu/cgi-bin/eos/eos_title.pl?callnum=DT73.I3P5_cop2 and https://archive.org/details/cu31924086199514.

Petrie, W. M. Flinders. *Ten Years' Digging in Egypt 1881–1891*, 2nd ed., rev. (London: The Religious Tract Society, 1893), 124–25.

Pinches, T. "Sumerian or Cryptography." *Journal of the Royal Asiatic Society* 32, no. 1 (January 1900): 75–96. The idea of Sumerian as a form of secret writing only used by priests, and never spoken, is finally rejected definitively.

Russon, Mary-Ann. "Tiny Egyptian Mummy Confirmed by CT Scan to be Baby's Remains." *International Business Times*, May 8, 2014, available online at http://www.ibtimes.co.uk/tiny-egyptian-mummy-confirmed-by-ct-scan-be-babys-remains-1447717.

Sauneron, S. "Le papyrus magique illustré de Brooklyn [Brooklyn Museum 47.218.156]." *Wilbur Monographs* 3 (New York: The Brooklyn Museum, 1970).

Schulman, Alan R. "The Ossimo Scarab Reconsidered." *Journal of the American Research Center in Egypt* 12 (1975): 15–18.

Schulman, Alan Richard. "Two scarab impressions from Tel Michal." *Tel Aviv* 5, no. 3–4 (1978): 148–51. In the acknowledgments section of this paper, Schulman wrote, "I should like to thank Dr. Z. Herzog for reminding me that I had looked at these items 8 years ago and for allowing me to revise what I had written then, since it in retrospect, was nonsense." I like people who can admit this sort of thing!

Woollaston, Victoria. "The mummified FOETUS: Scans reveal tiny ancient Egyptian sarcophagus contains the remains of a 16-week-old embryo," *Daily Mail*, May 9, 2014, available online at http://www.dailymail.co.uk/sciencetech/article-2624136/The-mummified-FOETUS-Scans-reveal-tiny-ancient-Egyptian-sarcophagus-contains-remains-16-week-old-baby.html.

关于古希腊

Bakker, Egbert J. *A companion to the Ancient Greek Language* (Chichester, U.K.: Wiley-Blackwell, 2010).

Boardman, John. " 'Reading' Greek Vases?" *Oxford Journal of Archaeology* 22, no. 1 (2003): 109–14. This paper argues that the inscriptions on the vases were meant to be read aloud. But how to read gibberish? Boardman's explanation is this:

As for the many nonsense inscriptions on vases; were these simply an invitation to improvisation on the part of the owner displaying them, a form of Hellenic *karaoke*? All seem to be either decorative or making a false pretense to literacy. They are part of the look of the vase, nothing to do with reading.

Clark, A. J., Maya Elston, and Mary Louise Hart. *Understanding Greek Vases: A Guide to Terms, Styles, and Techniques* (Los Angeles: Getty Publications, 2002).

Clifford, Kathleen Elizabeth. "Lingering Words: A Study of Ancient Greek Inscriptions on Attic Vases." (master's thesis, Florida State University Electronic Theses, Treatises and Dissertations, Paper 3580, 2007). This thesis is a very nice (gentle) introduction to the topic of inscriptions on Attic vases. It can be read and enjoyed by a layperson. I suggest that any nonspecialist wishing to delve deeper begin here, even though the possibility of encryption is not addressed.

Cook, R. M. *Greek Painted Pottery* (London: Routledge, 1997).

Gardthausen, Viktor Emil. *Griechische Palaeographie* (Leipzig, Germany: Veit, 1913).

Grammenos, Dimitris. "Abba-Dabba-Ooga-Booga-Hoojee-Goojee-Yabba-Dabba-Doo: Stupidity, Ignorance & Nonsense as Tools for Nurturing Creative Thinking." *Proceedings CHI EA '14, CHI '14, Extended Abstracts on Human Factors in Computing Systems* (New York: ACM, 2014), 695–706, available online at https://www.ics.forth.gr/_publications/dgrammenos_sin_FINAL.pdf. This paper includes an image of the Memnon pieta as an example of a nonsense inscription. The author wrote:

In Ancient Greek pottery inscriptions are sometimes seemingly meaningless combinations of letters. Up to now, the prevailing explanation was that these were made by illiterate vase-painters either to imitate the decorative effect of literate inscriptions or, to give the impression that they were literate. But, "nonsense inscriptions" often coexist with others that do make sense. Furthermore, there are too many of them (about 1/3 of vases in the Corpus of Attic Vase Inscriptions). Recently, Mayor et al. came to a groundbreaking conclusion. There is evidence that (at least some of them) constitute names and words of "barbarian" tongues transliterated into Greek. Sometimes experts label what they cannot understand as nonsense, while in reality the distance between nonsense and sense is just a matter of standpoint. After all, our everyday lives are full of nonsense and there is so much of it that we rarely even notice. For example, in June 2010, almost half of the earth's population spent at least one minute watching 22 grown-up guys kicking around an inflated piece of plastic (i.e., FIFA World Cup South Africa). In 2006, "no. 5, 1948" a painting by Jackson Pollock showing colored paint drizzles was sold for $140m, the highest sum ever been paid for a painting. And so on.

Hunt, A. S. "A Greek Cryptogram." *Proceedings of the British Academy* 15 (1929): 1–10. This easy-to-read paper presents a Greek ciphertext in which the letters were turned half over or modified in other small ways to disguise the writing. No knowledge of Greek is needed to follow this ten-page paper, as an English translation of the ciphertext is provided. Unfortunately, an approximate date for the ciphertext examined (Michigan cryptographic papyrus) is not given.

Hyman, Malcolm D. "Of Glyphs and Glottography." *Language & Communication* 26 (2006): 231–49.

Immerwahr, Henry R. *Attic Script: A Survey*, Oxford Monographs on Classical Archaeology (Oxford, U.K.: Clarendon Press, 1990).

Immerwahr, Henry R. "Observations on Writing Practices in the Athenian Cera-micus." *Studies in Greek Epigraphy and History in Honor of Stephen V. Tracy* (Ausonius Éditions Études 26), eds. G. Reger, F. X. Ryan, and T. F. Winters (Pessac, France: Ausonius Éditions, 2010), 107–22.

Immerwahr, Henry R. Corpus of Attic Vase Inscriptions, http://www2.lib.unc.edu/dc/attic/about.html.

This website contains the following quotes:

> I should add at this point that we are dealing with a total of over 100,000 known Attic vases (a rough guess), of which (again roughly) one in ten has some inscription. There are certainly over 10,000 inscribed Attic vases. Rudolf Wachter, who is working on a project based on this corpus, conjectures that the number is about 4,000; my guess is that he did not include nonsense inscriptions and perhaps not graffiti made by users (as against inscriptions written in the workshop). Four thousand is probably about right for meaningful decorative inscriptions. In fact, Sir John Beazley, in 1947, advised me not to undertake this collection because I would not finish it. He was right, in the sense that my collection is not complete, although with well over 8,000 items it is extensive.
>
> I made the decision that the so-called nonsense inscriptions (of which there are thousands) had to be included, for the corpus was to be of service not only to linguists but also to archaeologists, and the nonsense inscriptions are a part of the ornamentation of the vase, which varies from workshop to workshop.

Note his reluctance to include them—it sounds like he'd really rather not.

> The next form this project takes will be Wachter's *AVI* (note that it is no longer called a corpus), and it will also be in electronic form which will make it possible to update it continuously. This is especially important for vases, for new material is constantly produced by excavation and tomb robbery.

"Inscriptions," University of Oxford, Classical Art Research Centre and the Beazley Archive, http://www.beazley.ox.ac.uk/tools/pottery/inscriptions/.

Leighton, Albert C. "Secret Communication among the Greeks and Romans." *Technology and Culture* 10, no. 2 (April 1969): 139–54.

Lorber, Fritz. *Inschriften auf Korinthischen Vasen. Archaologisch-Epigraphische Untersuchungen zur Korinthischen Vasenmalerei im 7 end 6 JH. V. CHR.*, Deutsches Archäologisches Institut, Archäologische Forschungen 6 (Berlin: Gebr. Mann Verlag, 1979). This catalog includes nonsense inscriptions.

Mayor, Adrienne, John Colarusso, and David Saunders. "Making Sense of Nonsense Inscriptions Associated with Amazons and Scythians on Athenian Vases," Version 2.0, Princeton/Stanford Working Papers in Classics, July 2012.

Mayor, Adrienne, John Colarusso, and David Saunders. "Making Sense of Nonsense Inscriptions Associated with Amazons and Scythians on Athenian Vases." *Hesperia: The Journal of the American School of Classical Studies at Athens* 83, no. 3 (July–September 2014): 447–93.

Nock, A. D. "A Greek Cryptogram by Arthur S. Hunt." *The Classical Review* 43, no. 6 (December 1929): 238.

Pappas, A. "More than Meets the Eye: The Aesthetics of (Non)sense in the Ancient Greek Symposium." *Aesthetic Value in Classical Antiquity* (*Mnemosyne* Suppl. 305), eds. I. Sluiter and R. M. Rosen (Leiden, Netherlands: Brill, 2012), 71–111. Note: The Symposia were celebratory drinking parties for men. The author gives an odd explanation for the purpose of nonsense inscriptions.

Plutarch's Life of Lysander, sec. 19, available online at http://www.gutenberg.org/files/14114/14114-h/14114-h.htm#LIFE_OF_LYSANDER.

Polybius. *Histories*, Book 10 is relevant here; see sec. 45–47. Available online at http://www.gutenberg.org/files/44126/44126-h/44126-h.htm.

Sironen, Erkki. "Edict of Diocletian and a Theodosian Regulation at Corinth." *Hesperia: The Journal of the American School of Classical Studies at Athens* 61, no. 2 (April–June 1992): 223–26.

Steiner, Ann. *Reading Greek Vases* (Cambridge, U.K.: Cambridge University Press, 2007). Steiner argues that some nonsense inscriptions could be ridiculing the accents of foreigners.

Suess, Wilhelm. "Ueber antike Geheimschreibemethoden und ihr Nachleben." *Philologus* LXXVIII (June 1922): 142–75.

Sulzer, A. B. "Making Sense of Nonsense Inscriptions: Orality, Literacy and the Greek Dipinti on Four Vases in the Arthur M. Sackler Museum" (senior thesis, Harvard University, 2003).

Thomsen, Megan Lynne. "Herakles Iconography on Tyrrhenian Amphorae" (master's thesis, University of Missouri–Columbia, 2005).

Vanderpool, E. "An Unusual Black-Figured Cup." *American Journal of Archaeology* 49, no. 4 (October 1945): 436–40.

Vergano, Dan. "Amazon Warriors' Names Revealed Amid 'Gibberish' on Ancient Greek Vases." *National Geographic*, September 22, 2014, available online at http://news.nationalgeographic.com/news/2014/09/140923-amazon-greek-vase-translations-science/. This brief article presents the work of Mayor et al. in a very entertaining way.

关于维京人

I've included some references that deal with controversial decipherments. I'm simply providing the titles; it's up to you to decide whether or not the claims are correct.

Barnes, Michael P. *Runes: A Handbook* (Suffolk, U.K. and Rochester, NY: Boydell Press, 2012). Despite having a chapter devoted to "Cryptic inscriptions and cryptic runes," Barnes is dismissive of the so-called nonsense inscriptions.

Braunmüller, Kurt. *Der Maltstein: Versuch einer Deutung* (Berlin: Walter de Gruyter, 1992).

Eliasson, Stig. " 'The letters make no sense at all . . .': språklig struktur i en 'obegriplig' dansk runinskrift?" *Nya perspektiv inom nordisk språkhistoria. Föredrag hållna vid ett symposium i Uppsala 20–22 januari 2006*, ed. Lennart Elmevik, Acta Academiae Regiae Gustavi Adolphi, 97, 45-80 (Uppsala, Sweden: Kungl. Gustav Adolfs Akademien för svensk folkkultur, 2007). This paper describes the Sørup runestone, Rundata ID DR 187, in Denmark, which Eliasson interprets as possibly representing Basque in his other paper cited here.

Eliasson, Stig. "When 'chance' explanations turn tenuous: Basque verbal morphology, argument marking, and an allegedly nonsensical Danish runic inscription," available online at http://www.orientalistik.uni-mainz.de/robbeets/verbalmorphv8/_Media/abs-eliasson2.pdf.

Franksen, Ole. *Mr. Babbage's Secret: The Tale of a Cypher and APL* (Englewood Cliffs, NJ: Prentice-Hall, 1985).

Gordon, Cyrus, and Roy Bongartz. "Stone Inscription Found in Tennessee Proves that America was Discovered 1500 Years before Columbus." *Argosy* (January 1971): 23–27. Gordon was a World War II cryptanalyst.

Landsverk, O. G. *Ancient Norse Messages on American Stones* (Glendale, CA: Norseman Press, 1969).

Landsverk, O. G. "Cryptography in runic inscriptions." *Cryptologia* 8, no. 4 (1984): 302–19.

Ljosland, Ragnhild. "Pondering Orkney's runic inscriptions." *The Orcadian*, June 4, 2015, p. 21, available online at http://www.uhi.ac.uk/en/research-enterprise/cultural/centre-for-nordic-studies/mimirs-well-articles/pondering-orkneys-runic-inscriptions. Ljosland wrote:

> New light was also shed on the metal pendant found by James Barrett's excavation team on the Brough of Deerness. The runes on this pendant look very strange, a bit like Christmas trees with wild-growing branches. Could they be in the tree-like cipher code, like some of the Maeshowe inscriptions, which are also found in Anglo-Saxon manuscripts (as shown by Aya Van Renterghem at the conference)? Yet, interpreting the Deerness runes as such didn't bring up any sensible text.
>
> But resulting from a closer study of the Deerness pendant, Sonia Pereswetoff-Morath discovered that the runes are actually complex ligatures made from combinations of ordinary runes. She has not yet interpreted the whole text, but was able to give us a few words, involving talk of a "large payment" or "big secret", and a command to "go away". It was great to hear that the runes are actually meaningful, and not just nonsense-runes as we first thought. It's also exciting to have another runic object whose function seems to have been as an amulet, in addition to the bear's or seal's tooth amulet from the Brough of Birsay, which was also discussed at the conference. Birsay itself was identified in Jan Ragnar Hagland's talk as a centre of runic literacy, along with Orphir.

Meijer, Jan. "Corrections in Viking Age Rune-Stone Inscriptions." *Arkiv för nordisk filologi* 110 (January 1995): 77–83, available online at http://journals.lub.lu.se/ojs/index.php/anf/article/viewFile/11543/10639.

Mongé, A., and O. G. Landsverk. *Norse Medieval Cryptography in Runic Carvings* (Glendale, CA: Norseman Press, 1967). Mongé's skills as a cryptanalyst are discussed briefly in chapter 7.

Page, Raymond Ian. *Runes* (*Reading the Past*) (Berkeley, CA, and Los Angeles: University of California Press/British Museum, 1987). Examples of gibberish appear on p. 29 (the Lindholm "amulet," listed as DR 261 in Rundata) and p. 58 (the Hunterston brooch).

Stevenson, Robert B. K. "The Hunterston Brooch and Its Significance." *Medieval Archaeology* 18 (1974): 16–42. The odd inscription is dismissed in this paper.

Syversen, Earl. *Norse Runic Inscriptions with Their Long-Forgotten Cryptography* (Sebastopol, CA: Vine Hill Press, 1979). This book has the lovely dedication "In memory of my true friend ALF MONGE who solved the enigma of Medieval Norse Runic Cryptography."

Thompson, Claiborne W. "Nonsense Inscriptions in Swedish Uppland." *Studies for Einar Haugen*, ed. E. S. Firchow et al. (The Hague, Netherlands: Mouton, 1972), 522–34.

Whittaker, Helène. "Social and Symbolic Aspects of Minoan Writing." *European Journal of Archaeology* 8, no. 1 (2005): 29–41. This wide-ranging paper talks about the Phaistos disc and the Arkalochori axe, among other challenges, but also includes Scandinavian bracteates bearing unintelligible runic inscriptions. Whittaker concludes, "It seems obvious that these bracteates were made by illiterate goldsmiths for customers who were themselves illiterate and unable to check the genuineness of the writing." I have no idea why this is supposed to seem obvious.

第3章

Adams, Byron, ed. *Edward Elgar and His World*, Bard Music Festival (Princeton, NJ: Princeton University Press, 2007).

All About Elgar, A comprehensive guide to the man, his music and the organisations which support him, available online at http://www.elgar.org/welcome.htm.

Anderson, Martin. Code-breaker, civil servant, musicologist and Shakespeare scholar, available online at http://www.ericsams.org/index.php/eric-sams/100-code-breaker-civil-servant-musicologist-and-shakespeare-scholar.

Anderson, Robert. *Elgar*, The Master Musicians (New York: Schirmer Books, 1993).

Buckley, Robert J. *Sir Edward Elgar* (London: Ballantyne, Hanson & Co., 1905).

Centro Studi Eric Sams—the Eric Sams Archive, http://www.ericsams.org/.

Centro Studi Eric Sams, Eric Sams: on Cryptography, Solutions to ciphers, esssays, articles, reviews, http://www.ericsams.org/index.php/on-cryptography has seventeen pieces on crypto by Eric Sams.

Elgar, Edward. *My Friends Pictured Within* (London: Novello & Co Ltd., 1946).

Elgar, Edward, and Jerrold Northrup Moore. *Edward Elgar: Letters of a Lifetime* (Oxford, U.K.: Oxford University Press, 1990).

Fardon, Michael. *"Dear Carice . . ." Postcards from Edward Elgar to His Daughter* (Worcester, U.K.: Osbourne Books, 1997).

Fiske, Roger. "The Enigma: A Solution." *Musical Times* 110, no. 1521 (November 1969): 1124–26.

Grimley, Daniel M., and Julian Rushton, eds. *The Cambridge Companion to Elgar*, Cambridge Companions to Music (Cambridge, U.K.: Cambridge University Press, 2004).

Jones, Kevin. "The Puzzling Mr. Elgar." *New Scientist* 184, no. 2479 (December 25, 2004): 56.

Kennedy, Michael. *Portrait of Elgar*, 3rd ed. (Oxford, U.K.: Clarendon Press, 1987), Clarendon paperback edition 1993.

Kennedy, Michael. *The Life of Elgar*, Musical Lives (New York: Cambridge University Press, 2004).

Kent, Christopher. *Edward Elgar: A Thematic Catalogue and Research Guide,* Routledge Music Bibliographies, 2nd ed. (New York: Routledge, 2013).

Kenyon, Nicholas, introduction. *Elgar: An Anniversary Portrait* (London: Continuum, 2007).

Kolodin, Irving. "What is Enigma?" *Saturday Review*, February 2, 1953, pp. 53, 55, and 71.

Kruh, Louis. "Still Waiting to be Solved: Elgar's 1897 Cipher Message." *Cryptologia* 22, no. 2 (April 1998): 97–98.

Macnamara, Mark. "The Artist of the Unbreakable Code, Composer Edward Elgar still has cryptographers playing his tune." *Nautilus* no. 6, October 17, 2013, http://nautil.us/issue/6/secret-codes/the-artist-of-the-unbreakable-code.

McVeagh, Diana. *Elgar the Music Maker* (Woodbridge, U.K.: Boydell Press, 2007).

Messenger, Michael. *Edward Elgar: An Illustrated Life of Sir Edward Elgar (1857–1934)* (Princes Risborough, U.K.: Shire Library, 2005).

Moore, Jerrold Northrup. *Edward Elgar: A Creative Life* (Oxford, U.K.: Oxford University Press, 1984).

Palmer, Jean. *The Agony Column Codes & Ciphers* (Bedfordshire, U.K.: Bright Pen, 2005).

Parrott, Ian. "Music and Cipher." *The Musical Times* 109, no. 1508 (October 1968): 920–21.

Parrott, Ian. *Elgar* (London: J. M. Dent and Sons Ltd., and New York: Farrar, Straus and Giroux, Inc., 1971).

Powell, Mrs. Richard. *Edward Elgar: Memories of a Variation* (London: Oxford University Press, 1937).

Powell, Mrs. Richard. *Edward Elgar: Memories of a Variation*, 4th ed., rev. and ed. Claud Powell, with an addendum by Jerrold Northrop Moore (Aldershot, Hants, U.K.: Scolar Press, 1994).

Rushton, Julian. *Elgar: Enigma Variations*, Cambridge Music Handbooks (Cambridge, U.K.: Cambridge University Press, 1999).

Sams, Eric. "Elgar's Cipher Letter to Dorabella." *The Musical Times* 111, no. 1524 (February 1970): 151–54.

Sams, Eric. "Musical Cryptography." *Cryptologia* 3, no. 4 (October 1979): 193–201.

Sams, Eric. "Cracking the Historical Codes." *Times Literary Supplement*, February 8, 1980, p. 154.

Sams, Eric. "Elgar and Cryptology," letter to the editor, *The Musical Times* 125, no. 1695 (May 1984): 251.

Sams, Eric. "Cryptanalysis and Historical Research." *Archivaria* 21 (Winter 1985–1986).

Sams, Eric. "Elgar's Enigmas." *Music & Letters* 78, no. 3 (August 1997): 410–15.

Santa, Charles Richard, and Matthew Santa. "Solving Elgar's Enigma." *Current Musicology* 89 (2010).

Schooling, John Holt. "Secrets in Cipher IV. From the Time of George II to the Present Day." *The Pall Mall Magazine* 8, no. 36 (April 1896): 608–18. This article contains the cipher that Elgar broke on p. 618.

Schridde, Christian. "The Dorabella Cipher (Part 3)," July 31, 2013, available online at http://numberworld.blogspot.co.uk/2013/07/the-dorabella-cipher-part-3.html.

Sypher, F. J. "Eric Sams, Cryptography and the Voynich Manuscript," a pamphlet, New York, 2011, available online at http://www.ericsams.org/index.php/a-portrait-of-f-j-sypher/278-eric-sams-cryptography-and-the-voynich-manuscript.

Tatlow, Ruth. *Bach and the Riddle of the Number Alphabet* (Cambridge, U.K.: Cambridge University Press, 1991). This volume investigates claims of Bach's use of cryptography.

Tibbetts, John C. "John C. Tibbetts Interviews Eric Sams," available online at http://www.ericsams.org/index.php/interview-by-john-c-tibbetts.

Trowell, B. "Elgar's Marginalia." *The Musical Times* 125, no. 1693 (March 1984): 139–41, 143. Trowell looked at the notebook pages where Elgar played with a cipher and noted

> The presence of the sentence 'DO YOU GO TO LONDON TOMORROW', which contains no E (the commonest letter in English by far), with the figure '23' (for the total of letters) and the remark '9 Os', strongly suggests that the Dorabella message has been deliberately constructed in order to defeat the solver's normal resource of analyzing the frequency-count of the letters of the alphabet in English.

Eric Sams rejected this notion, saying that it would still have been solved quickly. See Sams, Eric. "Elgar and Cryptology," letter to the editor, *The Musical Times* 125, no. 1695 (May 1984): 251.

Young, Percy M. *Elgar O. M. A Study of a Musician* (London: Collins, 1955).

第4章

America's Most Wanted. December 1998. Also in the companion magazine, *Manhunter.*

Anonymous. "Are They Closing in on Zodiac?" *Detective Cases* (April 1974).

Butterfield, Michael. *Zodiac Death Machine: The Untold Story of the Unsolved Zodiac*, to appear. For now, there's the website http://zodiackillerfacts.com/.

Covino, Joseph, Jr. *San Francisco's Finest: Gunning for the Zodiac* (Walnut Creek, CA: Epic Press, 2012).

Crowley, Kieran. *Sleep My Little Dead: The True Story of the Zodiac Killer* (New York: St. Martin's Paperbacks, 1997). This is about a copycat killer in New York, not the original Zodiac. The author signed my copy "To Craig—What's Your Sign?" It gave me a chill—thanks!

Dell, Jessica. The Zodiac Killer Annotated Bibliography, http://jsscdell7.wordpress.com/2011/04/22/the-zodiac-killer-annotated-bibliography-project-3/ (mostly websites).

Doss, Diane. *NEXUS*, date unknown, but not long before August 20, 1985. Doss found Penn's evidence "voluminous and overwhelming."

Fraley, Craig. *Zodiac Killer Final Thoughts*, CreateSpace Independent Publishing Platform, 2014.

Francis, Carmen. "Zodiac Casts a Strangler's Shadow." *Startling Detective* (March 1970): 18–21, 62–64.

Graysmith, Robert. Letter to Waltz, May 25, 1979, National Cryptologic Museum files.

Graysmith, Robert. *Zodiac* (New York: St. Martin's/Marek, 1986). This is the best book on the Zodiac case. It's creepy and was made into a movie that's also creepy.

Graysmith, Robert. *Zodiac Unmasked: The Identity of America's Most Elusive Serial Killer Revealed* (New York: Berkley Books, 2002).

Haugen, Brenda. *The Zodiac Killer: Terror and Mystery* (Mankato, MN: Capstone Point Books, 2010)—True Crime—96 pages. This is a young adult book.

Hodel, Steve, with Ralph Pezzullo. *Most Evil: Avenger, Zodiac, and the Further Serial Murders of Dr. George Hill Hodel* (New York: Dutton, 2009).

Holt, Tim. "The Men Who Stalk the Zodiac Killer." *San Francisco Magazine*, April 1974.

Jordan, John Robert. Hunter among the Stars: A Critical Look at the Zodiac Killer as Serial Killer, Occultist, and Speller (CreateSpace Independent Publishing Platform, 2011).

Kelleher, Michael D., and David Van Nuys. *"This Is the Zodiac Speaking": Into the Mind of a Serial Killer* (Westport, CT: Praeger Publishers, 2002).

Lafferty, Lyndon E. *The Zodiac Killer Cover-Up: AKA The Silenced Badge* (Vallejo, CA: MANDAMUS, 2012).

Lowall, Gene. "Zodiac California's Blood-Thirsty Phantom." *Argosy*, September 1970.

Montgomery, John. "Your Daughter May Be Next." *Inside Detective*, January 1969.

O'Hare, Michael. "Confessions of a Non-Serial Killer. Conspiracy theories are all fun and games until you become the subject of one." *Washington Monthly*, May/June 2009, available online at http://www.washingtonmonthly.com/features/2009/0905.ohare.html.

Oranchak, David. http://www.zodiackillerciphers.com/ (my favorite Zodiac website).

Oswell, Douglas. *The Unabomber and the Zodiac* (Douglas Evander Oswell, 2007).

Penn, Gareth. "The Calculus of Evil." *MENSA Bulletin*, no. 288 (July/August 1985): 5–8, 25. At the end of this piece, it was noted that "A different form of this story appeared serially in *The Ecphorizer*, then edited by Richard Amyx." *MENSA Bulletin* chose not to include references that specifically identified Penn's suspect, although Penn had already gone public with his accusation, according to the editor's note. Penn's address was provided for readers who wanted more information. In response to such a query, the writer would not receive a name, but rather enough information so that the suspect's identity could be found with the help of a library with a "half-decent reference collection", according to Gareth Penn, from a response to queries, August 20, 1985. As I stated in the main text of this chapter, I think Penn's suspect is innocent and has, in fact, been victimized by Penn's accusation.

Penn, Gareth. Form response to queries, August 20, 1985, 8 pp.

Penn, Gareth. *Times 17: The Amazing Story of the Zodiac Murders in California and Massachusetts 1966–1981* (New York: Foxglove Press, 1987). Penn seems to have put much more thought into the killings than Zodiac himself. It seems to me that almost all of the patterns he picked up on are coincidental, but I do like the radian theory.

Schillemat, Brandon. "Massachusetts Man Says He's Cracked Zodiac Killer Code," July 21, 2011, available online at http://belmont-ca.patch.com/articles/massachusetts-man-says-hes-cracked-zodiac-killer-code; comments can be found at http://www.fark.com/comments/6408574/70565655#c70565655.

Smith, Dave. "Zodiac Killer: Is He Still at Large?" *Coronet*, October 1973.

Stephens, Hugh. "Has the Zodiac Killer Trapped Himself?" *Front Page Detective*, February 1970.

Stephens, Hugh. "He Wants Slave Girls Waiting for Him in Paradise." *Front Page Detective*, September 1975.

Stewart, Gary L., and Susan Mustafa. *The Most Dangerous Animal of All: Searching for My Father . . . and Finding the Zodiac Killer* (New York: Harper, 2014).

Symons, C. *Solving the Zodiac: The Zodiac Killer Case Files* (CreateSpace Independent Publishing Platform, 2009).

Weissman, Jerry. *Zodiac Killer* (Los Angeles: Pinnacle Books, 1979).

Williams, Bryan. "The Zodiac Killings: California's No. 1 Murder Mystery." *True Detective* 95, no. 4 (August 1971): 12–17, 65–67.

Yancey, Diane. *The Case of the Zodiac Killer*, Crime Scene Investigations (Detroit: Lucent Books, 2008).

第5章

Anderson, Jeanne. "Breaking the BTK Killer's Cipher." *Cryptologia* 37, no. 3 (July 2013). This is a nice example of a killer's cipher for which the solution was found.

Anderson, Jeanne. "Kaczynski's Ciphers." *Cryptologia* 39, no. 3 (July 2015). Again, killer ciphers that were solved. This time, because the key was kept with them.

Anonymous. "Police Close Susan Powell Case, Offer New Details." *USA Today*, May 20, 2013, available online at http://www.usatoday.com/story/news/nation/2013/05/20/susan-powell-case/2344681/.

Beattie, Robert. *Nightmare in Wichita: The Hunt for the BTK Strangler* (New York: New American Library, 2005). This book covers the noncryptologic aspects of this case.

Bennett, Donald H. "An Unsolved Puzzle Solved." *Cryptologic Spectrum* (Spring 1980). Available online at https://www.nsa.gov/public_info/_files/cryptologic_spectrum/unsolved_puzzle.pdf.

Bennett, Donald H. "An Unsolved Puzzle Solved." *Cryptologia* 7, no. 3 (July 1983): 218–34.

Carlisle, Nate. "Susan Powell Case Closed, Files Are Opened." *The Salt Lake Tribune*, May 21, 2013, available online at http://www.sltrib.com/sltrib/news/56338157-78/powell-susan-police-valley.html.csp.

Daniel, Mac, and Diane Allen. "Killings, Suicide Baffle Authorities. Husband Left Note; Motive a Mystery." *The Boston Globe*, February 5, 2004, 3rd ed., p. B1. This article was posted to Bruce Schneier's blog.

Davis, William F. "The Debosnys Murder Case." Undated, but after 1961. Davis served as deputy sheriff in Essex County. The report on the case was a joint effort carried out by Davis and his wife, according to Cheri L. Farnsworth, *Adirondack Enigma: The Depraved Intellect and Mysterious Life of North Country Wife Killer Henry Debosnys* (Charleston, SC: The History Press, 2010), 51.

"Debosnys Suspected of Another Murder." *Elizabethtown Post*, August 10, 1882.

Farmer, Tom. "Suicide Note Leaves Motive for Saugus Killings a Mystery." *The Boston Herald*, February 5, 2004, p. 14.

Farnsworth, Cheri L. *Adirondack Enigma: The Depraved Intellect and Mysterious Life of North Country Wife Killer Henry Debosnys* (Charleston, SC: The History Press, 2010).

Hodgkins, James D. "The Copiale Cipher: An Early German Masonic Ritual Unveiled." *The Scottish Rite Journal* (March/April 2012): 4–8, available online at http://scottishrite.org/about/media-publications/journal/article/the-copiale-cipher-an-early-german-masonic-ritual-unveiled/.

Knight, Kevin, Beáta Megyesi, and Christiane Schaefer. "The Copiale Cipher." *Proceedings of the 4th Workshop on Building and Using Comparable Corpora*, 49th Annual Meeting of the Association for Computational Linguistics, Portland, OR, June 24, 2011 (Stroudsburg, PA: Association for Computational Linguistics, 2011) 2–9.

Kourofsky, Niki. *Adirondack Outlaws: Bad Boys and Lawless Ladies*, Bedside Readers Series (Helena, MT: Farcountry Press, 2015). The Henry Debosnys case is detailed on pp. 71–78.

Lohr, David. "Susan Powell's Parents Search for Clues in Father-in-Law's Former Home." May 27, 2014, *Huffington Post*, available online at http://www.huffingtonpost.com/2014/05/27/susan-powell-missing-clues_n_5398147.html.

Mahon, Tom, and James J. Gillogly. *Decoding the IRA* (Cork, Ireland: Mercier Press, 2008).

Meguid, Halia. Best known for her singing on television's *Doctor Who*, Halia indicated online that she's at work on a novel about the Henry Debosnys case. See Halia Meguid, http://haliameguid.tumblr.com/post/91617594118/pls-read-if-you-are-an-ingenious-codebreaker-or.

Morris, Brent. "Fraternal Cryptography." *Cryptologic Spectrum* (Summer 1978).

Morris, Brent. "Fraternal Cryptography, Cryptographic Practices of American Fraternal Organizations." *Cryptologia* 7, no. 1 (January 1983): 27–36.

Morris, S. Brent. *The Folger Manuscript: The Cryptanalysis and Interpretation of an American Masonic Manuscript* (Ft. Meade, MD, 1992).

Morris, S. Brent. The Folger Manuscript, a lecture, available online at http://www.themasonictrowel.com/articles/manuscripts/manuscripts/folger_manuscript/the_folger_manuscript_lecture.htm.

Olsen, Greg, and Rebecca Morris. *If I Can't Have You: Susan Powell, Her Mysterious Disappearance, and the Murder of Her Children* (New York: St. Martin's Press, 2014).

Rule, Ann. *Fatal Friends, Deadly Neighbors and Other True Cases*, Ann Rule's Crime Files: Vol. 16 (New York: Pocket Books, 2012). Pages 1–156 of this book are devoted to the Powell case. No mention is made of encryption.

Ryba, Jim. "What Is He Saying?" *The Phoenician*, Summer 2006, p. 26.

Schmeh, Klaus. Codeknacker auf Verbrecherjagd, Folge 4: Der Maskenmann, http://scienceblogs.de/klausis-krypto-kolumne/2014/03/12/codeknacker-auf-verbrecherjagd-folge-4-der-maskenmann/, March 12, 2014.

Schmeh, Klaus. http://scienceblogs.de/klausis-krypto-kolumne/2015/02/23/die-ungeloesten-codes-des-mutmasslichen-frauenmoerders-henry-debosnys-teil-1/, February 23, 2015.

Schmeh, Klaus. Die ungelösten Codes des mutmaßlichen Frauenmörders Henry Debosnys (Teil 2), http://scienceblogs.de/klausis-krypto-kolumne/2015/02/25/die-ungeloesten-codes-des-mutmasslichen-frauenmoerders-henry-debosnys-teil-1-2/, February 25, 2015.

Schneier, Bruce. Handwritten Real-World Cryptogram, Schneier on Security, https://www.schneier.com/blog/archives/2006/01/handwritten_rea.html.

Wagner, Ingo."Verschlüsselt und nicht knackbar, Polizei scheitert an Festplatten des 'Maskenmanns.' " *Focus Magazine*, January 23, 2012. Available online at http://www.focus.de/panorama/welt/verschluesselt-und-nicht-knackbar-polizei-scheitert-an-festplatten-des-maskenmanns_aid_705621.html.

第 6 章

关于萨默顿男子

Abbott, Derek. Cipher Cracking, https://www.eleceng.adelaide.edu.au/personal/dabbott/wiki/index.php/Cipher_Cracking. Scroll down to the section titled "See Also." This section has a number of useful links, including "Primary source material on the Taman Shud Case" and "Secondary source material on the Taman Shud Case." Each of these leads to many useful documents.

Andrew, Christopher. *Defend the Realm: The Authorized History of MI5* (New York: Alfred A. Knopf, 2009). See this for background on possibly relevant Cold War spy cases.

Anonymous. "Dead Man Found Lying on Somerton Beach." *The News*, Adelaide, December 1, 1948, p. 1. Available online at http://trove.nla.gov.au/ndp/del/article/129897161.

Anonymous. "Body Found on Beach." *The Advertiser*, Adelaide, December 2, 1948, p. 3.

Anonymous. " 'Dead' Man Walks into Police HQ." *The News*, Adelaide, December 2, 1948.

Anonymous. "Somerton Beach Body Mystery." *The Advertiser*, Adelaide, December 4, 1948.

Anonymous. "Luggage as Clue to Beach Body." *The News*, Adelaide, January 12, 1949.

Anonymous. "Attempts to Solve Somerton Mystery Deepen Mystery." *Adelaide Truth*, June 25, 1949, p. 6.

Anonymous. "Police Test Book for Somerton Body Clue." *The Advertiser,* Adelaide, July 26, 1949, p. 3. Available online at http://trove.nla.gov.au/ndp/del/article/36677872.

Ashton, Lucy F. "Logical thinking," Letter to the Editor, *The Advertiser*, November 15, 1995, p. 14.

Balint, Ruth. "The Somerton Man an Unsolved History." *Cultural Studies Review* 16, no. 2 (September 2010): 159–78, http://epress.lib.uts.edu.au/journals/index.php/csrj/index.

The Body on the Beach: The Somerton Man—Taman Shud Case, http://brokenmeadows.hubpages.com/hub/The-Mystery-of-the-Somerton-Man-Taman-Shud-Case This is a great article with some rare pictures.

Bouda, Simon. *Crimes that Shocked Australia* (Sydney, Australia: Bantam Books, 1991). This book contains an incorrect version of Somerton Man's cipher.

Campbell, MacGregor. "Unbreakable: Somerton Man's Poetic Mystery." *New Scientist*, May 26, 2011.

Castello, Renato. "New Twist in Somerton Man Mystery as Fresh Claims Emerge." *The Advertiser, Sunday Mail*, South Australia, November 23, 2013, available online at http://www.adelaidenow.com.au/news/south-australia/new-twist-in-somerton-man-mystery-as-fresh-claims-emerge/story-fni6uo1m-1226766905157.

Clegg, Edward (His Honour Judge Clegg QC). *Famous Australian Murders* (Sydney, Australia: Angus and Robertson, 1975.

Clemow, Matt. " 'Poisoned in SA'—Was He a Red Spy?" *Sunday Mail*, November 7, 2004, 76–77. Portions reproduced in Feltus's book, pp. 186–87.

Coupe, Stuart, and Julie Ogden, eds. *Case Reopened* (Sydney, Australia: Allen and Unwin, 1993). This anthology contains a Kerry Greenwood short story, where an unknown man is smuggling for Ireland. It makes use of an incorrect version of the cipher from Simon Bouda's *Crimes that Shocked Australia* (Sydney, Australia: Bantam Books, 1991). The short story was reprinted in Greenwood's nonfiction *Tamam Shud: The Somerton Man Mystery* on pp. 177–216; p. 176 offers an introduction.

Dash, Mike. "The Body on Somerton Beach." Smithsonian.com, August 12, 2011, available online at http://www.smithsonianmag.com/history/the-body-on-somerton-beach-50795611/.

Debelle, Penelope. "A Body, a Secret Pocket and a Mysterious Code. Can the Riddle be solved?" *The Advertiser, Weekend Magazine,* August 1, 2009, pp. W14–W16.

Feltus, Gerald (Gerry) Michael. *The Unknown Man: A suspicious death at Somerton Beach* (Richmond, Australia: Hyde Park Press, 2010). Also see www.theunknownman.com. Feltus was born in 1943 and grew up near the area where the unknown man was found. He joined the South Australian police force in 1964, but knew about the case before becoming a police officer. In 1975, as a member of the Major Crimes Squad, the old unsolved case came under his purview. It wasn't as cold of a case as the intervening decades might make it seem, as periodic media attention led to new leads. Feltus left the Major Crimes Squad in 1979, but returned in 1992 (the name had changed to the Major Crime Task Force). He began devoting some of his spare time to the case (just like Graysmith did with Zodiac). He retired from the police force in 2004 (bio taken from pp. 10–11).

Feltus, Gerry. The Unknown Man, http://theunknownman.com/. This is the website for the book, and one can find useful updates and other information not in the book here, as well.

Ferguson, John. "After Discovery of Body on Somerton Beach in 1948 . . . Mystery Remains on the 'Unknown Man.' " *The News,* May 15, 1987, p. 5.

Fife-Yeomans, Janet. "The Man with No Name." *The Weekend Australian Magazine*, September 15–16, 2001. Portions of this article were reproduced in Feltus's book, pp. 185–86.

Gibson, Candy. "Students Aim to Crack 60-Year-Old Mystery." *Adelaidean* 18, no. 3 (May 2009): 9.

Grace, Kynton. "South Australia's *X-Files*: Part 2—The Somerton Man Mystery and the Secrets of Adelaide's Tunnels." *The Advertiser*, June 9, 2014, available online at http://www.theaustralian.com.au/news/south-australias-xfiles-part-2-the-somerton-man-mystery-and-the-secrets-of-adelaides-tunnels/story-e6frg6n6-1226941307664.

Greenwood, Kerry. *Tamam Shud: The Somerton Man Mystery* (Sydney, Australia: NewSouth Publishing, University of New South Wales Press Ltd., 2012).

Greenwood, Kerry. "Riddle on the Sands." *The Sydney Morning Herald*, November 28, 2012, edited extract from *Tamam Shud: The Somerton Man Mystery* by Kerry Greenwood, available online at http://www.smh.com.au/national/riddle-on-the-sands-20121119-29kwz.html.

"Inquest into the death of a body located at Somerton on 1.12.48" [refers to December 1, 1948] 1949, pdf available online at http://www.eleceng.adelaide.edu.au/personal/dabbott/tamanshud/inquest1949ocr.pdf.

"Inquest into the death of a body located at Somerton on 1.12.48" [refers to December 1, 1948] 1958, pdf available online at http://www.eleceng.adelaide.edu.au/personal/dabbott/tamanshud/inquest1958.pdf.

Jory, Rex. "The Dead Man Who Sparked Many Tales." *The Advertiser*, December 1, 2000, p. 18.

King, Stephen. *The Colorado Kid* (New York: Hard Case Crime, 2005). Although a reader might get the sense that this book was inspired by the Somerton Man case, King said that it was not. This novel inspired the TV series *Haven* (2010).

Lewes, Jacqueline Lee. "30-Year-Old Death Riddle Probed in New Series." *TV Times,* August 19, 1978, pp. 20–21.

Loftus, Tom. "The Somerton Body Mystery." *The News*, December 1–2, 1982. Portions reproduced in Feltus's book, pp. 197–200.

MacGregor Campbell. "Killer Codes." *New Scientist,* May 21, 2011, pp. 40–45.

Maguire, Shane. "Death Riddle of a Man with No Name." *The Advertiser*, March 9, 2005, p. 28.

Orr, Stephen. "Riddle of the End." *The Sunday Mail*, January 11, 2009, pp. 71–76.

Pedley, Derek. "Detective Still on Trail 47 Years On." *The Advertiser*, November 4, 1995, p. 14.

Pelling, Nick. "Sorry, the unknown man is (very probably) not H. C. Reynolds . . .." March 15, 2013, http://www.ciphermysteries.com/2013/03/15/sorry-but-the-unknown-man-is-almost-certainly-not-h-c-reynolds.

Phillips, John Harber. "So When That Angel of the Darker Drink." *Criminal Law Journal* 18, no. 2 (April 1994): 108–10.

Schmeh, Klaus. *Nicht zu Knacken* [Impossible to Crack] (Munich, Germany: Carl Hanser Verlag, 2012). This German language book is a survey of unsolved ciphers. Chapter 7 is devoted to Somerton Man. No English edition is available.

Steadwell. Somerton Man: A True Mystery "Down Under" with *Casablanca* Intrigue, August 23, 2011, http://blogcritics.org/culture/article/somerton-man-a-true-mystery-down/#ixzz1esR3uG12.

Taman Shud Case, http://en.wikipedia.org/wiki/Taman_Shud_Case.

Trove, digitised newspapers and more, http://trove.nla.gov.au/newspaper. A large number of Australian newspaper articles on Somerton Man are available here.

Turner, Jeff. "Beach Keeps Its Grim Secret." *The Advertiser*, December 7, 1998, p. 19. Portion reproduced in Feltus's book, p. 200.

Watkins, Emily. "Is British Seaman's Identity Card Clue to Solving 63-Year-Old Beach Body Mystery?" *The Advertiser, Sunday Mail,* Adelaide, November 20, 2011, pp. 4–5, available online at http://www.adelaidenow.com.au/is-british-seamans-identity-card-clue-to-solving-63-year-old-beach-body-mystery/story-e6frea6u-1226200076344.

视　频

All of the videos can be found at the indicated YouTube pages, but the first three immediately below are also linked from the webpage https://www.eleceng.adelaide.edu.au/personal/dabbott/wiki/index.php/Primary_source_material_on_the_Taman_Shud_Case.

The Somerton Beach Mystery, *Inside Story*, Australian Broadcasting Corporation, 1978. Online in three parts: http://www.youtube.com/watch?v=nnPqlYPQ9lY, http://www.youtube.com/watch?v=605V1-03r1Y, and http://www.youtube.com/watch?v=ieczsZRQnu8.

Tamam Shud—The Time Traveler that Died Before He Was Born! http://www.youtube.com/watch?v=zV3tfUoJyr4. No time travel is actually suggested in this video!

Taman Shud Case, *Stateline Report*, Australian Broadcasting Corporation, March 27, 2009, and May 1, 2009. Online in two parts: http://www.youtube.com/watch?v=GIUP-wVw6ok and http://www.youtube.com/watch?v=GNgsA1aHNHA.

The segment "The Somerton Man" from *60 Minutes* (Australia) is available at http://sixtyminutes.ninemsn.com.au/article.aspx?id=8759245. If you want to watch the entire program (including unrelated segments) go to http://www.youtube.com/watch?v=iy5s5oF3uB8. This program aired on November 24, 2013. For "Extra Minutes," providing an extended interview with Roma and Rachel Egan discussing the reasons they want the Somerton Man exhumed, go to http://www.youtube.com/watch?v=pJKMnX4WHSA.

Derek Abbott and Matthew Berryman have led several student projects on Somerton Man. The YouTube videos associated with this work are listed here:

Somerton Man Code Investigation. http://www.youtube.com/watch?v=YrjOvbk6QVI (2009) Students Andrew Turnbull and Denley Bihari drew the same conclusion that I did.

Somerton Man Code Investigation. http://www.youtube.com/watch?v=rFsFpSBGhQw (2010) Students Kevin Ramirez and Michael Lewis-Vassallo.

Project Video. http://www.youtube.com/watch?v=KoDEF-FDLso (2011) Students Steven Maxwell and Patrick Johnson. The audio portion of this one doesn't work.

Code Cracking: Who Murdered the Somerton Man? http://www.youtube.com/watch?v=jWE7xl9LiOw&feature=youtu.be (2012) Students Aidan Duffy and Thomas Stratfold.

Code Cracking: Who Murdered the Somerton Man? http://www.youtube.com/watch?v=JX4bt7VuJQs (2013) Students Lucy Griffith and Peter Varsos. These students tried other languages, but found that English had the best match. They also conducted mass spectral analysis on one of Somerton Man's hairs and learned that the amount of lead in his system decreased toward his time of death.

关于保罗·鲁宾

Some of the newspaper articles cited here were found in clipped form in a scrapbook in the uncataloged portion of the National Cryptologic Museum's library. Others were obtained as photocopied clippings from an FBI file on the case held at the National Archives and Records Administration. Hence, full bibliographic details were not always apparent. Journalists were hardly ever credited by name in these articles.

"Airport Death Called Murder by Ominsky." Philadelphia, January 21, 1953.

Apt, Jay. "Note Baffles Decoders in Poison Death." *Daily News*, Philadelphia, January 22, 1953, p. 2. Story continues on page 19, but retains the headline.

"Body at Airport Is Identified as NYU Student." Philadelphia, January 21, 1953.

"Body Found at Airport, Coded Note on Abdomen." *The Philadelphia Inquirer*, January 21, 1953, p. 1. This story continues on p. 18, col. 3, under the headline "Mystery Death at City Airport."

"Body with Code Note Found Here." *Daily News*, Philadelphia, p. 2, January 20, 1953.

"Code Death Mystery Still Baffles FBI." January 23, 1953.

"Code Message Found Taped to Dead Man." *Washington Times-Herald*, January 21, 1953, p. 35.

"Code Note Taped to Body in Ditch." *The Evening Bulletin*, Philadelphia, January 20, 1953, p. 1, col. 2.

"Coded Note, Poison Deepen Mysteries of Airport Death." *Philadelphia Daily News* 28, no. 253, January 21, 1953.

"Codes Seen Key in Airport Death." January 22, 1953.

"Coroner Hints Poison Slaying of Student." *The Knickerbocker News*, Albany, NY, January 22, 1953, p. 14-B.

"Cyanide Caused Death at Airport." *The Evening Bulletin*, Philadelphia, January 21, 1953, p. 1.
"Cyanide Killed Cloak-and-Dagger Airport Victim." (big headline, references story on p. 3)
"Dad Identifies Poisoned Youth as N.Y. Student." *Washington Times-Herald*, January 22, 1953, p. 3.
"Dead Youth with Message on Body was NY Pupil." *Buffalo Courier-Express*, January 22, 1953, p. 2.
"Death of Student." *Shamokin News-Dispatch*, Shamokin, PA, January 23, 1953, p. 2.
" 'Dulles' Name Taped to Body of 'Suicide.' " *Washington Post*.
"FBI Investigates Possibilities of Murder in Mysterious Cyanide-Death of Student." *Schenectady Gazette*, January 22, 1953.
"Hazing Studied in Airport Death." *The Philadelphia Inquirer*, January 28, 1953, p. 5.
"Life on the Newsfronts of the World, Student Leaves a Coded Clue to his Death, a Soldier Plants a Flag and Flu Virus Sweeps the World." *Life*, February 2, 1953, p. 30.
"Man Poisoned, Paper in Code Taped to Body; FBI Called In." *Washington Star*, p. A-26.
"Murder Hinted in Cyanide Death." *The Washington Daily News*, January 22, 1953.
"Murder Hinted in Youth's Death." *Long Island Star-Journal*, January 22, 1953, p. 4.
"Mystery Man at Airport Is Found Dead of Poison." *The Philadelphia Inquirer* 248, no. 21, January 21, 1953, p. 1. Story continues on p. 37.
"Mystery Man Cyanide Clue Hints Murder."
"Mystery of Body, Code Note Deepens in Airport Death." January 21, 1953.
"Mystery Shrouds Death by Cyanide." *The Philadelphia Inquirer*, March 25, 1953, p. 19.
"Note Taped to Body at Airport Bears Symbols, Word 'Dulles.' " January 20, 1953.
"Philly Jury Finds Boro Student Died from Cyanide Poisoning." *Brooklyn Eagle*, March 25, 1953, p. 4.
"Poison Victim Identified as N.Y. Youth." *The Philadelphia Inquirer*, January 22, 1953, final city ed., p.1, illustrated on p. 3 and continued on p. 13.
"Sift Cyanide Death of Boro Collegian." *Brooklyn Eagle*, January 22, 1953, 7 star ed., p. 1.
"Student Identified as Cyanide Victim." *New York Times*, January 22, 1953.
"Test Tube Found at Site of Airport Ditch Death." *The Philadelphia Inquirer*, January 23, 1953, p. 28.
"Test Tube Near Code Body Yields no Clue." January 23, 1953.
"Verdict Remains Open in Mysterious Death Philadelphia." *The Evening Sun*, Hanover, PA, March 25, 1953, p. 9.
"Youth Dabbled in Cryptography." *Philadelphia Evening Bulletin*, January 22, 1953.
Many of the newspaper articles cited above are available here as pdfs: http://fultonhistory.com/Fulton.html.

关于里基·麦考密克

Anthony, Shane. "From 1999: Body Found in Field Puzzles Police." *St. Louis Post-Dispatch*, July 2, 1999, available online at http://www.stltoday.com/news/article_bcc02074-5b1a-11e0-b199-0017a4a78c22.html.
Castigliola, Angelo. "Ricky McCormick FBI Letters Decoded." *Angelo on Security*, March 30, 2011, http://www.castigliola.com/index.php?option=com_content&task=view&id=123&Itemid=1 (a proposed solution).
Cryptanalysts Part 2: Help Solve an Open Murder Case, http://www.fbi.gov/news/stories/2011/march/cryptanalysis_032911.
Help Break the Code, https://forms.fbi.gov/code.

Howard, Trisha L. "Store Manager Gets 38-Year Sentence in Killing of Customer." *St Louis Post-Dispatch*, September 5, 2002. This is the place to start if you want to follow the continuing adventures of Baha Hamdallah.

Kessler, Ronald. *The Secrets of the FBI* (New York: Crown Publishers, 2011). This book has some useful general information beginning on p. 257.

Mann, Jennifer. "FBI Wants Public to Help Crack Code in St. Charles County Cold Case." *St. Louis Post-Dispatch*, March 31, 2011, available online at http://www.stltoday.com/news/local/crime-and-courts/fbi-wants-public-to-help-crack-code-in-st-charles/article_10af026c-e2d1-59f1-92e5-0576d36e2ffc.html.

Ricky McCormick's encrypted notes. http://en.wikipedia.org/wiki/Ricky_McCormick's_encrypted_notes.

Schmeh, Klaus. Top-25 der ungelösten Verschlüsselungen—Platz 7: Der Mord an Ricky McCormick, Klausis Krypto Kolumne, http://scienceblogs.de/klausis-krypto-kolumne/2013/08/29/top-25-der-ungelosten-verschlusselungen-platz-7-der-mord-an-ricky-mccormick/. This piece is in German, although some use comments appear in English.

Tritto, Christopher. "Code Dead: Do the Encrypted Writings of Ricky McCormick Hold the Key to His Mysterious Death?" *Riverfront Times*, St. Louis, June 14, 2012, available online at http://www.riverfronttimes.com/2012-06-14/news/ricky-mccormick-code-mysterious-death-st-louis/.

Tweedie, Neil. "Calling All Codebreakers . . . ." *The Telegraph*, April 7, 2011, available online at http://www.telegraph.co.uk/news/uknews/crime/8432893/Calling-all-codebreakers....html. This article provides various experts' opinions on the cipher.

第 7 章

Anonymous. "£20 to Solve Riddle." *Psychic News*, July 15, 1950, p. 4. This piece is about Wood's cipher.

Anonymous. "£20—If You Can Beat the Spirits to It!" *Two Worlds* 63, no. 3277 (September 16, 1950): 923. This piece is about Wood's cipher.

Anonymous. "S. P. R. Member's Survival Test Amplified." *Two Worlds* (December 2, 1950). In this piece, Wood referred to the "substantial prize" he offered.

Anonymous. "After Death." *Daily Express*, August 15, 2015. This piece is about Wood's cipher.

Augustine, Keith. "The Case against Immortality." *Skeptic Magazine* 5, no. 2 (1997), available online in an expanded form at http://infidels.org/library/modern/keith_augustine/immortality.html.

Berger, Arthur S. "Better than a Gold Watch: The Work of the Survival Research Foundation." *Theta* 10 (1982): 82–84.

Berger, Arthur S. "Death Comes Alive." *Journal of Religion and Psychical Research* 5 (1982): 139–47.

Berger, Arthur S. "Project: Unrecorded Information." *Christian Parapsychologist* 4 (1982): 159–61.

Berger, Arthur S. "Letter to the Editor." *Journal of the Society for Psychical Research* 52 (1983): 156–57.

Berger, Arthur S. "Experiments with False Keys." *Journal of the American Society for Psychical Research* 78, no. 1 (January 1984): 41–54.

Berger, Arthur S. "The Development and Replication of Tests for Survival." *Parapsychological Journal of South Africa* 5 (1984): 24–35.

Berger, Arthur S. *Aristocracy of the Dead: New Findings in Postmortem Survival* (London and Jefferson, NC: McFarland & Co., 1987).

Berger, Arthur S. "Reincarnation by the Numbers: A Criticism and Suggestion." *Reincarnation: Fact or Fable?* Arthur S. Berger and Joyce Berger, eds. (London: The Aquarian Press, 1991), 221–33.

Carroll, R. T. "How Not to Conduct Scientific Research." (review of *The Afterlife Experiments* by Gary Schwartz), The Skeptic's Dictionary, available online at http://skepdic.com/refuge/afterlife.html.

Feely, Joseph Martin. *Electrograms from Elysium: A Study on the Probabilities in Postmortuary Communication through the Electronics of Telepathy and Extrasensory Perception, Including the Code of Anagrams in the Purported Sender's Name* (New York, 1954).

Fox, Margalit. "Ian Stevenson Dies at 88; Studied Claims of Past Lives." *New York Times*, February 18, 2007, available online at http://www.nytimes.com/2007/02/18/health/psychology/18stevenson.html?_r=0.

Gillogly, James J., and Larry Harnisch. "Cryptograms from the Crypt." *Cryptologia* 20, no. 4 (October 1996): 325–29.

Kellock, Harold. *Houdini: His Life-story, from the Recollections and Documents of Beatrice Houdini* (New York: William Heinemann, 1928).

Leighton, A. C. "Has Dr. Thouless Survived Death?" *Cryptologia* 10, no. 2 (April 1986): 108–9.

Levin, Michael. "Encryption Algorithms in Survival Evidence." *The Journal of Parapsychology* 58, no. 2 (June 1994): 189–95.

MacDonald, Leo. "Rebuttal to Keith Augustine's Look at the Tests Done by Ian Stevenson," February 3, 2009, available online at http://paranormalandlifeafterdeath.blogspot.com/2009/02/rebuttal-to-keith-augustines-look-at.html.

Martin, Michael, and Keith Augustine. *The Myth of an Afterlife: The Case against Life After Death* (Lanham, MD: Rowman & Littlefield Publishers, 2015).

Mongé, A. "Solution of a Playfair Cipher." *Signal Corps Bulletin* 93 (November–December 1936). Reprinted in W. F. Friedman. *Cryptography and Cryptanalysis Articles* 1 (Laguna Hills, CA: Aegean Park Press, 1976) and in Brian J. Winkel. "A Tribute to Alf Mongé." *Cryptologia* 2, no. 2 (1978): 178–85.

Polidoro, Massimo. "The Day Houdini (Almost) Came Back from the Dead." *Skeptical Inquirer* 36, no. 2 (March/April 2012), available online at http://www.csicop.org/si/show/the_day_houdini_almost_came_back_from_the_dead.

Roggo, D. Scott. "Parapsychology—Its Contributions to the Study of Death." *Omega Journal of Death and Dying* 5, no. 2 (July 1974): 99–113.

Salter, W. H. "F.W.H. Myers' Posthumous Message." Proceedings *of the Society for Psychical Research* 52 (1958): 1–32 (a sealed envelope message).

Schmeh, Klaus. *Nicht zu knacken* [Impossible to Crack] (Munich, Germany: Carl Hanser Verlag, 2012). Chapter 10 of Schmeh's book covers Thouless's ciphers.

Schmeh, Klaus. "Parapsychologische Verschlüsselungsexperimente, Tote verraten keine Geheimwörter—bisher jedenfalls." *Skeptiker* 3 (2012): 2–9.

Schmeh, Klaus. "Skurriles Experiment: Verraten Tote Geheimwörter?" *Telepolis*, May 13, 2013, available online at http://www.heise.de/tp/artikel/38/38934/1.html.

Schwartz, Gary E. R., and L. G. S. Russek. "Testing the Survival of Consciousness Hypothesis: The Goal of the Codes." *Journal of Scientific Exploration* 11, no. 1 (1997): 79–88.

Schwartz, Gary E. R., and Linda G. S. Russek. "Celebrating Susy Smith's Soul: Preliminary Evidence for the Continuance of Smiths Consciousness after Her Physical Death." *Journal of Religion and*

Psychical Research 24, no. 2 (April 2001): 82–91, available online at http://www.innerknowing.net/research.html.

Schwartz, Gary E., with William L. Simon, foreword by Deepak Chopra. *The Afterlife Experiments, Breakthrough Scientific Evidence of Life After Death* (New York: Pocket Books, 2002).

Smith, Susy (pen name of Ethel Elizabeth Smith). Introduction by Gary E. R. Schwartz and Linda G. S. Russek. *The Afterlife Codes: Searching for Evidence of the Survival of the Human Soul* (Charlottesville, VA: Hampton Roads Publishing Company, Inc., 2000).

Stevenson, Ian. "The Combination Lock Test for Survival." *Journal of the American Society for Psychical Research* 62 (1968): 246–54.

Stevenson, Ian. "Further Observations on the Combination Lock Test for Survival." *Journal of the American Society for Psychical Research* 70 (1976): 219–29.

Stevenson, Ian, Arthur T. Oram, and Betty Markwick. "Two Tests of Survival After Death: Report on Negative Results." *Journal of the Society for Psychical Research* 55, no. 815 (April 1989): 329–36.

Thouless, Robert H. "A Test of Survival." *Proceedings of the Society for Psychical Research* 48 (July 1948): 253–63.

Thouless, Robert H. "Additional Notes on a Test of Survival." *Proceedings of the American Society for Psychical Research* 48 (1948): 342–43.

Tribbe, Frank C. "The Tribbe/Mulders Code." *Journal of the Academy of Religion and Psychical Research* 3, no. 1 (January 1980): 44–46.

Winkel, Brian J. "A Tribute to Alf Mongé." *Cryptologia* 2, no. 2 (April 1978): 178–85.

Wood, T. E. "A Further Test for Survival." *Proceedings of the Society for Psychical Research* 49 (1950): 105–6.

第 8 章

Bauer, Craig. *Secret History: The Story of Cryptology* (Boca Raton, FL: Chapman and Hall/CRC, 2013).

Bauer, Craig, and Elliott Gottloeb. "Results of an Automated Attack on the Running Key Cipher." *Cryptologia* 29, no. 3 (July 2005): 248–54.

Bauer, Craig, and Christian N. S. Tate. "A Statistical Attack on the Running Key Cipher." *Cryptologia* 26, no. 4 (October 2002): 274–82.

Bellovin, Steven M. "Frank Miller: Inventor of the One-Time Pad." *Cryptologia* 35, no. 3 (July 2011): 203–22.

Byrne, John F. *Silent Years: An Autobiography with Memoirs of James Joyce and Our Ireland* (New York: Farrar, Straus and Young, 1953).

Byrne, John, Cipher A. Deavours, and Louis Kruh. "Chaocipher Enters the Computer Age When Its Method Is Disclosed to *Cryptologia* Editors." *Cryptologia* 14, no. 3 (July 1990): 193–98.

Calof, Jeff, Jeff Hill, and Moshe Rubin. "Chaocipher Exhibit 5: History, Analysis, and Solution of *Cryptologia*'s 1990 Challenge." *Cryptologia* 38, no. 1 (January 2014): 1–25.

Friedman, William F. "The Cryptanalyst Accepts a Challenge." *The Signal Corps Bulletin* 103 (January–March 1939).

Griffing, Alexander. "Solving the Running Key Cipher with the Viterbi Algorithm." *Cryptologia* 30, no. 4 (October 2006): 361–67. This is the best paper on the topic of breaking running key ciphers.

"Historical Survey of Strip Cipher Systems." This is available from NARA; NSA Historical Collections 190/37/7/1, NR 3525 CBRK24 12957A 19450000.

"History of Army Strip Cipher, SRH-366." This is available from NARA; RG 0457: NSA/CSS Finding Aid A1, 9020 U.S. Navy Records Relating to Cryptology 1918–1950 Stack 190 Begin Loc 36/12/04 Location 1-19.

Kruh, Louis. "The Genesis of the Jefferson/Bazeries Cipher Devices." *Cryptologia* 5, no. 4 (October 1981): 193–208.

Kruh, Louis (under his ACA pen name, MEROKE). "The M-94 Test Messages." *The Cryptogram* XLVIII, no. 6 (July–August 1982): 4–5, available online at http://www.prc68.com/I/M94TM.htm.

Kruh, Louis (under his ACA pen name, MEROKE). "A 77-Year-Old Challenge Cipher." *The Cryptogram* (March/April 1991): 7.

Kruh, Louis. "A 77-Year-Old Challenge Cipher." *Cryptologia* 17, no. 2 (April 1993): 172–74. Note: Mauborgne is misspelled in this paper, and the reference to the paper in *The Cryptogram* leads one to look in early 1992 issues, when it is actually March/April 1991.

Kruh, Louis. "Riverbank Laboratory Correspondence, 1919 (SRH-50)." *Cryptologia* 19, no. 3 (July 1995): 236–46.

Mauborgne, Joseph O. *Practical Uses of the Wave Meter in Wireless Telegraphy* (New York: McGraw-Hill Book Co., 1913).

Mauborgne, Joseph O. *An Advanced Problem in Cryptography and Its Solution* (Fort Leavenworth, KS: Press of the Army Services Schools, 1914). A second edition appeared in 1918.

Mauborgne, Joseph O. *Data for the Solution of German Ciphers: Also a Diagram of Cipher Analysis* (Fort Leavenworth, KS: Army Service School Press, 1917).

Mauborgne, Joseph O. "One Method of Solution of the Schooling 'Absolutely Indecipherable' Cryptogram." *The Signal Corps Bulletin* 104 (April–June 1939): 27–40.

Mauborgne, Joseph O. "Reminiscences of Joseph Oswald Mauborgne: Oral History." 1971 (held in Columbia University Library's rare book collection).

Smoot, Betsy Rohaly. "Parker Hitt's First Cylinder Device and the Genesis of U.S. Army Cylinder and Strip Devices." *Cryptologia* 39, no. 4 (October 2015): 315–21. It was long believed that Hitt came up with his version of the cipher wheel in 1913. This paper shows that it was 1912. So, it shouldn't surprise anyone if Mauborgne's version turns out to be a little older than believed as well!

Wesencraft, Fenwick (under his ACA pen name TRIO). "Solutions of the M-94 Test Messages." *The Cryptogram* XLVIII no. 8 (November–December 1982): 6–7, available online at http://www.prc68.com/I/M94S.htm.

第9章

关于达加佩耶夫密码

Barker, Wayne G. "The Unsolved d'Agapeyeff Cipher." *Cryptologia* 2, no. 2 (April 1978): 144–47.

d'Agapeyeff, Alexander. *Codes and Ciphers* (London: Oxford University Press, 1939).

d'Agapeyeff, Alexander. *Maps* (London: Oxford University Press, 1942). Some online sources claim that this book appeared before d'Agapeyeff's *Codes and Ciphers*, but that's not correct.

d'Agapeyeff cryptogram revisited, http://www.rodinbook.nl/dagapeyeff.html.

Lann, Jew-Lee Irena. D'Agapeyeff's Code: A New Breakthrough Leads to A New Paradigm, May 25, 2009, available online at www.thekryptosproject.com/tjp/release.doc.

Rugg, Gordon, and Gavin Taylor. A very British mystery: The case of the D'Agapeyeff Cipher, https://hydeandrugg.wordpress.com/2013/07/02/a-very-british-mystery-the-case-of-the-dagapeyeff-cipher/ July 2, 2013, last accessed May 19, 2015.

Rugg, Gordon, and Gavin Taylor. A very British mystery, part 2: The D'Agapeyeff Cipher and the first edition, https://hydeandrugg.wordpress.com/2013/07/17/a-very-british-mystery-part-2-the-dagapeyeff-cipher-and-the-first-edition/.

Rugg, Gordon, and Gavin Taylor. A very British mystery, part 3: The D'Agapeyeff Cipher's Table of Contents, https://hydeandrugg.wordpress.com/2013/07/25/a-very-british-mystery-part-3-the-dagapeyeff-ciphers-table-of-contents/, July 25, 2013, last accessed May 19, 2015.

Rugg, Gordon, and Gavin Taylor. A very British mystery, part 4: Quiet Bodies, https://hydeandrugg.wordpress.com/2013/08/12/a-very-british-mystery-part-4-quiet-bodies/, August 12, 2013, last accessed May 19, 2015.

Rugg, Gordon, and Gavin Taylor. A very British mystery, part 5: Gavin Finds a Typo, https://hydeandrugg.wordpress.com/2013/08/16/a-very-british-mystery-part-5-gavin-finds-a-typo/, August 16, 2013, last accessed May 19, 2015.

Rugg, Gordon, and Gavin Taylor. A very British mystery, part 6: A possible solution? Probably not . . .. As of this writing, this installment of Rugg and Taylor's series has not yet appeared online. Rugg shared a draft version of it with me.

Shulman, David (under his ACA pen name AB STRUSE). "The D'Agapeyeff Cryptogram: A Challenge." *The Cryptogram* (April/May 1952): 39–40, 46.

Shulman, David (under his ACA pen name AB STRUSE). "D'Agapeyeff Cipher: Postscript." *The Cryptogram* (March/April 1959): 80–81.

Van Zandt, Armand. His proposed solution has been archived at https://web.archive.org/web/20131207173351/http://www.gather.com/viewArticle.action?articleId=281474981054022.

关于克里普托斯雕塑

Andrews, Robert M. "Sculpture with Code Poses Mystery at the CIA." *Los Angeles Times*, April 28, 1991, p. A34, available online at http://articles.latimes.com/1991-04-28/news/mn-1468_1_mystery-sculpture.

Bauer, Craig P. *Secret History: The Story of Cryptology* (Boca Raton, FL: Chapman and Hall/CRC, 2013).

Bauer, Craig, Gregory Link, and Dante Molle. "James Sanborn's *Kryptos* and the Matrix Encryption Conjecture." *Cryptologia* 40, no. 6 (November 2016), 541–52. .

Bauer, Craig, and Katherine Millward. "Cracking Matrix Encryption Row by Row." *Cryptologia* 31, no. 1 (January 2007). A good attack on matrix encryption is described in this paper, but it requires that the numerical assignments for the letters be known. Since publication, other authors have improved this attack.

Brown, Dan. *The Da Vinci Code* (New York: Doubleday, 2003).

Brown, Dan. *The Lost Symbol* (New York: Doubleday, 2009).

Central Intelligence Agency. *Central Intelligence Agency Employee Bulletin*. June 1, 1988, released in response to a Freedom of Information Act (FOIA) request.

Central Intelligence Agency. *Central Intelligence Agency Employee Bulletin*. May 17, 1989, released in response to a Freedom of Information Act (FOIA) request.

Central Intelligence Agency. "The Puzzle at CIA Headquarters: Cracking the Courtyard Crypto." *Studies in Intelligence* 43, no. 1 (1999): 11 pages. Available online at http://www2.gwu.edu/~nsarchiv/NSAEBB/NSAEBB431/docs/intell_ebb_010.PDF.

Dunin, Elonka. *Kryptos*, http://elonka.com/kryptos/.

Ellis, David, reported by Daniel S. Levy. "The Spooks' Secret Garden." *Time*, March 18, 1991.

Gillogly, James (under his ACA pen name, Scryer). "The Kryptos Sculpture Cipher: A Partial Solution." *The Cryptogram* 65, no. 5 (September–October 1999): 1–7.

Gillogly, James (under his ACA pen name, Scryer). "Kryptos Clue." *The Cryptogram* (January–February 2011): 11.

Hill, Lester S. "Cryptography in an Algebraic Alphabet." *The American Mathematical Monthly* 36, no. 6 (1929): 306–12.

Markoff, John. "C.I.A.'s Artistic Enigma Yields All but Final Clue." *New York Times*, June 16, 1999, p. A24. Available online at http://www.nytimes.com/library/tech/99/06/biztech/articles/16code.html. This piece gets the artist's name wrong in a caption, calling him David Sanborn, instead of James Sanborn (he actually prefers Jim). The article also errs in attributing the sculpture's dedication to October 1990 instead of November 1990.

National Security Agency. "CIA KRYPTOS Sculpture—Challenge and Resolution." *United States Government—Memorandum*, June 9, 1993. Available online at https://docs.google.com/file/d/0B7G1aFZQuZtXRmRkcmhkNGtqQ2c/edit.

Overbey, J., W. Traves, and J. Wojdylo. "On the Keyspace of the Hill Cipher." *Cryptologia* 29, no. 1 (January 2005): 59–72.

Schwartz, John. "Clues to Stubborn Secret in C.I.A.'s Backyard." *New York Times*, November 20, 2010.

关于蝉 3301

Anonymous. *How Hackers Changed the World* (BBC Documentary) https://www.youtube.com/watch?v=jwA4tSwmS-U , last accessed March 3, 2015.

Bell, Chris. "The Internet Mystery That Has the World Baffled." *The Telegraph* (November 25, 2013), http://www.telegraph.co.uk/technology/internet/10468112/The-internet-mystery-that-has-the-world-baffled.html.

Bell, Chris. "Cicada 3301 Update: The Baffling Internet Mystery Is Back." *The Telegraph* (January 7, 2014), http://www.telegraph.co.uk/technology/internet/10555088/Cicada-3301-update-the-baffling-internet-mystery-is-back.html.

Dailey, Timothy. *The Paranormal Conspiracy: The Truth about Ghosts, Aliens, and Mysterious Beings* (Minneapolis: Chosen Books, 2015). Chapter 10 is titled "The Mystery of Cicada 3301."

Eriksson, Joel. "Cicada 3301," *ClevCode*, http://www.clevcode.org/cicada-3301/. Scroll down to get to a clear explanation of the beginning of the 2012 Cicada 3301 puzzle.

Ernst, Douglas. "Secret Society Seeks World's Brightest: Recruits Navigate 'Darknet' Filled with Terrorism, Drugs." *The Washington Times*, November 26, 2013, available online at http://www.washingtontimes.com/news/2013/nov/26/secret-society-seeks-worlds-smartest-cicada-3301-r/.

Ethliel. The Forum > Technical Corner > Hidden message in image, Two Cans and String, http://twocansandstring.com/forum/technical/4123/.

Grothaus, Michael. Meet the man who solved the mysterious cicada 3301 puzzle, http://www.fastcompany.com/3025785/meet-the-man-who-solved-the-mysterious-cicada-3301-puzzle.

Kushner, David. "Cicada: Solving the Web's Deepest Mystery." *Rolling Stone,* no. 1227 (January 15, 2015): 52–59, available online at http://www.rollingstone.com/culture/features/cicada-solving-the-webs-deepest-mystery-20150115.

NPR Staff. "The Internet's Cicada: A Mystery without an Answer." *All Things Considered*, National Public Radio, January 5, 2014. Text and audio recording also available at link, http://www.npr.org/2014/01/05/259959632/the-internets-cicada-a-mystery-without-an-answer.

Nursall, Kim. "Cicada 3301—The hunt continues in 2014." *Toronto Star*, January 10, 2014, available online at http://www.thestar.com/life/2014/01/10/cicada_3301_the_hunt_continues_in_2014.html.

Øverlier, Lasse, and Paul Syverson. "Locating Hidden Servers." *IEEE Symposium on Security and Privacy* (May 2006).

Uncovering Cicada, http://uncovering-cicada.wikia.com/wiki/Uncovering_Cicada_Wiki. This is an extensive site that covers all of the challenges to date.

Vincent, James. "Masonic Conspiracy or MI6 Recruitment Tool? Internet Mystery Cicada 3301 Starts Up Again." *Belfast Telegraph*, July 1, 2014, available online at http://www.belfasttelegraph.co.uk/technology/masonic-conspiracy-or-mi6-recruitment-tool-internet-mystery-cicada-3301-starts-up-again-29896340.html.

关于圣殿骑士团密码

Amland, Bjoern, and Sarah DiLorenzo. "Lawyer: Norway Suspect Wanted a Revolution." July 24, 2011, available online at http://news.yahoo.com/lawyer-norway-suspect-wanted-revolution-100757635.html.

Belfield, Richard. *The Six Unsolved Ciphers: Inside the Mysterious Codes that Have Confounded the World's Greatest Cryptographers* (Berkeley, CA: Ulysses Press, 2007). Chapter 4 is titled "Shugborough: The Shepherd's Monument." I chose not to include an in-depth treatment of this cipher (and many others) to prevent the present volume from growing to thousands of pages.

Defalcouss, Liath. "Jahbulonian 'Sons of Fallen' Oddity Data." *Cybercomopolotian*, https://cybercosmopolitan.wordpress.com/p-c-c-t-jahbulonian-website/jahbulonian-sons-of-fallen-oddity-data/.

Herstein, Olivia. "Still Standing—As One." *Viking* 108, no. 12 (December 2011): 16–20, 22, 24. This article details Breivik's attack and the aftermath. It includes the visit of Norway's prime minister, Jens Stoltenberg, to a mosque and his quote, "We will be one community, across religion, ethnicity, gender, and rank."

Jahbulon, https://en.wikipedia.org/wiki/Jahbulon.

Norway Terror Attacks Fast Facts, CNN Library, July 17, 2015, available online at http://www.cnn.com/2013/09/26/world/europe/norway-terror-attacks/.

Ramsden, Dave. *Unveiling the Mystic Ciphers: Thomas Anson and the Shepherd's Monument Inscription* (CreateSpace, 2014).

"The True Sons of the Fallen Are Back—Mysterious Website." Before It's News, December 6, 2013, http://beforeitsnews.com/strange/2013/12/the-true-sons-of-the-fallen-are-back-mysterious-website-2453196.html.

关于比尔密码

Aaron, Frank H. "Historical Facts Supporting Beale." *Proceedings of the Fourth Beale Cypher Association,* 1986 (Warrington, PA: Beale Cipher Association, 1987).

Anonymous. "Believers Still Searching for the Beale Treasure." *Bedford Bulletin-Democrat*, August 31, 1967.
Anonymous. " 'Buford County' Still Attracts Buried Treasure Hunters." *Bedford Bulletin-Democrat*, August 1, 1968.
Anonymous. "Newspaper Reports Cipher of Bedford Treasure Broken." *Roanoke Times*, April 20, 1972.
Anonymous. "Beale's Treasure Tale Revived." *The News*, April 21, 1972.
Anonymous. "Using Computers to Hunt Beale Treasure." *Bedford Bulletin-Democrat*, May 4, 1972.
Anonymous. "Beale Treasure Termed Hoax." *Staunton Leader*, February 5, 1974.
Anonymous. "The Second Beale Cipher Sympoisum—Call for Papers." *Cryptologia* 3, no. 3 (1979): 191.
Anonymous. "Treasure Hunter Freed on Bond." *Roanoke World News*, January 13, 1983.
Anonymous. "Retrial Is Ordered for Woman Charged with Disturbing Graves." *Richmond Times Dispatch*, February 24, 1983.
Anonymous. "Follow Guidelines When Treasure Hunting." *Lynchburg News*, July 14, 1985.
Anonymous. "Many Still Seek Beale Treasure." *Lynchburg News*, July 14, 1985.
Atwell, Albert. "The Mystery of Beale's Treasure Solved." (Ridgeway, VA: A. L. Atwell, 1990), 36 pp.
Barnes, Raymond. "Famed 1822 Beale Treasure Led Roanoke Brothers to Futile Hunt 66 Years Ago." *Roanoke World News*, June 1, 1963.
Bauman, Ken Andrew. *The National (Beale) Treasure at Red Knee* (Pittsburgh: RoseDog Books, 2007).
Beale Cypher Association. *Proceedings of the Third Beale Cipher Symposium* (Medfield, MA: Beale Cypher Association, 1981).
Beale Cypher Association. *The Beale Ciphers in the News* (Medfield, MA: Beale Cypher Association, 1983).
Bechtel, Stefan. "Solid Gold Mystery." *Southern World* (July–August 1980).
Boegli, Jacques S. "Madison-Beale-Hite Connection." *Proceedings of the Fourth Beale Cypher Association,* 1986 (Warrington, PA: Beale Cipher Association, 1987).
Brooks, Dorothy S. "Story of Buried Treasure has Disappointing Ending." *Lynchburg Daily Advance*, May 8, 1970.
Burchard, Hank. "Legendary Treasure Quests." *The Washington Post*, October 5, 1984.
Burleson, Bill. "Trio Hunts Famed Beale Treasure." *Roanoke Times & World News*, May 8, 1962.
Burleson, Bill. "Bedford County's Buried Treasure Lures Hunters." *Lynchburg Daily Advance*, May 19, 1962.
Chan, Wayne S. "Key Enclosed: Examining the Evidence for the Missing Key Letter of the Beale Cipher." *Cryptologia* 32, no. 1 (January 2008): 33–36.
Clayton, Stan. *Beale Treasure Map to Cipher Success* (Peterborough, U.K.: FastPrint Publishing, 2012).
Daniloff, Ruth. "A Cipher's the Key to the Treasure in Them Thar Hills." *Smithsonian Magazine* (April 1981).
"Death of James B. Ward." *The Lynchburg News*, May 17, 1907.
Easterling, E. J. *In Search of a Golden Vault: The Beale Treasure Mystery* (Roanoke, VA: Avenel, 1995).
Gillogly, James. "The Beale Cipher: A Dissenting Opinion." *Cryptologia* 4, no. 2 (April 1980): 116–19.
Greaves, Richard H. "One Letter, One Enclosure Subject: The Beale Treasure." private publication, 1986.
Greaves, Richard H. "Subject: The Beale Treasure," advertisement in *Lynchburg News & Daily Advance*, September 21, 1986.
Hart, George L., Sr. The Beale Papers, Manuscript to Roanoke Library, 1952, 1964.

Hinson, Larry C. *Secret Mission of Thomas Jefferson Beale: Intrigue and Hidden Treasure—With Beale Code 3 Solved* (Denver: Outskirts Press, Inc., 2011).

Hohmann, Robert E. "Beale Code No. 3 Deciphered." *True Treasure Mag.* (March–April 1973).

Holst, Per A. *Handbook of the Beale Ciphers* (Medfield, MA: Beale Cypher Association, 1980).

Innis, Pauline B. "The Beale Fortune." *Argosy Magazine* (August 1964).

Innis, Pauline B. and Walter Dean Innis. *Gold in the Blue Ridge, the True Story of the Beale Treasure* (Washington, DC: R. B. Luce, 1973).

Jolley, Boyd M. "Has Beale's Fabulous Treasure Been Found?" *Treasure Mag.* (August 1982).

Jolley, Boyd M. "Circle Tightens Further on Beale Treasure." *Treasure Magazine* (December 1982).

Kendall, Ray. *Solved: The T. J. Beale Treasure Code of 1822* (Birmingham, AL: Colonial Press, 1998).

Kennedy, Joe. "And the Treasure Hunt Continues." *Roanoke Times*, August 11, 1982.

Kenny, Tom. 30 Million Dollar Beale Treasure Hoax, Private, 1990.

King, John C. "A Reconstruction of the Key to Beale Cipher Number Two." *Cryptologia* 17, no. 3 (July 1993): 305–18. In this paper, King gave a more thorough analysis of the unlikely patterns that Gillogly found arose in Cipher 1, when the Declaration of Independence was applied as the key.

Kruh, Louis. "Reminiscences of a Master Cryptologist." *Cryptologia* 4, no. 1 (January 1980): 45–50. The master cryptologist was Frank Rowlett. These reminiscences include a paragraph where Rowlett talks about the Beale and Swift ciphers, as well as another similar one of which he couldn't recall many details. He believed all three to be hoaxes.

Kruh, Louis. "Beale Society Material: Book Reviews." *Cryptologia* 6, no. 1 (January 1982): 39.

Kruh, Louis. "A Basic Probe of the Beale Cipher as a Bamboozlement." *Cryptologia* 6, no. 4 (October 1982): 378–82. [This is a transcript of a slide-illustrated talk delivered by the author at the Third Beale Cipher Symposium, held September 12, 1981, in Arlington, VA.]

Kruh, Louis. "The Beale Cipher as a Bamboozlement—Part II." *Cryptologia* 12, no. 4 (October 1988): 241–46.

Leighton, Albert C., and Stephen M. Matyas. "Search for the Key Book to Nicholas Trist's Book Ciphers." *Cryptologia* 7, no. 4 (October 1983).

Masters, Al. "Has the Beale Treasure Code Been Solved?" *True Treasure Mag.* (September–October 1968).

Mateer, Todd D. "Cryptanalysis of Beale Cipher Number Two." *Cryptologia* 37, no. 3 (July 2013): 215–32.

Matyas, Stephen M. *The Beale Ciphers: Containing Research Findings, and Documents and Data, As Well As Several Predictions about How The Ciphers Were Constructed* (Poughkeepsie, NY: Published privately, 1966).

McCartney, Sean. *Breaking the Beale Code: The Treasure Hunters Club Book 2* (Ogden, UT: Mountainland Publishing, Inc., 2011).

Nelson, Carl W., Jr. Historical & Analytical Studies in Relation to the Beale Cyphers, Proprietary, 1970.

Nickell, Joe. "Uncovered—The Fabulous Silver Mines of Swift and Filson." *Filson Club History Quarterly* LIV (1980): 325–45. This paper details an earlier fictional story that has much in common with the Beale Papers. Did it inspire the unknown author?

Nickell, Joe. "Discovered, The Secret of Beale's Treasure." *The Virginia Magazine of History and Biography* 90, no. 3 (July 1982): 310–24.

Nickell, Joe, and John F. Fischer. *Mysterious Realms: Probing Paranormal, Historical, and Forensic Enigmas* (Buffalo, NY: Prometheus Books, 1992), 53–67.

Nicklow, Douglas. "Beale's Buried Treasure." *RUN* (August 1984): 48–57.

Ostler, Reinhold. "100 Millionen! Das Gold in der Höhle." *Bild am Sonntag* (Berlin: Axel Springer AG, December 1, 1985).

Poe, Edgar Allan. "The Gold Bug." 1843. This short story has some features in common with the Beale papers. In particular, consider the style of the beginnings. "The Gold Bug" may have served as inspiration for a hoax. First saw print in two installments in *Philadelphia Dollar Newspaper*, June 21 and June 28, 1843.

Price, Steve. "Bedford Treasure Hunt Goes On and On." *Lynchburg News*, August 20, 1967.

Ray, Richard. "Silent for Years, Famous THer Reveals New Clues to Famous Cache." *Treasure Search Mag.* (January–February 1987).

Rubin, Robert. "Jail Is at the End of Treasure-Hunter's Rainbow." *Roanoke Times and World News*, January 15, 1983.

Smith, Becca C. *Alexis Tappendorf and the Search for Beale's Treasure*, The Alexis Tappendorf Series, Book 1, Kindle ed. (Red Frog Publishing, 2012).

Timm, John W. *Mystery Treasure* (Bedford, VA: Tracy Book Co., 1973).

Viemeister, Peter. *The Beale Treasure: New History of a Mystery* (Bedford, VA: Hamilton's, 1997).

Ward, James B. *The Beale Papers* (Lynchburg, VA: Virginian Book and Print Job, 1885).

Williams, Clarence R. The Beale Papers, Library of Congress, Legislative Reference Service Memorandum, April 26, 1934.

Yancey, Dwayne. "Buried Treasure in the Blue Ridge." *Commonwealth* (September 1980).

关于芬恩及其他宝藏密码

Charroux, Robert. *Trésors du monde, trésors de France, trésors de Paris: Enterrés, emmurés, engloutis* (Paris: Fayard, 1972).

de La Roncière, Charles. *Le Flibustier mystérieux, histoire d'un trésor cache* (Paris: Masque, *1934*).

Deutschmann, Jennifer. "Forrest Fenn: Hunt for Buried Treasure Is 'Out of Control.' " *Inquisitr* (April 28, 2015), available online at http://www.inquisitr.com/2050321/forrest-fenn/.

Fenn, Forrest. *The Thrill of the Chase, A Memoir* (Santa Fe, NM: One Horse Land & Cattle Co., 2010). The relevant excerpt is available online at https://www.oldsantafetradingco.com/assets/book-previews/thrill-of-the-chase.pdf.

Fenn, Forrest. *Too Far to Walk* (Santa Fe, NM: One Horse Land & Cattle Co., 2013).

Ford, Dana. "California Couple Strikes Gold after Finding $10 Million in Rare Coins." CNN, February 26, 2014, available online at http://www.cnn.com/2014/02/25/us/california-gold-discovery/.

Goldsmith, Margie. "Well Over $1 Million in Buried Treasure: Find It!" *Huffington Post*, February 18, 2011, updated February 27, 2013, available online at http://www.huffingtonpost.com/margie-goldsmith/over-1-million-in-buried-_b_822894.html.

Lammle, Rob. Get Rich Quick: 6 People Who Accidentally Found a Fortune, *mental_floss*, February 26, 2014, available online at http://mentalfloss.com/article/22449/get-rich-quick-6-people-who-accidentally-found-fortune.

McCurley, Kevin. Cryptograms on Gold Bars from China, International Association for Cryptologic Research, http://www.iacr.org/misc/china/. This is, sadly, the best reference for this story. It has pictures of all of the bars, but very little information beyond that.

Thomas, Athol. *Forgotten Eden: A View of the Seychelles Islands in the Indian Ocean* (London and Harlow, U.K.: Longmans, Green and Co. Ltd., 1968).

第10章

Enăchiuc, V. *Rohonczy Codex* (Bucharest, Romania: Editura Alcor, 2002). A claimed solution to the Rohonc Codex.

Gould, Stephen Jay. *Time's Arrow, Time's Cycle: Myth and Metaphor in the Discovery of Geological Time* (Cambridge, MA: Harvard University Press, 1987).

Gould, Stephen Jay. "James Hampton's Throne and the Dual Nature of Time." *Smithsonian Studies in American Art* 1, no. 1 (Spring 1987): 46–57.

Gould, Stephen Jay. "Nonoverlapping Magisteria." *Natural History* 106 (March 1997): 16–22.

Hampton, James, http://fortean.wikidot.com/james-hampton.

Hampton, James writings [ca. 1950–1964], Archives of American Art, http://www.aaa.si.edu/collections/james-hampton-writings-7162.

Hartigan, Lynda Roscoe. *James Hampton: The Throne of the Third Heaven of the Nations' Millennium General Assembly* (Boston: Museum of Fine Arts, October 19 (1975?)–February 13, 1976). This essay was originally published for a 1976 exhibition of James Hampton's work.

Hartigan, Lynda Roscoe. *The Throne of the Third Heaven of the Nations' Millennium General Assembly* (Montgomery, AL: Montgomery Museum of Fine Arts, 1977).

Ingalls, Helen. "James Hampton's Throne of the Third Heaven of the Nations' Millennium General Assembly," https://www.youtube.com/watch?v=oabZVM-o2IQ. This is a video of a lecture titled "James Hampton's *Throne*: All That Glitters Is Not Gold," delivered by Helen Ingalls, of the Lunder Conservation Center of the Smithsonian American Art Museum, under whose care the objects came in 1988.

Kerckhoffs, Auguste. "La Cryptographie Militaire." *Journal des science militaires* (Paris: Baudoin, 1883), vol. IX, 5–83.

Láng, Benedek. "Why Don't We Decipher an Outdated Cipher System? The Codex of Rohonc." *Cryptologia* 34, no. 2 (April 2010): 115–44.

Marshall, Steve. "St James the Janitor." *Fortean Times* no. 150 (2001). Available online at http://web.archive.org/web/20020125163459/http://www.forteantimes.com/articles/150_jamesjanitor.shtml.

Nyíri, A. "Megszólal 150 év után a Rohonci-kódex? [After 150 Years, the Rohonc Codex Starts to Speak?]." *Theologiai Szemle* 39 (1996): 91–98. A claimed solution to the Rohonc Codex.

Rugg, Gordon. Home page at Keele University, http://www.keele.ac.uk/scm/staff/academic/drgordonrugg/.

Rugg, Gordon. The Penitentia Manuscript, http://www.scm.keele.ac.uk/research/knowledge_modelling/km/people/gordon_rugg/cryptography/penitentia/index.html.

Rugg, Gordon. The Ricardus Manuscript, http://www.scm.keele.ac.uk/research/knowledge_modelling/km/people/gordon_rugg/cryptography/ricardus_manuscript.html.

Rugg, Gordon. "Visualising Structures in Ancient Texts." *Search Visualizer*, Nov. 16, 2012, https://searchvisualizer.wordpress.com/2012/11/16/visualising-structures-in-ancient-texts/.

Rugg, Gordon. "The 'Genesis Death Sandwich' Story." *Search Visualizer*, Feb. 21, 2013, https://searchvisualizer.wordpress.com/2013/02/21/the-genesis-death-sandwich-story/.

Schmeh, Klaus. Klaus Schmeh's List of Encrypted Books, http://scienceblogs.de/klausis-krypto-kolumne/klaus-schmehs-list-of-encrypted-books/.

Schmeh, Klaus. "The Voynich Manuscript: The Book Nobody Can Read." *Skeptical Inquirer* 35.1 (January/February 2011), available online at http://www.csicop.org/si/show/the_voynich_manuscript_the_book_nobody_can_read/_1.

Schmeh, Klaus. "The Pathology of Cryptology—A Current Survey." *Cryptologia* 36, no. 1 (January 2012): 14–45.

Schmeh, Klaus. "Neue Scans zeigen den Codex Rohonci in seiner ganzen Schönheit," April 25, 2015, http://scienceblogs.de/klausis-krypto-kolumne/2015/04/25/neu-scans-zeigen-den-codex-rohonci-in-seiner-ganzen-schoenheit/.

Schmeh, Klaus. "Encrypted Books: Mysteries that Fill Hundreds of Pages." *Cryptologia* 39, no. 4 (October 2015): 1–20.

Singh, Mahesh Kumar. "Rohonci Kódex [The Codex of Rohonc]." *Turán* 6 (2004): 9–40. Also, there is more in 1 (2005). A claimed solution to the Rohonc Codex.

Stallings, Dennis J. The Secret Writing of James Hampton, African American Sculptor, Outsider Artist, Visionary, http://ixoloxi.com/hampton/index.html.

Stamp, Mark. Hamptonese, http://www.cs.sjsu.edu/faculty/stamp/Hampton/hampton.html.

Walsh, Mike. The Miracle of St. James Hampton, *Expresso Tilt*, available online at http://www.missioncreep.com/tilt/hampton.html.

第 11 章

Anonymous. "Signal from Mars. Professor Pickering Saw Bright Lights Upon That Planet." *Toledo Weekly Blade*, January 17, 1901, p. 5. Available online at https://news.google.com/newspapers?nid=1350&dat=19010117&id=uO8SAAAAIBAJ&sjid=cP4DAAAAIBAJ&pg=5259,940727&hl=en.

Anonymous. "Scientific World Stirred by Possibility of Communication from Mars." *The Monthly Evening Sky Map* XIV, no. 159 (March 1920). Available online at Google Books.

Anonymous. "Mars Is Signaling to Us, Says Marconi." *The Monthly Evening Sky Map* XV, no. 179 (November 1921). Available online at Google Books.

Anonymous. "Asks Air Silence When Mars is Near. Prof. Todd Obtains Official Aid from Washington Despite Doubts of Its Efficacy." *New York Times*, August 21, 1924, available online at http://theartpart.jonathanmorse.net/tag/david-peck-todd/.

Anonymous. "Weird 'Radio Signal' Film Deepens Mystery of Mars." *Washington Post*, August 27, 1924, available online at http://www.shorpy.com/node/12482.

Anonymous. "Seeks Sign from Mars in 30-foot Radio Film, Dr. Todd Will Study Photograph of Mysterious Dots and Dashes Recently Recorded." *New York Times*, August 28, 1924, available online at http://theartpart.jonathanmorse.net/tag/david-peck-todd/.

Bracewell, Ronald. "What to Say to the Space Probe When It Arrives." *Horizon*, January 1977.

Callimahos, Lambros D. "Communication with Extraterrestrial Intelligence." Published by the National Security Agency, pp. 79–86 and 109. I haven't been able to find the name of the journal, but a copy of the paper is available online at https://www.nsa.gov/public_info/_files/ufo/communication_with_et.pdf. NSA also published it somewhere with the page numbers 107–15. See https://www.nsa.gov/public_info/_files/tech_journals/communications_extraterrestrial_intelligence.pdf. The paper was reprinted in *Cryptologic Spectrum* 5, no. 2 (Spring 1975): 4–10, available online at https://www.nsa.gov/public_info/_files/cryptologic_spectrum/communications_with_extraterrestrial.pdf. Few papers get published three times!

Cameron, A. G. W. *Interstellar communication: A Collection of Reprints and Original Contributions* (New York: W.A. Benjamin, 1963).

Campaigne, Howard. "Key to the Extraterrestrial Messages." *NSA Technical Journal* XIV, no. 1 (Winter 1969): 13–23, available online at https://www.nsa.gov/public_info/declass/tech_journals.shtml.

Campbell, MacGregor. "Unbreakable: The MIT Time-Lock Puzzle." *New Scientist* no. 2813 (May 27, 2011).

Darling, David. "Green Bank Conference (1961)." *Encyclopedia of Science*, http://www.daviddarling.info/encyclopedia/G/GreenBankconf.html.

Darling, David. "Todd, David Peck (1855–1939)." *Encyclopedia of Science*, http://www.daviddarling.info/encyclopedia/T/Todd.html.

Ehman, Jerry R. "The Big Ear Wow! Signal What We Know and Don't Know about It after 20 Years." original draft completed: September 1, 1997. Last revision: February 3, 1998, available online at http://www.bigear.org/wow20th.htm.

Fort, Charles. *New Lands* In *The Complete Books of Charles Fort*, with an introduction by Tiffany Thayer (Omnibus ed.) (New York: Henry Holt and Co., 1941), chap. 32, p. 494. Available online in single-volume edition at http://www.resologist.net/landsei.htm—see part 2, chap. 20.

Gardner, Martin. Mathematical Games, "Thoughts on the Task of Communication with Intelligent Organisms on Other Worlds." *Scientific American* 213, no. 2 (August 1, 1965): 96–100. This paper also appears in Martin Gardner, *Martin Gardner's 6th Book of Mathematical Diversions from Scientific American* (Chicago: The University of Chicago Press, 1971) as chap. 25, "Extraterrestrial Communication." pp. 253–62. As Gardner typically did with his collected articles, an addendum is provided at the end of this piece. In this instance, it takes the form of a pair of letters related to the piece that appeared in the January 1966 issue of *Scientific American*. However, all they discuss is the reality, or not, of the canals on Mars. Not relevant to our interest here. In the body of the paper, an Arecibo-style communication scheme is discussed, but instead of primes it uses 100 bits to form a 10 × 10 square. Gardner points out, "Indeed, this is the technique by which pictures are now transmitted by radio as well as the basis of television-screen scanning." Of course, that was done at a much higher resolution.

Gregg, Justin. "Dolphins Aren't as Smart as You Think." *The Wall Street Journal*, December 18, 2013. Available online at http://www.wsj.com/articles/SB10001424052702304866904579266183573854204.

Kraus, John. "We Wait and Wonder." *Cosmic Search* 1, no. 3 (Summer 1979), available online at http://www.bigear.org/CSMO/PDF/CS03/cs03p31.pdf. This paper is on the Wow! signal.

Lowell, Percival. *Proceedings of the American Philosophical Society* 40, no. 167 (December 1901): 166–76. This was reprinted, with the addition of many illustrations, in the journal *Popular Astronomy* 10 (April 1902): 185–94. Available online through Google Books and http://adsabs.harvard.edu/full/1902PA.....10..185L.

Lunan, Duncan. "Spaceprobe from Epsilon Bootes." *Spaceflight*, British Interplanetary Society, 1973. This paper presented another claim of a signal being received. The Lunan references that follow continue the story.

Lunan, Duncan. "Long-Delayed Echoes and the Extraterrestrial Hypothesis." *Journal of the Society of Electronic and Radio Technicians* 10, no. 8 (September 1976).

Lunan, Duncan. *The Mysterious Signals from Outer Space* (New York: Bantam Books, 1977).

Lunan, Duncan. "Epsilon Boötis Revisited." *Analog Science Fiction and Fact* 118, no. 3 (March 1998).

Morton, Ella. "Messages to the Universe: A Short History of Interstellar Communication." *Slate*,

November 14, 2014, available online at http://www.slate.com/blogs/atlas_obscura/2014/11/14/the_arecibo_message_and_other_interstellar_communication_attempts.html.

Oliver, Bernard M. "Interstellar Communication." *Interstellar Communication: A Collection of Reprints and Original Contributions*, ed., A. G. W. Cameron (New York: W.A. Benjamin, 1963), 294–305.

Piper, H. Beam. "Omnilingual." *Astounding Science Fiction*, February 1957, available online at http://www.gutenberg.org/files/19445/19445-h/19445-h.htm. "To translate writings, you need a key to the code—and if the last writer of Martian died forty thousand years before the first writer of Earth was born . . . how could the Martian be translated . . .?" I won't give away the punchline, but I will reveal that it isn't mathematics that provides the first break.

Rivest, Ron, Adi Shamir, and Len Adleman. "On Digital Signatures and Public-Key Cryptosystems." MIT/LCS/TM-82, Massachusetts Institute of Technology, Laboratory for Computer Science, Cambridge, MA, 1977. There was soon a title change to "A Method for Obtaining Digital Signatures and Public-Key Cryptosystems," but the date is the same for both. This report later appeared in *Communications of the ACM* 21, no. 2 (1978): 120–26, with the latter title.

Rivest, Ronald L. Description of the LCS35 Time Capsule Crypto-Puzzle, April 4, 1999, available online at http://people.csail.mit.edu/rivest/lcs35-puzzle-description.txt.

Rivest, Ronald L., Adi Shamir, and David A. Wagner. "Time-Lock Puzzles and Timed-Release Crypto." Revised March 10, 1996, available online at http://theory.lcs.mit.edu/~rivest/RivestShamirWagner-timelock.ps.

RSA Secret-Key Challenge, http://en.wikipedia.org/wiki/RSA_Secret-Key_Challenge. This page details other challenges put forth by RSA Labs between 1997 and 2007.

Steele, Bill. "It's the 25th Anniversary of Earth's First (and Only) Attempt to Phone E.T." *Cornell News* (November 12, 1999), available online at http://web.archive.org/web/20080802005337/http://www.news.cornell.edu/releases/Nov99/Arecibo.message.ws.html.

Wooster, Harold (moderator), Paul J. Garvin, Lambros D. Callimahos, John C. Lilly, William O. Davis, and Francis J. Heyden. "Communication with Extraterrestrial Intelligence." *IEEE Spectrum* 3, no. 3 (March 1966).

图片来源

Fig. 1.1 From Bolton, Henry Carrington, *The Follies of Science at the Court of Rudolph II 1576–1612*, Milwaukee: Pharmaceutical Review Publishing Co., 1904

Fig. 1.2 From the National Library of Sweden, MS A 148, with permission

Fig. 1.3 From Yale, Beinecke Rare Book & Manuscript Library, with permission

Fig. 1.4 From Yale, Beinecke Rare Book & Manuscript Library, with permission

Fig. 1.5 From Yale, Beinecke Rare Book & Manuscript Library, with permission

Fig. 1.6 From Yale, Beinecke Rare Book & Manuscript Library, with permission

Fig. 1.7 From Yale, Beinecke Rare Book & Manuscript Library, with permission

Fig. 1.8 From Yale, Beinecke Rare Book & Manuscript Library, with permission

Fig. 1.10 From Reddy, Sravana and Kevin Knight, "What We Know about the Voynich Manuscript," *Proc. ACL Workshop on Language Technology for Cultural Heritage, Social Sciences, and Humanities*, 2011, with permission

Fig. 1.11 From Serafini, Luigi, *Codex Seraphinianus*, 2 vols., Milan, Italy: Franco Maria Ricci, 1981

Fig. 1.12 From Pelcl, František Martin, *Abbildungen böhmischer und mährischer Gelehrten und Künstler: nebst kurzen Nachrichten von ihren Leben und Werken*, Prag: Bey Wolffgang Gerle, 1773–1782. Rare Book Division, Department of Rare Books and Special Collections, Princeton University Library

Fig. 1.13 From Linda Hall Library, LHL Digital Collections, http://lhldigital.lindahall.org/cdm/ref/collection/color/id/4014, with permission

Fig. 1.14 From file RMC2007_1065, Carl A Kroch Library, Cornell University, with permission

Fig. 1.15 Image from "The Erwin Tomash Library on the History of Computing," www.cbi.umn.edu/hostedpublications/Tomash/, used with permission

Fig. 1.16 Michał Wojnicz (later known as Wilfrid Michael Voynich) c. 1885, Central Archives, Moscow

Fig. 1.17 From Newbold, William Romaine, *The Cipher of Roger Bacon*, edited with foreword and notes by Roland Grubb Kent, Philadelphia: University of Pennsylvania Press; London, H. Milford, Oxford University Press, 1928

Fig. 1.18 From Newbold, William Romaine. *The Cipher of Roger Bacon*, edited with foreword and notes by Roland Grubb Kent, Philadelphia: University of Pennsylvania Press; London, H. Milford, Oxford University Press, 1928

Fig. 1.19	From Newbold, William Romaine. *The Cipher of Roger Bacon*, edited with foreword and notes by Roland Grubb Kent, Philadelphia: University of Pennsylvania Press; London, H. Milford, Oxford UniversityPress, 1928)
Fig. 1.20	From Yale, Beinecke Rare Book & Manuscript Library, with permission
Fig. 1.21	From Yale, Beinecke Rare Book & Manuscript Library, with permission
Fig. 1.22	From "Something More Than a Secretary," *Christian Science Monitor*, Sept. 30, 1924
Fig. 1.23	From Kraus, Hans Peter, *A Rare Book Saga*, New York: G.P. Putnam's Sons, 1978. Courtesy of Mary Ann Kraus Folter
Fig. 1.24	From D'Imperio, M. E., *The Voynich Manuscript—An Elegant Enigma*, National Security Agency, 1976
Fig. 1.25	Courtesy of the National Cryptologic Museum
Fig. 1.26	From D'Imperio, M. E., *The Voynich Manuscript—An Elegant Enigma*, National Security Agency, 1976
Fig. 1.27	Courtesy of the National Cryptologic Museum
Fig. 1.28	D'Imperio, M. E., *The Voynich Manuscript—An Elegant Enigma*, National Security Agency, 1976
Fig. 1.29	From Rugg, Gordon, "An Elegant Hoax? A Possible Solution to the Voynich Manuscript," *Cryptologia*, Vol. 28, No. 1, January 2004, with permission
Fig. 1.30	Courtesy of the National Cryptologic Museum
Fig. 2.1	© The Egypt Centre, Swansea University, with permission
Fig. 2.3	Courtesy of Adrienne Mayor
Fig. 2.4	From Vanderpool, E., "An Unusual Black-Figured Cup," *American Journal of Archaeology*, Vol. 49, No. 4, pp. 436–440 (fig. 1), October 1945. Courtesy of the Archaeological Institute of America and the *American Journal of Archaeology*
Fig. 2.5	From Vanderpool, E., "An Unusual Black-Figured Cup," *American Journal of Archaeology*, Vol. 49, No. 4, pp. 436–440 (fig. 3), October 1945. Courtesy of the Archaeological Institute of America and the *American Journal of Archaeology*
Fig. 2.6	From Riksantikvarieämbetet / Swedish National Heritage Board, with permission
Fig. 2.7	From Franksen, O. I., *Mr. Babbage's Secret: The Tale of a Cypher—and APL*, Englewood Cliffs, NJ: Prentice Hall, 1984
Fig. 2.8	From Franksen, O. I., *Mr. Babbage's Secret: The Tale of a Cypher—and APL*, Englewood Cliffs, NJ: Prentice Hall, 1984
Fig. 2.9	From Riksantikvarieämbetet / Swedish National Heritage Board, with permission
Fig. 2.11	From Riksantikvarieämbetet / Swedish National Heritage Board, with permission
Fig. 2.14	From Franksen, O. I., *Mr. Babbage's Secret: The Tale of a Cypher—and APL*, Englewood Cliffs, NJ: Prentice Hall, 1984
Fig. 2.15	From Wimmer, L. F. A. *De danske runemindesmærker*, København: Gyldendal, 1914.
Fig. 2.16	From Riksantikvarieämbetet / Swedish National Heritage Board, with permission
Fig. 2.17	From Riksantikvarieämbetet / Swedish National Heritage Board, with permission
Fig. 2.18	From Riksantikvarieämbetet / Swedish National Heritage Board, with permission
Fig. 2.19	From Riksantikvarieämbetet / Swedish National Heritage Board, with permission
Fig. 2.20	From Riksantikvarieämbetet / Swedish National Heritage Board, with permission
Fig. 2.21	From Riksantikvarieämbetet / Swedish National Heritage Board, with permission

Fig. 2.22 From Riksantikvarieämbetet / Swedish National Heritage Board, with permission
Fig. 2.23 From Riksantikvarieämbetet / Swedish National Heritage Board, with permission
Fig. 2.24 From Riksantikvarieämbetet / Swedish National Heritage Board, with permission
Fig. 2.25 From Riksantikvarieämbetet / Swedish National Heritage Board, with permission
Fig. 3.1 From Elgar, Edward, *My Friends Pictured Within*, Novello & Co Ltd, 1946
Fig. 3.2 From Elgar, Edward, *My Friends Pictured Within*, Novello & Co Ltd, 1946
Fig. 3.3 From Powell, Mrs. Richard, *Edward Elgar: Memories of a Variation*, London: Oxford University Press, 1937
Fig. 3.4 Redrawn from Elgar, Edward and Jerrold Northrup Moore, *Edward Elgar: Letters of a Lifetime*, Oxford University Press, 1990
Fig. 3.5 Reproduced by kind permission of the Elgar Will Trust and the Elgar Birthplace Museum
Fig. 3.6 Reproduced by kind permission of the Elgar Will Trust and the Elgar Birthplace Museum
Fig. 3.7 Reproduced by kind permission of the Elgar Will Trust and the Elgar Birthplace Museum
Fig. 3.8 Reproduced by kind permission of the Elgar Will Trust and the Elgar Birthplace Museum
Fig. 3.9 From Schooling, John Holt, "Secrets in Cipher IV. Form the time of George II to the present day," *The Pall Mall Magazine*, Vol. 8, No. 36, April 1896
Fig. 3.10 From Schooling, John Holt, "Secrets in Cipher IV. Form the time of George II to the present day," *The Pall Mall Magazine*, Vol. 8, No. 36, April 1896
Fig. 3.11 From Schooling, John Holt, "Secrets in Cipher IV. Form the time of George II to the present day," *The Pall Mall Magazine*, Vol. 8, No. 36, April 1896
Fig. 3.12 Reproduced by kind permission of the Elgar Will Trust and the Elgar Birthplace Museum
Fig. 3.13 Reproduced by kind permission of the Elgar Will Trust and the Elgar Birthplace Museum
Fig. 3.14 Reproduced by kind permission of the Elgar Will Trust and the Elgar Birthplace Museum
Fig. 3.15 Reproduced by kind permission of the Elgar Will Trust and the Elgar Birthplace Museum
Fig. 3.16 Reproduced by kind permission of the Elgar Will Trust and the Elgar Birthplace Museum
Fig. 3.17 Reproduced by kind permission of the Elgar Will Trust and the Elgar Birthplace Museum
Fig. 3.18 Reproduced by kind permission of the Elgar Will Trust and the Elgar Birthplace Museum
Fig. 3.19 Reproduced by kind permission of the Elgar Will Trust and the Elgar Birthplace Museum
Fig. 3.20 Reproduced by kind permission of the Elgar Will Trust and the Elgar Birthplace Museum
Fig. 3.21 Reproduced by kind permission of the Elgar Will Trust and the Elgar Birthplace Museum
Fig. 3.22 Reproduced by kind permission of the Elgar Will Trust and the Elgar Birthplace Museum
Fig. 3.23 Reproduced by kind permission of the Elgar Will Trust and the Elgar Birthplace Museum
Fig. 3.24 From Powell, Mrs. Richard, *Edward Elgar: Memories of a Variation*, London: Oxford University Press, 1937
Fig. 3.25 From the collection of the author
Fig. 4.1 From Graysmith, Robert, *Zodiac*, New York: St. Martin's / Marek, 1986
Fig. 4.2 From Graysmith, Robert, *Zodiac*, New York: St. Martin's / Marek, 1986
Fig. 4.3 From Graysmith, Robert, *Zodiac*, New York: St. Martin's / Marek, 1986
Fig. 4.4 The cover of *Argosy*, September 1970
Fig. 4.5 Retrieved from http://zodiackillerfacts.com/feed/
Fig. 4.6 Courtesy of the National Cryptologic Museum
Fig. 4.7 From www.zodiackillerfacts.com/radian.htm

Fig. 4.8 From www.zodiackillerfacts.com/radian.htm
Fig. 4.9 From www.zodiackillerfacts.com/radian.htm
Fig. 4.10 Modified from www.zodiackillerfacts.com/radian.htm
Fig. 5.1 From http://cipherfoundation.org/modern-ciphers/scorpion-ciphers
Fig. 5.2 From http://cipherfoundation.org/modern-ciphers/scorpion-ciphers
Fig. 5.3 From http://coldcasecameron.com/wp-content/uploads/2014/02/John-Walsh-Zodiac-Killer-Letter.pdf
Fig. 5.4 From http://coldcasecameron.com/wp-content/uploads/2014/02/John-Walsh-Zodiac-Killer-Letter.pdf
Fig. 5.5 From http://coldcasecameron.com/wp-content/uploads/2014/02/John-Walsh-Zodiac-Killer-Letter.pdf
Fig. 5.6 From *The Phoenician*, Summer 2006, p. 26
Fig. 5.7 From http://www.thephoenixsociety.org/puzzles/special_puzzle.htm
Fig. 5.8 From http://scienceblogs.de/klausis-krypto-kolumne/2014/03/12/codeknacker-auf-verbrecherjagd-folge-4-der-maskenmann/
Fig. 6.1 Redrawn from Feltus, Gerald (Gerry) Michael, *The Unknown Man: A Suspicious Death at Somerton Park*, Richmond, Australia: Hyde Park Press, 2010, with permission
Fig. 6.2 From Kerry Greenwood, *Tamam Shud: The Somerton Man Mystery*, Sydney, Australia: NewSouth Publishing, University of New South Wales Press Ltd., 2012, with permission
Fig. 6.3 Retrieved from https://www.eleceng.adelaide.edu.au/personal/dabbott/wiki/index.php/The_Taman_Shud_Case_Coronial_Inquest#Exhibit_C.2.
Fig. 6.4 From Feltus, Gerald (Gerry) Michael, *The Unknown Man: A Suspicious Death at Somerton Park*, Richmond, Australia: Hyde Park Press, 2010, with permission
Fig. 6.5 From the South Australia Police Historical Society
Fig. 6.6 From Feltus, Gerald (Gerry) Michael, *The Unknown Man: A Suspicious Death at Somerton Park*, Richmond, Australia: Hyde Park Press, 2010, with permission
Fig. 6.7 From the Australian police
Fig. 6.12 From *The Phoencian*, Winter 2012–13
Fig. 6.13 From "The Body on the Beach: The Somerton Man–Taman Shud Case," http://hubpages.com/education/The-Mystery-of-the-Somerton-Man-Taman-Shud-Case
Fig. 6.14 Redrawn from "The Body on the Beach: The Somerton Man–Taman Shud Case," http://hubpages.com/education/The-Mystery-of-the-Somerton-Man-Taman-Shud-Case
Fig. 6.15 From the Australian police
Fig. 6.16 Retreived from https://en.wikipedia.org/wiki/File:SomertonManEars.jpg
Fig. 6.17 From "Perth Poet-Suicide Chose Omar Verse as His Epitaph," *Mirror*, (Perth, WA), August 25, 1945, http://trove.nla.gov.au/ndp/del/article/76017408
Fig. 6.18 From "Perth Poet-Suicide Chose Omar Verse as His Epitaph," *Mirror*, (Perth, WA), August 25, 1945, http://trove.nla.gov.au/ndp/del/article/76017408
Fig. 6.19 Redrawn from "The Body on the Beach: The Somerton Man–Taman Shud Case," http://hubpages.com/education/The-Mystery-of-the-Somerton-Man-Taman-Shud-Case

Fig. 6.20 From collection of the author
Fig. 6.21 From *Information Bulletin, the Monthly Magazine of the Office of US High Commissioner*
Fig. 6.22 *for Germany*, March 1953, https://commons.wikimedia.org/wiki/File:Dr._James_B_Conant_1953_Berlin.jpeg
Fig. 6.25 From *Galaxy Science Fiction*, February 1953
Fig. 6.23 Courtesy of the National Cryptologic Museum
Fig. 6.24 From http://www.investigatingcrimes.com/unsolved-ciphers-hiding-murder
Fig. 6.26 -mysteries/, accessed February 27, 2016.
Fig. 6.29 From the FBI
Fig. 6.27 From the FBI
Fig. 6.28 From the Master and Fellows of Corpus Christi College, Cambridge, with permission
Fig. 7.2 Courtesy of Paragon House
Fig. 7.4 From Berger, Arthur S., *Aristocracy of the Dead: New Findings in Postmortem Survival*, London and Jefferson, NC: McFarland & Co., 1987
Fig. 7.5 Courtesy of Jim Tucker
Fig. 8.1 Courtesy of Dirk Rijmenants
Fig. 8.2 Courtesy of the National Cryptologic Museum
Fig. 8.3 Courtesy of Nicholas Gessler
Fig. 8.4 Courtesy of the National Cryptologic Museum
Fig. 8.5 Courtesy of the National Cryptologic Museum
Fig. 9.1 From d'Agapeyeff, Alexander, *Codes and Ciphers*, London: Oxford University Press, 1939
Fig. 9.2 From d'Agapeyeff, Alexander, *Codes and Ciphers*, London: Oxford University Press, 1939
Fig. 9.3 From d'Agapeyeff, Alexander, *Codes and Ciphers*, London: Oxford University Press, 1939
Fig. 9.4 From the collection of the author
Fig. 9.5 Courtesy of James Sanborn
Fig. 9.6 From https://www.cia.gov/about-cia/headquarters-tour/kryptos/KryptosPrint.pdf
Fig. 9.7 From https://www.cia.gov/about-cia/headquarters-tour/kryptos/KryptosPrint.pdf
Fig. 9.8 From https://stilljane7.files.wordpress.com/2015/02/kryptos_transcript.jpg
Fig. 9.9 From the secretive group Cicada 3301, retrieved from http://uncovering-cicada.wikia.com/wiki/What_Happened_Part_1_(2012)
Fig. 9.10 From the secretive group Cicada 3301, retrieved from http://i.imgur.com/m9sYK.jpg
Fig. 9.11 From the secretive group Cicada 3301, retrieved from http://www.reddit.com/r/a2e7j6ic78h0j/
Fig. 9.12 From the secretive group Cicada 3301, retrieved from http://uncovering-cicada.wikia.com/wiki/What_Happened_Part_1_(2012)
Fig. 9.13 From the secretive group Cicada 3301, retrieved from http://uncovering-cicada.wikia.com/wiki/What_Happened_Part_1_(2012)

Fig. 9.14 Retrieved from http://www.davidkushner.com/article/cicada-solving-the-webs-deepest-mystery/

Fig. 9.15 From the secretive group Cicada 3301, retrieved from http://i.imgur.com/vjuNp.jpg

Fig. 9.16 Retrieved from https://web.archive.org/web/20101020055419/http://pccts.com/

Fig. 9.17 Retrieved from http://jahbulonian.byethost7.com/

Fig. 9.18 Retrieved from http://jahbulonian.byethost7.com/

Fig. 9.19 Courtesy of Andrew Baker

Fig. 9.20 Courtesy of Andrew Baker

Fig. 9.21 Retrieved from http://jahbulonian.byethost7.com/

Fig. 9.23 Retrieved from http://jahbulonian.byethost7.com/666partsofhell.html

Fig. 9.24 Modified from http://jahbulonian.byethost7.com/666partsofhell.html

Fig. 9.25 From Ward, James B. *The Beale Papers*, Lynchburg, VA: Virginian Book and Print Job, 1885

Fig. 9.26 Courtesy of Addison Doty

Fig. 9.27 From Fenn, Forrest, *Too Far to Walk*, Santa Fe, NM: One Horse Land & Cattle Co., 2013, with permission

Fig. 9.28 Courtesy of the IACR

Fig. 9.29 Courtesy of the IACR

Fig. 9.30 Courtesy of the IACR

Fig. 9.31 From de la Roncière, Charles, *Le Flibustier mystérieux, histoire d'un trésor cache*, Paris, 1934

Fig. 9.34 From http://www.findagrave.com/cgi-bin/fg.cgi?page=gr&GRid=8179530

Fig. 10.2 Courtesy of Gordon Rugg

Fig. 10.3 Courtesy of Gordon Rugg

Fig. 10.4 Courtesy of Gordon Rugg

Fig. 10.5 Courtesy of Gordon Rugg

Fig. 10.6 Courtesy of Gordon Rugg

Fig. 10.7 Courtesy of Gordon Rugg

Fig. 10.8 Courtesy of Gordon Rugg

Fig. 10.9 Courtesy of Gordon Rugg

Fig. 10.10 Courtesy of Oliver Knörzer

Fig. 10.11 Courtesy of Oliver Knörzer

Fig. 10.12 Courtesy of Oliver Knörzer

Fig. 10.13 Courtesy of Oliver Knörzer

Fig. 10.14 Courtesy of Oliver Knörzer

Fig. 10.15 From the secretive group Cicada 3301, retrieved from http://uncovering-cicada.wikia.com/wiki/File:0.jpg

Fig. 11.1 Photograph by Pach Bros, 1908, from the United States Library of Congress, Prints and Photographs division.

Fig. 11.2 From the David Peck Todd papers, 1862–1939 (inclusive), Manuscripts and Archives, Yale University, with permission

Fig. 11.3 Retrieved from http://www.americaspace.com/2014/06/03/of-alien-life-and-intelligence-are-we-ready-for-contact-part-3/